Cell Culture: Techniques and Applications

Cell Culture: Techniques and Applications

Editor: Ted Myers

STATES
ACADEMIC PRESS
www.statesacademicpress.com

States Academic Press,
109 South 5th Street,
Brooklyn, NY 11249, USA

Visit us on the World Wide Web at:
www.statesacademicpress.com

ISBN: 978-1-63989-098-9 (Hardback)

Cataloging-in-Publication Data

Cell culture : techniques and applications / edited by Ted Myers.
p. cm.
Includes bibliographical references and index.
ISBN 978-1-63989-098-9
1. Cell culture. 2. Cytology--Technique. 3. Cytology. 4. Cultures (Biology). I. Myers, Ted.
QH585.2 .C45 2022
571.638--dc23

Table of Contents

Preface

Over the recent decade, advancements and applications have progressed exponentially. This has led to the increased interest in this field and projects are being conducted to enhance knowledge. The main objective of this book is to present some of the critical challenges and provide insights into possible solutions. This book will answer the varied questions that arise in the field and also provide an increased scope for furthering studies.

The process of growing cells outside their natural environment is called cell culture. After isolating them from the living tissue, the cells are grown under controlled conditions. They can be grown using an artificial substrate or in a free floating culture medium. Cell culture finds extensive applications in the fields of tissue engineering, molecular biology, cellular agriculture, nanomaterials assessment, drug discovery, regenerative medicines and vaccines. From theories to research to practical applications, case studies related to all contemporary topics of relevance to this field have been included in this book. The various studies that are constantly contributing towards advancing technologies and evolution of this field are examined in detail. This book is a vital tool for all researching or studying techniques and applications of cell culture as it gives incredible insights into emerging trends and concepts.

I hope that this book, with its visionary approach, will be a valuable addition and will promote interest among readers. Each of the authors has provided their extraordinary competence in their specific fields by providing different perspectives as they come from diverse nations and regions. I thank them for their contributions.

Editor

1

In Vitro Toxicity Testing of Nanomaterials

Ansie Martin and Angshuman Sarkar

Abstract

Cell culture-based techniques are employed world-wide to assess risks to optimize design and handling of materials. Nano-scale particles particularly pose a challenge to the predictability of toxicity. This is due to the interplay of several factors including size, shape, surface properties, micro-environment, charge, electronic species etc., that affect the degree of cellular stress. Numerous assays could be employed to understand the nature of toxicity. Some of these include cytotoxicity, metabolic kinetics, modulation in cellular morphology. Various models such as primary cell culture and cell lines can also be employed that best suits a study. Signaling cascades could also be monitored in better understanding cellular responses holistically. Fluorescence microscopy can further be attempted in studying spatial and temporal variations of bio-markers. Ultimately any plethora of dose-response data, need to be streamlined for developing toxicity evaluation protocols that are accurate, cost effective and time saving.

Keywords: nanoparticles, toxicity, signaling, cell culture, *in vitro*

1. Introduction

Nanoparticles are particles that have at least one dimension less than a 100 nm [1]. Manufactured nanoscale particles find themselves in multitudes of applications such as industrial, food, cosmetic, medical, computing and so on. Distinctive properties such as increased surface area and enhanced reactivity largely fuel nanoingenuity and the net result is an exponential raise in the manufacture and use of nanoparticles. As such human exposure has also risen at an alarming rate. Deleterious effects of this exposure include pulmonary distress, hypertension, cardiovascular damage, irritation of the otolaryngological tissue, impairment of reproductive functions, neural injuries and blood disorders. Long term exposure can also lead to epigenetic alterations and cancer.

Benefits of nanoscale materials are numerous and the market for nanotechnology is enormous owing to the large turnout for profits. The only ethical way forward to manage the harmful effects of nanoparticles is in educated design with innovations, applications and in ethical disposal of nanowastes [2]. Understanding the mechanisms of toxicity attributed by nanoparticles is crucial to this endeavor. Some tools to assess dose and time dependent toxicity have already been developed. Toxicity can either be directly- cell death or the impairment of normal functions in the cell. There are different ways to test toxicity as there are different routes of exposure to the toxic substance.

Conventionally 'acute toxicity testing' has been carried out on a model organism, with each candidate, being tested for a single dose and a single exposure time. Following the exposure, biochemical and histological changes observed from different tissue samples of the dead animal were documented for analysis. The determination of LD_{50}; administration dose with 50% lethality thus required the sacrifice of many animals. LD_{50} has long been used the comparative standard in assessing the degree of toxicity. Several skin sensitivity tests and lymph analysis have also been performed to assess immunogenic potential of an exposing agent. There are other function-based toxicity tests to study effects on reproduction, mutagenic potential, neural management and embryonic development.

The advantage of cell-based toxicity assays is that large numbers of experiments can be conducted to screen the exponential dose-time combinations of exposure [3]. It is greatly time and cost effective as compared to *in vivo* testing. And ethical concerns of animal sacrifice and need for elaborate and regulated laboratories are avoided. No doubt, *in vitro* and *in vivo* results tend to vary depending on the case study, but the wider use of *in vitro* studies allows for only a fraction of the most promising outcomes to be further evaluated with live animal testing. Again, this alleviates many concerns associated with the using animal testing as the first line of investigation.

2. Types of cell culture

Cell culture broadly denotes maintaining cell population, enabling both growth and propagation of cells. There are different types of cell culture methods. The primary culture involves desegregation of cells from the mother tissue by application of enzymatic or shear processes. Cells are transferred to a sterile system with favorable media for growth. This is usually done in glass or plastic containments such as flasks, petri plates, dishes and so on. Primary cultures are often heterogeneous, that is they are a mixed collection of different cell types that are present in the source tissue. Depending on the application requirements, the mixed pool can directly be considered for a study or the different cell populations may be identified by examining biomarkers and further sorted to obtain a culture of a single type of cells. Primary cultures are often further classified into adherent and suspension cultures. Adherent cultures are anchorage dependent [4]. They require surface support for normal proliferation. Adherent cultures are grown in containers coated with a basal polymeric protein matrix such as lysine.

When normal cells are isolated for a period, detached from the surrounding extracellular matrix, it leads to growth arrest and even the induction of anoikis. The cell-cell contact [5] in the extracellular matrix provides a local niche that provides the necessary growth factors, cytokines and integrin binding molecules that favors cell survival and growth. On the other hand, cells growing as a monolayer in a culture dish, also experience a growth arrest when they have crowded and have no more space to spread out. Cells stop proliferating once they fill out the culture dish. As such routine sub culturing or passaging is necessary, to not only maintain the proliferating pool of cells but also to keep them healthy. Tissue from most organs are candidates for adherent cultures.

Other than primary, there is also the suspension type cultures. These cells do not require a support matrix to grow. Examples include cells of the hematopoietic system. From an industrial perspective, suspension cultures are easier to maintain and implement in a large scale set up. Optimization and quantification of various culture parameters are quicker, and this helps efficient development of protocol for production. Like passaging in adherent cultures, suspension cultures need dilution for further growth and propagation. Cell population in a culture is finite and a function of the concentration of cells in the medium. Constant agitation is required to avoid flocculation of cells. Although do not require enzymatic and mechanical detachment as in adherent cultures. Adherent cultures best suit cytological studies while suspension cultures enhance bulk protein production.

Although primary cells retain genetic integrity of the source tissue, there is a limit to its life span and proliferative ability. It varies between donor tissues, requires optimization of culture conditions and is time consuming to grow. The other alternative to primary are continuous cell lines. They are primary cells transformed through subsequent culturing. Transformation can occur naturally or be induced through chemical and viral means. They greatly benefit from their ease of culturing methods. They are characterized for markers and often are available with well-established protocols of handling and propagation. They enable quicker enable quicker biochemical and cellular analysis of mammalian cells. A large number of experiments can be conducted and repeated to add accuracy to experimental evaluation. This is highly desirable for research and industrial applications. Antibody production, screening of toxic compounds and gene expression studies can all be achieved in a time and cost-efficient manner. Drawbacks however include disparity of investigations with *in vivo* systems. Thus, any use of continuous cell line as a model of study, needs to be followed by *in vivo* analysis and medical trials before consideration for implementation.

3. Cell culture practices

All cultures face challenges of contamination through contact and lack of appropriate sterility techniques. Bacterial and fungal contaminations cloud the culture and shift the pH. pH imbalances also occur due to presence of incorrect salts, bicarbonate buffering and gaseous tension. pH changes may at times result in media precipitation, although this may also be the case for contamination with detergent phosphate used for cleaning the culture vessels and equipment. Contamination with magnesium and calcium ions can lead to cell clumping and lysis particularly in a suspension culture. Increased duration of enzyme treatment such as trypsin not only results in subsequent cell adherence issues but may adversely affect cell's survivability due impairment of membrane integrity. Cell death can ultimately be induced by several parameters such as fluctuation in the conditioning temperature, CO_2, repeated freeze and thawing of cells, bad cryopreservation techniques, production of toxic metabolites in the culture media etc. [6]. Thus, proper handling and care are vital to cell culture.

Some common cell culture techniques are exercised to ensure maintaining an aseptic environment for culture [7]. These include routine cleaning of designated rooms and facilities for removal of dust and grease. Use of chlorine and xylenol-based disinfectants for open surfaces is another precaution commonly employed. Decontamination of culture premises by fumigation using potassium permanganate and formaldehyde is carried out as per requirement such as the instances of biological contamination. Steam sterilization of all glass and plastic ware, particularly those used for actual culturing within the biosafety cabinet is undertaken. 70% ethanol or isopropyl alcohol is commonly used to swab hands and wipe surfaces during handling. Use of single open glass flame and sterile tissue rolls also supplements this purpose. Double autoclaved and ion free water is to be used to prepare all sterile solutions. Filtration with at least a 0.22-micron pore sized filter is advised to remove biological contaminants in heat labile substances such as trypsin and antibiotic/antimycotic agents. Use of powder free and sterile oil-resistant gloves is recommended where necessary. The laminar air flow system should house a high efficiency particulate air flow system (HEPA filter). A good HEPA filter must be able to retain and remove at least 99% of particles of 0.3 micron in diameter, suspended in the penetrating air. The maximum speed of the filter should not exceed 0.025 m/s as low speed penetration achieves maximum filtration capacity. As a user dependent precaution, it is recommended, to avoid talking during handling, to prevent generation of contaminant carrying aerosols. Proper planning of experiments routes in better execution. Experiments need to be performed as quickly as possible avoiding all unnecessary steps especially those that involve physical contact with the culture.

Once the experiments are conducted, the right and efficient disposal of culture waste is also equally crucial for ethical reasons. Use of 70% alcohol, isopropyl alcohol, sodium hypochlorite (bleach) and autoclaving can all be employed depending upon the material.

Another crucial step to cell culture is the use of ideal media and storage conditions for preservation of cells. Cryopreservation either by storing vials containing cells in −80°C or liquid nitrogen is commonly followed. Components of freezing media are usually serum and dimethyl sulfoxide. Cryovials can be snap frozen by adding them directly to storage conditions or by gradual incubations with decreasing temperatures. The latter is particularly preferred for sensitive cells.

Some safety measures to be considered while dealing with unauthenticated source is to complete quarantine procedures. Particularly new samples need to be tested for mycoplasma, bacterial and fungal contamination. Unless absolutely required, use of antibiotics is not recommended as they may lead to development of resistant strains and may put stress the cultured cells. Sub culturing needs to be done around 80% confluency, to avoid effects of growth arrest by contact inhibition. Cells need to be routinely frozen and revived. Otherwise a continuous culture that runs for months especially for transformed cells has risks of picking up uncharacterized mutations. Care needs to be taken that all reagents used for cell culture are within the recommended shelf life. All equipment need to be well calibrated and safety cabinets need to be tested for efficacy. Water baths need to be routinely cleaned to avoid contamination. Always sterile water needs to be used in water baths. All work surfaces need to be free of clutter. There must be minimal cardboard packaging if at all required. A splash proof apron and eye protection are to be utilized where necessary.

4. Cell culture-based assays to evaluate toxicity associated with nanoparticles

Nanoparticle formulations to be administered in *in vitro* experiments need to be prepared with care. To avoid inhaling aerosolized nanoparticle, an appropriate pollution mask needs to be worn while handling nanoparticles. It is best to sonicate or vortex and add nanoparticles to allow for dispersion in the nanoscale. This process is called charging. If charging is not done properly, nanoparticle may aggregate and present themselves as micro range particles to the experimental set up. Another precaution to avoid aggregation is to use stock solutions with least possible nanoparticle concentration. Incubation with nanoparticles need to be followed with appropriate washing steps to remove as much particles as possible that have adhered to the cell surface. This prevents interference to the downstream processing of the cell.

One of the first investigations in understanding the effects of any exposing agent is to conduct a cell proliferation assay. Cellular viability, a synonym term is the number of healthy cells in a sample. Viability monitored as a function of dose and time provides information on cell death and hence is also a measure of cytotoxicity [8]. Different parameters indicate the viability of the cell and thus can be used to quantify the cytotoxicity [9]. Some of the methods used to study nanoparticle exposure on cells are described as follows. The dye exclusion methods are a preliminary test based on permeation of dye in dead or dying cells owing to loss of membrane integrity. Trypan blue, eosin and propidium iodide can all be implemented in the dye exclusion test [10]. Dyes are added to cell suspensions and appropriate volumes are loaded onto a counter to aid determination of live cells. Neubauer hemocytometer, a manual counting slide, has been conventionally used to count cells.

Another cell viability assessment is through the documenting the metabolic or enzymatic activity of viable cells. These cells convert the substrate to a colored or fluorescent product and as cell death increases, the degree of this conversion also lacks behind. Examples of this type of tests include protease activity assay and reduction of tetrazolium and resazurin salts [11]. The protease viability assay includes the use of glycylphenylalanyl-aminofluorocoumarin [12]. Abbreviated as GF-AFC, it's a recently developed marker. It permeated live cells and is acted upon by cytoplasmic aminopeptidase. This results in cleavage of glycine and phenyl-alanine amino acid, releasing AFC (aminofluorocoumarin). AFC generates fluorescent signals corresponding to the number of live cells.

MTT [13] (3-(4,5-dimethylthiazol-2-yl)-2,5-diphenyltetrazolium bromide) a positively charged tetrazolium dye readily penetrates the cell. It is converted to the colored formazan product by specific NADH dependent mitochondrial enzymes. Formazan crystals accumulate as insoluble precipitates inside the cell. Cells are lysed, and these crystals are solubilized through combinations of various reagents such as detergent, DMSO, SDS, acidified isopropanol, dimethylformamide, etc. Absorbance readings are then documented, where maximum absorbance is proportional to higher survival. There are some negatively charges tetrazolium dyes such as MTS, XTT, and WST-1. These do not readily penetrate the cell. However, they can still be used for the assay by incorporating an intermediate electron acceptor.

Resazurin (7-hydroxy-10-oxidophenoxazin-10-ium3-one) is particularly used to assay for mitochondrial dysfunction. Resazurin, a deep blue colored complex is reduced to resorufin, a pink colored complex by the enzymes in the inner mitochondrial membrane. Use of resazurin is inexpensive and more sensitive than using tetrazolium dyes.

ATP cell assay is yet another type of evaluation for cell viability. It overrides any incubation with live cell population. It is quicker, more sensitive and less prone to artifacts. Cells with damaged membranes cannot synthesize ATP and endogenous ATPases rapidly depletes cellular ATP concentration. Cells are lysed through detergent activity for this assay, in the presence of ATPase inhibitors to stabilize the total ATP content. Firefly or shrimp derived luciferase acts on substrate luciferin to generate light in the presence of ATP. Higher intensity of light is proportional to the number of live cells.

Sulforhodamine B (SRB) is a fluorescent dye [14]. It is bright pink in color. This aminoxanthene dye binds to cellular protein under mildly acidic conditions. Extraction in a basic environment is proportional to cell mass and thus an indicator of cell viability. Clonogenic cell survival assay investigates the cell's capability for propagation [15]. Treated sample of cells are plated and colonies stained and counted. DNA synthesis cell proliferation assay involves incubating cells with 3H-thymidine. Proliferating cells incorporates this radioactive tracker. Proliferation and thus cell viability can be measured by a scintillation counter. A non-radioactive alternative such as 5-bromo-2′-deoxyuridine (BrdU) can also be implemented. Although this includes an additional step of binding with BrdU-specific antibody and probably a secondary antibody before the colorimetric estimation [16]. A developed 5-ethynyl-2′-deoxyuridine (EdU) can be detected by a fluorescent azide through a Cu(I)-catalyzed cycloaddition reaction. This fast and sensitive method has additional advantages over BrdU assay [17], which are lack of sample fixation and preservation of DNA structure [18]. Raman micro spectroscopy detects variations in Raman bands associated with O–H stretching in water [19]. These variations can non-destructively correspond with ionic concentrations in the intracellular and extracellular fluids. This can be useful in determining the rate and direction of ionic transport. Loss of membrane integrity is dead and dying cells are often associated with leaching of ions, which can be detected and quantified by this technique.

5. Structural tracking as a function of nanoparticle treatment

Morphology modulations can be studied through documenting the changes in the cell as a dose and time dependent function of the exposing agent, using an inverted microscope [20]. Merged with Hoechst stained pictures, can provide additional information on the nuclear morphology. This blue fluorescent dye is membrane permeable and binds to DNA. Hoechst staining has also been successfully used to track changes in the nuclear integrity, thus documenting the progress of cell fate. Necrosis, apoptosis and cell enlargement have been successfully shown as consequences of different nanoparticle exposure in various cell-based studies.

Phalloidin, a mushroom toxin, with a high affinity for F-actin can be used to track changes in actin dynamics [21]. Fluorescent probes are bound to phalloidin and thus help visualize the actin networks. Gap junctions in neuronal cells can be imaged by using biotinylated dextrans. Biotinylated dextran amines can be introduced in neural cell cultures by pressure injections [22]. Further they can be visualized by avidin conjugated horse radish peroxidase with a metal enhanced diaminobenzidine reaction.

Biocytin hydrazide, an aldehyde-based fixative, is another transneuronal tracer. It detects glycoconjugates and can be used to trace neuronal projections and visualize gap junctions. Biocytin hydrazide [23] in turn is detected by a fluorescent dye conjugated to streptavidin. Cholera toxin subunit B is a protein commonly used to retrograde tracer [24]. It binds to glycosphingolipids in axonal membranes [25]. It can also be used to visualize retinofugal projections. Subunit B is non-toxic and aids internalization and transport of conjugates. Wheat germ agglutinin (WGA) is a lectin protein that binds to N-acetyl-D-glucosamine and sialic acid [26]. Neurons can endocytose WGA-HRP, thus they can be used as tracers as they cross through synapses. Isolectins from legumes *Griffonia simplicifolia* can be used to differentiate between neuronal subtypes [27]. The A subunit prefers N-acetyl-D-galactosamine end groups while the B subunit is selective for terminal a-D-galactosyl residues. Its therefore used as a vascular stain for study of adult neurogenesis. Carbocyanine is a lipophilic dye used to stain plasma membranes [28]. They are highly fluorescent in lipid bilayers while being weakly fluorescent in aqueous phase providing a strong contrast for visualization.

6. Evaluation of key signaling pathways in studying cell fate

Simple but conclusive evaluation of changes in expression of molecules is to conduct a reverse transcriptase polymerase chain reaction (RT-PCR) and carry out a western blot analysis. RT-PCR documents the changes in mRNA (transcriptional) levels, while western blot records changes at the protein (translational) level. Cells exposed to nanoparticles are harvested and thoroughly washed before downstream processing. RNA extracted is converted to cDNA and used as a template for PCR. Appropriate primers are designed for target amplification. Amplicons from the PCR are resolved using agarose gel electrophoresis and visualize by a UV illuminator using ethidium bromide staining.

For western blot analysis, protein is carefully extracted. Whole cell lysate (at least 40 μg) is resolved by SDS-PAGE [29]. Western blot is made by electrophoretic transfer on to membranes. Following transfer membranes are blocked with either non-fat dried milk or BSA. Overnight incubation with primary antibody is usually allowed. Post this, appropriate incubation is carried out with secondary antibody. Number of washes after each step, antibody dilution and incubation time all need to be optimized. Horse radish peroxidase conjugates are commonly used for color detection or chemiluminescent detection. However, chemiluminescent detection is much more sensitive than color detection. Expression of a target is normalized to an internal control such as β- Actin, to override experimental errors.

Different signaling cascade checkpoints and markers are commonly evaluated to understand the effect of the exposing agent on the cell through RT PCR and western blot analysis [30]. Some of these are highlighted. Akt, protein kinase B is involved in cell proliferation, transcription, migration and glucose metabolism. Caspases particularly, caspase 3 and caspase 9 indicate the progression of apoptosis. LC3B indicates the onset of autophagic processes. Hsp70 and Hsp90 indicate the heightened cellular responses of protein folding to external stresses. Expression of NFκB is an inflammatory response towards survival and cellular propagation. mTOR is directly or indirectly involved in the regulation of protein function.

7. Statistical analysis of experimental data

As important it is to design and execute meaningful investigations in understanding the effects of toxic agents on cellular functions, it is also important to do them with statistical validation. Experiment results need to be accurate and consistent within the framework of the defined system. Reproducibility of protocols and outcomes are important in formulating further scope for any research. In this endeavor, several statistical tools can be employed which include the determination of average, standard deviation, p value and range of error.

Especially with studying toxicity from nanoparticle exposure, the number of exposing agents is exponentially growing every year. As such it becomes time and cost consuming to exact the required number of experiments to test each particle for various routes of exposure. This necessitates the need for development of adequate and accurate prediction tools for toxicity. A raising technique today is the advent of quantitative structure activity-based relationship (QSAR) models [31]. A response curve is defined, and various physiochemical and biological parameters are gathered or determined and fitness to response is evaluated. For *in vitro* experiments the R^2 (correlation coefficient) of any such effort must be greater than 0.81. Such tools help screen large numbers of particles for their toxic outcomes. This ensure only the most promising leads are further evaluated with actual we laboratory experiments, in process, saving resources, time and funds.

8. Conclusion

Accurate risk assessment is an interdisciplinary approach, where acquiring information and processing is a never-ending ordeal. Therefore, an ethical yet smart approach is needed for studying risks associated with nanoparticles. Only the understanding of the mechanisms of toxicity can enable future endeavors towards strategies of safe nanomaterial design and stress recovery solutions.

In vivo investigations are largely failing to correlate with clinical trials for various reasons. Many pharmaceutical companies have been using rat, monkey and dog models to evaluate toxicity. These are time consuming and cost intensive with very little productivity in terms of actual drugs reaching the market. *In vitro* investigations have the advantage of being cost and time effective and reduces unethical animal sacrifices and cruelty. Any meaningful result can further be screened out for animal testing, if necessary.

Another advantage of *in vitro* models is that it can be developed into 3D cultures and organ on chip innovations to more accurately predict clinical outcomes [32]. Some recent publications show that the level of prediction is even higher than that of *in vivo* investigations, since there is the scope to study and test human tissue functions along with the mechanical fluidic motions that accurately mimic conditions in the human body.

Thus, it is highly intuitive to encourage young researchers to look at a problem holistically and design best possible routes towards solutions that may lead to implementation and relief. An appreciation of resources and time is vital to avoid wasting them in illogical endeavors. An open mind is to be inculcated that although is enriched with knowledge but chooses to not be limited by it.

Author details

Ansie Martin* and Angshuman Sarkar

*Address all correspondence to: ansiemartin@gmail.com

CMBL, Department of Biological Sciences, Birla Institute of Technology and Sciences, K K Birla Goa Campus, Sancoale, South Goa, India

References

[1] Riss TL, Moravec RA. Use of multiple assay endpoints to investigate the effects of incubation time, dose of toxin, and plating density in cell-based cytotoxicity assays. Assay and Drug Development Technologies. 2004;**2**(1):51-62

[2] Martin A, Sarkar A. Overview on biological implications of metal oxide nanoparticle exposure to human alveolar A549 cell line. Nanotoxicology. 2017;**11**(6):713-724

[3] Parasuraman S. Toxicological screening. Journal of Pharmacology & Pharmacotherapeutics. 2011;**2**(2):74

[4] Merten O-W. Advances in cell culture: Anchorage dependence. Philosophical Transactions of the Royal Society B. 2015;**370**(1661):20140040

[5] Abercrombie M. Contact inhibition and malignancy. Nature. 1979;**281**(5729):259

[6] Ramos TV et al. Standardized cryopreservation of human primary cells. Current Protocols in Cell Biology. 2014;**64**(1):A-3I

[7] Coté RJ. Aseptic technique for cell culture. Current Protocols in Cell Biology. 1998;**1**:1-3

[8] Hansen MB, Nielsen SE, Berg K. Re-examination and further development of a precise and rapid dye method for measuring cell growth/cell kill. Journal of Immunological Methods. 1989;**119**(2):203-210

[9] Mosmann T. Rapid colorimetric assay for cellular growth and survival: Application to proliferation and cytotoxicity assays. Journal of Immunological Methods. 1983;**65**(1-2):55-63

[10] Strober W. Trypan blue exclusion test of cell viability. Current Protocols in Immunology. 1997;**21**(1):A-3B

[11] Denizot F, Lang R. Rapid colorimetric assay for cell growth and survival: Modifications to the tetrazolium dye procedure giving improved sensitivity and reliability. Journal of Immunological Methods. 1986;**89**(2):271-277

[12] Niles AL et al. A homogeneous assay to measure live and dead cells in the same sample by detecting different protease markers. Analytical Biochemistry. 2007;**366**(2):197-206

[13] Bahuguna A et al. MTT assay to evaluate the cytotoxic potential of a drug. Bangladesh Journal of Pharmacology. 2017;**12**(2):115-118

[14] Orellana EA, Kasinski AL. Sulforhodamine B (SRB) assay in cell culture to investigate cell proliferation. Bio-Protocol. 2016;**6**(21)

[15] Munshi A, Hobbs M, Meyn RE. Clonogenic cell survival assay. In: Chemosensitivity. USA: Springer Humana Press; 2005. pp. 21-28

[16] Tada H et al. An improved colorimetric assay for interleukin 2. Journal of Immunological Methods. 1986;**93**(2):157-165

[17] Viggiani CJ, Knott SRV, Aparicio OM. Genome-wide analysis of DNA synthesis by BrdU immunoprecipitation on tiling microarrays (BrdU-IP-chip) in *Saccharomyces cerevisiae*. Cold Spring Harbor Protocols. 2010;**2010**(2) (pdb-prot 5385). http://cshprotocols.cshlp.org/content/2010/2/pdb.prot5385.long

[18] Bhaduri S, Ranjan N, Arya DP. An overview of recent advances in duplex DNA recognition by small molecules. Beilstein Journal of Organic Chemistry. 2018;**14**(1):1051-1086

[19] Puppulin L et al. Raman micro-spectroscopy as a viable tool to monitor and estimate the ionic transport in epithelial cells. Scientific Reports. 2017;**7**(1):3395

[20] Santimano C, Maria, et al. Zinc oxide nanoparticles cause morphological changes in human A549 cell line through alteration in the expression pattern of small GTPases at mRNA level. Journal of Bionanoscience. 2013;**7**(3):300-306

[21] Dancker P et al. Interaction of actin with phalloidin: Polymerization and stabilization of F-actin. Biochimica et Biophysica Acta (BBA)—Protein Structure. 1975;**400**(2):407-414

[22] Reiner A et al. Pathway tracing using biotinylated dextran amines. Journal of Neuroscience Methods. 2000;**103**(1):23-37

[23] Bayer EA, Ben-Hur H, Wilchek M. Biocytin hydrazide—A selective label for sialic acids, galactose, and other sugars in glycoconjugates using avidin-biotin technology. Analytical Biochemistry. 1988;**170**(2):271-281

[24] Trojanowski JQ, Gonatas JO, Gonatas NK. Horseradish peroxidase (HRP) conjugates of cholera toxin and lectins are more sensitive retrogradely transported markers than free HRP. Brain Research. 1982;**231**(1):33-50

[25] Lavail JH, Lavail MM. Retrograde axonal transport in the central nervous system. Science. 1972;**176**(4042):1416-1417

[26] Gonatas NK et al. Superior sensitivity of conjugates of horseradish peroxidase with wheat germ agglutinin for studies of retrograde axonal transport. Journal of Histochemistry & Cytochemistry. 1979;**27**(3):728-734

[27] Ernst C, Christie BR. Isolectin-IB4 as a vascular stain for the study of adult neurogenesis. Journal of Neuroscience Methods. 2006;**150**(1):138-142

[28] Heilemann M et al. Carbocyanine dyes as efficient reversible single-molecule optical switch. Journal of the American Chemical Society. 2005;**127**(11):3801-3806

[29] Sarkar A et al. Antagonistic roles of Rac and Rho in organizing the germ cell microenvironment. Current Biology. 2007;**17**(14):1253-1258

[30] Hynes NE et al. Signalling change: Signal transduction through the decades. Nature Reviews Molecular Cell Biology. 2013;**14**(6):393

[31] Puzyn T et al. Using nano-QSAR to predict the cytotoxicity of metal oxide nanoparticles. Nature Nanotechnology. 2011;**6**(3):175

[32] Huh D et al. A human disease model of drug toxicity-induced pulmonary edema in a lung-on-a-chip microdevice. Science Translational Medicine. 2012;**4**(159):159ra147-159ra147

2

2D vs. 3D Cell Culture Models for *In Vitro* Topical (Dermatological) Medication Testing

Arezou Teimouri, Pollen Yeung and Remigius Agu

Abstract

Due to ethical concerns regarding animal testing, alternative methods have been in development to test the efficacy and safety of pharmaceutical products and medications, specifically topical (dermatological) medications. Two-dimensional (2D) and three-dimensional (3D) skin cell cultures are examples of *in vitro* methods used as an alternative to animal testing. The first skin cells cultured were keratinocytes, a type of cell predominantly in the epidermal layer of the skin. However, with differences in skin characteristics and pathophysiology of different skin conditions, various skin cell cultures and models to better mimic these differences have been developed. These cell cultures include not only keratinocytes but also other skin cell types, such as fibroblasts, which are predominantly in the dermal layer of the skin, and certain immune cells and even melanocytes. To have a better understanding of the type of cell cultures used for testing dermatological products, this chapter aims to outline the differences between 2D and 3D skin cell cultures while considering the advantages and disadvantages of each culture. Different types of cell culture models used for wound healing and for inflammatory skin conditions such as psoriasis will also be discussed.

Keywords: cell culture, 2D cell culture, 3D cell culture, skin cell culture, *in vitro*, topical, dermatological, wound healing, psoriasis

1. Introduction

The growth of cells in an environment outside of an organism's body is referred to as a cell culture [1, 2]. Cells obtained from human tissues are often used for cell cultures to mimic the physiological and metabolic functions of humans for *in vitro* studies [3]. Cell cultures are maintained in controlled environments and under controlled temperatures for cells to maintain their ability to survive and to grow [3]. For this reason, a medium for the cells to obtain the required nutrients for their growth is essential [2].

Cells obtained directly from living tissues and then cultured are referred to as primary cell cultures [4]. These cells have a short time frame of survival and scientific use [4]. Cells that have been modified to survive indefinitely with continuous cell division are referred to as either continuous cell lines [1] or immortalized cell lines [2]. Cancer cells are an example of the type of cells that may be used to create cell lines due to their genetic mutations and ability to divide continuously [2].

The very first cell cultures developed were of human skin which allowed for a better understanding of the physiological functions of the skin [5]. Cell cultures are available today in two-dimensional (2D) and three-dimensional (3D) forms. Cell cultures that grow in controlled flat environments, such as a Petri dish, are 2D cell cultures [6]. Cell cultures that are 3D, however, combine and shape cells into a 3D form using a surrounding medium or specialized conditions to help maintain a 3D figure [7]. *In vitro* skin cell cultures developed initially were primarily 2D with keratinocytes as the primary cell types [8, 9]. In the epidermal layer, the main cell types are keratinocytes, while fibroblasts are the main cell types in the dermal layer of the skin [10]. It is thus evident that an *in vitro* model consisting of both keratinocytes and fibroblasts is required to better mimic the physiological functions of the human skin, especially with relation to the wound healing properties of the skin [5]. There are therefore ongoing efforts to develop various *in vitro* skin models that more closely resemble the skin's physiological functions and that will allow for testing of dermatological products [8]. To have a better understanding of the type of cell cultures used for testing dermatological products, this chapter aims to outline the differences between 2D and 3D skin cell cultures while considering the advantages and disadvantages of each culture. Different types of cell cultures used for wound healing and for inflammatory skin conditions such as psoriasis will also be discussed.

2. 2D cell cultures

2D cell cultures have been the main type of cell cultures used by scientists in various fields for decades [6]. As explained in the introduction, cell cultures that grow in controlled flat environments, such as a Petri dish, are 2D cell cultures [6]. To maintain cells outside of an organism's body, it is important to have the correct equipment and supplies to ensure the provision of an environment in which cells can survive and divide [2, 3]. Equipment involved typically includes a biosafety hood for maintaining the safety of cultures and of the individuals working with the cultures, a cell incubator for correct storage with a temperature of 37°C and 5% levels of CO_2, a refrigerator for culture media and other supplies, and flasks or Petri dishes for placement of cells, amongst other required equipment [2, 3].

Besides having the correct equipment for adequate handling and storage of cell cultures, the culture medium or environment in which cell culture grow is also important. The cell culture medium in 2D cell cultures initially consisted of blood plasma; however, over the years, synthetic medium with the right amount of nutrients and even antifungal and antibiotic ingredients have been developed [11]. Cell culture medium is expected to contain amino acids, a buffering system, vitamins, trace elements and more, to maintain cell viability [12]. Eagle's minimum essential medium (EMEM) and Dulbecco's modified Eagle's medium (DMEM) are two examples of cell culture media that have been developed [13]. Different cell types may require different culture media, and thus, the nutrient requirements may vary [13].

2.1. 2D skin cell cultures

The very first cell cultures developed were derived from human skin and were of primarily keratinocytes [5, 8, 9]. As the largest organ in the human body, [14] the skin consists of the epidermal and dermal layers [14, 15]. In the epidermal layer, the main cell types are keratinocytes, while fibroblasts are the main cell types in the dermal layer of the skin [10]. Initially, 2D skin cell cultures consisted primarily of keratinocytes [8, 9]. It was however determined that fibroblasts were essential for the growth of keratinocyte cultures [16] as fibroblasts are the cells responsible for secreting substances of the extracellular matrix such as collagen [17]. The primary fibroblast cell line, cultured by George Todardo and Howard Green, is known as the 3T3 cell line and was named based on its culturing method of transferring cultures every 3 days [16, 18]. The 3T3 cell line is typically used as a feeder layer in keratinocyte cultures [18]. HaCaT cells are a cell line of keratinocytes [19].

Prior to animal testing and clinical trials, traditionally, when testing pharmaceutical ingredients, the first step has been to test *in vitro* using 2D cell cultures [7]. This has however provided limitations with respect to testing as the results obtained from *in vitro* 2D cell culture studies have not translated to *in vivo* studies [7]. For this reason, *in vitro* 3D models have become increasingly popular as they are considered to more closely resemble *in vivo* processes [11, 20].

3. 3D cell cultures

Cell cultures that are 3D involve cells that are combined and shaped into a 3D form using a surrounding medium or specialized condition to help maintain the shape [7]. The equipment used for 2D cell cultures such as a biosafety hood, cell incubator with a temperature of 37°C and 95% O_2, refrigerator for culture medium storage and other supplies [2, 3] may also be used for 3D cell cultures. Supplies used for 3D cultures which differ from those in 2D cell cultures include the matrices, scaffolds, and proprietary plasticware for aggregate formation (e.g., AlgiMatrix, microplates, and multidishes) [21]. Other supplies used in 3D cell cultures may include cell reagents for determining cell health and viability, such as the live/dead Viability/Cytotoxicity Kit and CellTracker Deep Red Dye by ThermoFisher Scientific [21].

The techniques used in creating 3D cell cultures are typically divided into methods that use a scaffold and methods that do not (**Figure 1**) [22]. Methods that do not use a scaffold are typically considered to be a better representation of *in vivo* activity as the cells in this method aggregate on their own [22]. When cells come together and aggregate, they are referred to as

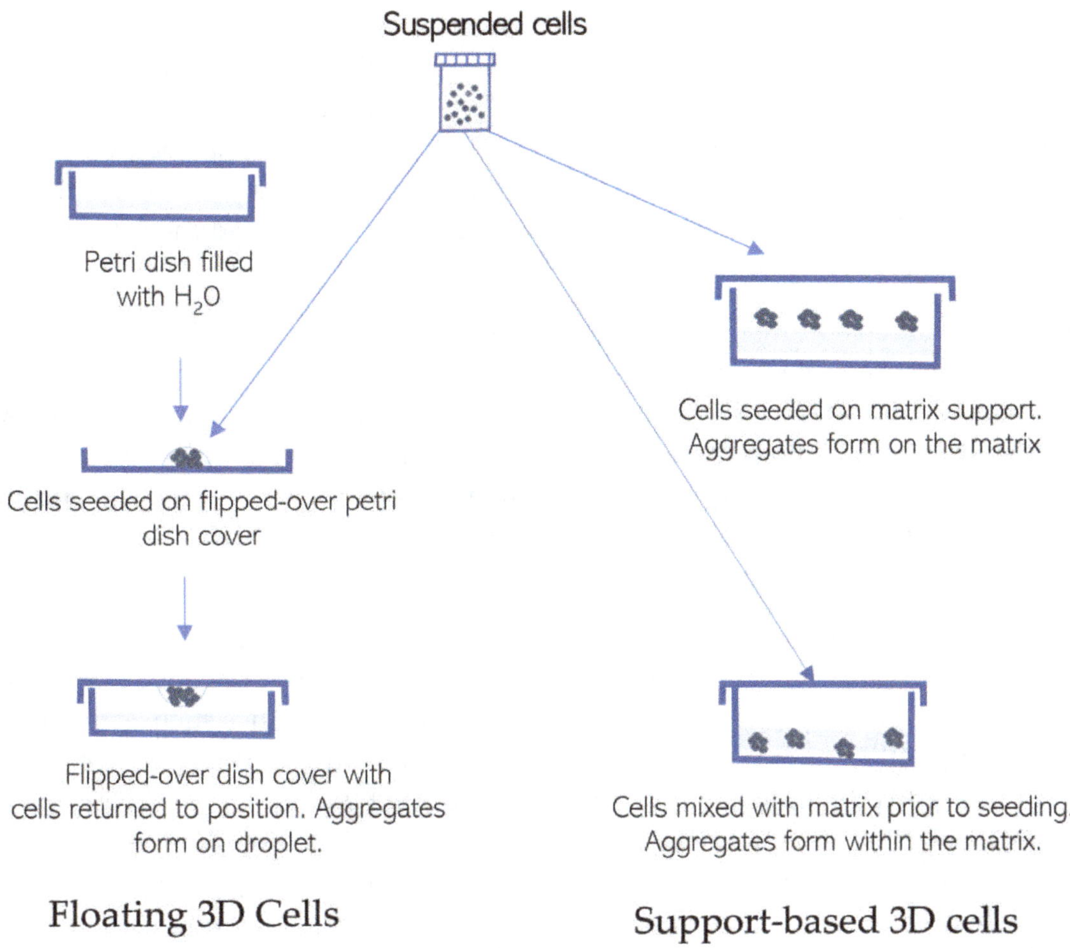

Figure 1. Illustration of general 3D tissue culture methods (floating and matrix-based). Commercially available well-plates and cell-support matrices are often used.

a spheroid in 3D cell cultures [22, 23]. Some of the methods used in 3D cell cultures that will be further explained include the hanging drop method, agitation-based approaches, forced-floating method, and the use of scaffolds [11, 24, 25]. Microfabricated 3D culture systems and bioprinting will also be discussed.

The hanging drop method provides a suspension of cells on a surface which is then inverted, allowing for the cells to hang and form cell-to-cell interactions and aggregate, forming spheroids [24, 25]. Another method is the utilization of continuous mixing methods in agitation-based approaches, which allows cells to aggregate to one another to form a 3D cell culture, as a container that is moving will not allow the cells to stick to the walls of that container [11]. Another method of preventing cells from sticking to the container in which they are placed in is using the forced-floating method which involves the placement of a nonadhesive coating such as poly-2-hydroxyethyl methacrylate (poly-HEMA) or agar, which will in turn force the cells to form interactions and aggregate as opposed to sticking to the surface of the container [11].

The use of scaffolds is another method of forming 3D cell cultures [11]. Scaffolds are materials with pores that allow cells to form aggregates [24]. Scaffolds can be prepared through different methods such as weaving and freeze drying [26] and are typically composed of gelatin, collagen, agarose, and fibronectin, amongst other materials [27]. The idea of scaffolds and matrices is to provide an imitation of the extracellular matrix, allowing for cells to grow and differentiate using these materials [27]. The extracellular matrix consists of essential substances, such as proteins, and surrounds cells in a fluid form that operates as a natural scaffold for cellular growth and differentiation [28, 29]. As discussed previously, fibroblasts in the body are the cells responsible for secreting extracellular matrix and the substances within it such as collagen [17].

Another type of 3D cell culture method is the microfabricated systems, also referred to as cells on chips, which uses microtechnology to provide cells with the required growth environment that can be easily controlled and uses microfluidics for providing nutrients [30]. Microfluidics is the handling of fluids within the microliters range [31]. Microfabricated systems have the advantage of being highly reproducible and provide a better control of the cell culture environment [30]. Another advantage of microfabricated systems is greater cell-to-extracellular matrix interactions while requiring less cell culture [32].

Bioprinting is another method of developing 3D cell cultures. Bioprinting involves the printing of 3D systems by placing living cells in layers [33]. Bioprinting involves both a pre-processing stage as well as the printing stage [34]. The use of imaging diagnostics such as magnetic resonance imagining (MRI), computed tomography (CT), and X-rays to scan an image of the organ or tissue of interest is the first step involved in bioprinting and is part of the pre-processing stage [33, 34]. Selecting the materials such as biomaterials (i.e., biopolymers) as well as cell selection is also part of the preprocessing stage [33, 34]. The printing stage is then completed using different methods such as inkjet printing [33, 35, 36]. Inkjet printing uses bioink for the process of bioprinting and may include the use of cells as the bioink [37, 38].

3.1. 3D skin cell cultures

As discussed previously, the results obtained from *in vitro* 2D monolayer cell culture models have not translated to *in vivo* successes [7]. Thus, alternative methods for testing dermatological products have been sought out. According to an article by Bergers et al., [39] there are various 3D models that exist for a number of different dermatological conditions such as psoriasis and other autoimmune diseases. Skin cell cultures that are 3D have the advantages of containing a stratum corneum and thus the ability of testing pharmaceutical products on the stratum corneum as a skin barrier [40]. Another advantage to the use of 3D skin cell cultures is the longer cell culture use, which is usually between 10 and 30 days [40]. More recent human skin equivalent (HSE) cultures can even be used for up to 20 weeks [41]. The main disadvantage to the use of 3D skin cell cultures however is the cost associated with developing these cultures [40]. The different types of 3D skin cell cultures that will be discussed in detail are histocultures, human skin equivalents, on-chip skin cultures, and pigmented cell cultures. A discussion on bioprinting of skin constructs is also included.

3.1.1. Histocultures

Histocultures are cultures of intact tissues that were developed to better mimic *in vivo* skin responses [42]. The process of hair growth is an example of the ability of histocultures to mimic *in vivo* processes, as this process occurs in histocultures of skin cells and allows for the testing of pharmaceutical products aimed at inhibiting or improving hair growth [42]. The process of histocultures involves growing skin tissues on a growth medium on its own or with the support of collagen [42, 43]. Research has shown that both epidermal and dermal cells as well as hair follicles maintain their functions and physiology in skin histocultures [44]. Histocultures have thus successfully been used as *in vitro* models for testing dermatological products, particularly with respect to toxicity screening [42, 44].

3.1.2. Human skin equivalents

In order to test pharmaceutical products on the skin *in vitro* and provide suitable skin replacement options for patients with various skin conditions, such as burn victims, HSEs were created [45]. HSEs are 3D cell culture models created from various human skin cells and materials that mimic the extracellular matrix [45] and are created as either epidermal equivalents, dermal equivalents or skin equivalents consisting of both layers [8, 45]. Skin equivalents that are used as replacement skin for patients, known as skin grafts, are useful in conditions where the skin cannot adequately heal on its own, such as in the case of a burn victim or chronic ulcers/wounds [10].

An example of the method involved in developing full thickness human skin equivalents includes the isolation of keratinocytes and fibroblasts from donor skin which are then cultured in medium at 37°C at 5% CO_2 [46]. Keratinocytes are then incubated for 2 days following their transfer onto a de-epidermized dermal equivalent and then placed at air-liquid interface for the development of full thickness human skin equivalent [46].

Currently available HSEs are primarily epidermal substitutes composed of keratinocytes in 3D cell culture models such as Epiderm™, SkinEthic RHE™, and EpiSkin™, which are used for pharmaceutical and cosmetic testing of skin irritation from topical products [45, 47, 48]. As these skin substitutes are derived of only the epidermal layer and primarily keratinocytes, it limits their use for testing of products related to particular types of skin conditions that involve the immune system, including testing of products for wound healing [47]. Full thickness models consisting of both the epidermal and dermal layers are thus beneficial [8]. One such model is the full thickness model developed by MatTeck Corporation, the EpidermFT™, which includes both normal human keratinocytes and fibroblasts cultured into a 3D model of several layers of the epidermis and dermis to more closely mimic human skin [49].

3.1.3. On-chip skin culture

Microfabricated systems, also referred to as cells on chips, are a type of 3D cell culture that uses microtechnology to provide cells with the required growth environment that can be easily controlled [30]. On-chip culturing also uses microfluidics for providing nutrients [30]. In order to mimic *in vivo* properties of the skin, an on-chip skin culture was created by Lee et al. [50] and is composed of a microfabricated cell culture of keratinocytes (HaCaT cells) and fibroblasts with microfluidic channels to mimic vasculature. This model provides an opportunity for dermatological medication testing on a microfabricated cell system as growth and differentiation of skin cells were made possible through this model [50]. Another on-chip skin culture model with the ability to mimic vasculature is the model developed by Mori et al. [51], which includes vascular channels attached to an external perfusion system and can be used to test vascular absorption of topical products.

A comparison of on-chip cell cultures to traditional transwell skin cultures was completed by Song et al., [52] in which a static (no flow) chip, dynamic (flow, based on a gravity flow system) chip, and transwell skin equivalents were compared. The comparison revealed that the static chip was not the ideal method as differentiation of skin did not occur and the epidermis was not attached by the end of 1 week [52]. It is thought that this is a result of lack of flow or perfusion in the static chip and thus insufficient flow of nutrients for the growth of the cells [52]. Thus an advantage to using microfabricated skin cell cultures vs. typical 3D cell cultures is the provision of microfluidics which can aid in the supply of essential nutrients for the growth and differentiation of skin cells as well as more closely mimicking *in vivo* processes such as vasculature [51, 52].

3.1.4. Pigmented skin cell cultures

A reconstructed pigmented human skin model was developed by Duval et al. [53]. Using a dermal equivalent with a steel ring on top, seeding of human keratinocytes and melanocytes was completed [53]. The culture developed a monolayer after a period of 1 week in medium after which it was exposed to air for another week to allow differentiation of keratinocytes [53]. This 3D model allows for a better understanding of the interactions between melanocytes, keratinocytes, and fibroblasts [53] and also allows for a pigmented human skin model that could potentially be used for testing of pharmaceutical products on pigmented skin.

3.1.5. Bioprinting of skin constructs

The process of bioprinting explained above is still the same process followed for bioprinting of skin constructs. An important consideration, however, in the pre-processing stage, is that the imaging equipment used ideally should be able to differentiate skin color [34]. Also, with respect to cell selection, keratinocytes are the primary cell types used for bioprinting of skin cells [34]. The advantages to using bioprinting of skin constructs includes greater accuracy in placement of cells and extracellular matrix as well as having the potential of imbedding vasculature in the skin construct as bioprinting of vasculature is also possible [54]. Skin constructs made through bioprinting are also considered to have great plasticity [54]. Skin bioprinting may also be used for developing 3D models for drug testing, such as diseased skin models, and are believed to provide more uniform models compared to manually developed skin models [55]. The main disadvantage of bioprinting for skin constructs is the high cost associated with its use [54].

4. Skin cell cultures for wound healing

Wound healing is a physiological process that consists of four phases: hemostasis, inflammation, proliferation, and remodeling [56]. In the first phase, after a wound injury, hemostasis, platelets are activated and migrated to the site of injury [57]. The second phase, inflammation, begins about 1 day postinjury and inflammatory mediators such as histamine are released, providing the typical traits of inflammation such as heat and swelling [57]. Proliferation is the phase in which granulation tissue forms at the site of injury and after which re-epithelization occurs [58]. In the last phase, remodeling of the tissue occurs to improve the strength of the skin tissue, which is only ever within 80% of the original tissue's strength [58]. The normal wound healing process could be affected, however, leading to chronic ulcers or excessive wound healing resulting in hypertrophic scars [59]. For this reason, topical agents to improve wound healing or reduce scarring may be of interest and as such *in vitro* testing models for wound healing will be discussed.

Both 2D and 3D skin cell cultures are available as wound healing models [60]. 2D wound healing models involve the creation of a site of injury in a monolayer of skin cells, either through mechanical or chemical means, in which cells then migrate to the site of injury [60, 61]. Cells in 2D monolayer cultures are thought to adhere to the flat environments, such as a Petri dish, in which they are cultured and will therefore migrate to areas of free space within the dish, an activity thought to mimic *in vivo* migration involved in cell differentiation [62]. One method of mechanical introduction of a wound to a 2D cell culture is through the scratch assay that utilizes materials such as pipette tips or needles to introduce a wound or scratch into the monolayer cell culture [63, 64]. Images of the wound are taken within set time frames to assess the migration of cells [63]. Typically, however, it is difficult to ensure wounds that are equal in size using this method [63], and thus for this reason, testing of pharmaceutical products on these types of cultures are not ideal. 2D skin cell cultures also lack essential functions that could mimic *in vivo* processes, such as immune functionality and blood perfusion [65]. 3D wound healing models have thus attempted to more closely mimic *in vivo* wound healing processes and are available as histocultures or HSEs.

Histocultures are cultures of intact tissues, consisting of more than one type of skin cell, such as neutrophils and other cells involved in wound healing, and are thus able to better mimic *in vivo* skin responses of wound healing [42, 43]. HSEs, on the other hand, are 3D cell culture models created from various human skin cells and materials that mimic the extracellular matrix [45] and are created as either epidermal equivalents, dermal equivalents or skin equivalents consisting of both layers [8, 45]. Some examples of 3D wound healing skin cell cultures and HSEs are described below.

A 3D wound healing skin equivalent developed by Herman et al. [66] included capillary endothelial cells which were capable of creating a similar formation to *in vivo* capillaries in the skin. This wound healing model involved several different cells, such as keratinocytes and epithelial cells, and used 3D matrices composed of Matrigel™ and collagen [66]. As angiogenesis is involved in the process of wound healing [66], this model is an excellent example of improved 3D wound healing cell culturing protocols to more closely mimic *in vivo* wound healing processes.

Another example of a 3D skin cell culture model that could be used for wound healing was reported by Sidgwick et al. [67]. This model allows for the immersion of biopsies in Williams E culture media with the epidermal layer of the skin uncovered and uses whole tissue biopsies in 24-well cell plates with transwell inserts [67]. This model allows application of topical medications to the uncovered epidermal layer and is thus useful for testing of pharmaceutical products, including those used for transdermal drug delivery [67].

The skin substitutes derived of only the epidermal layer and primarily keratinocytes, however, are limited in their use for testing of products related to particular types of skin conditions that involve the immune system, including the testing of products for wound healing [47]. Full thickness models consisting of both the epidermal and dermal layers are thus beneficial [8]. One such model is the full thickness model developed by MatTeck Corporation, the EpidermFT™ which includes both normal human keratinocytes and fibroblasts cultured into a 3D model of several layers of the epidermis and dermis to more closely mimic human skin [49]. This kit can be purchased with a wound healing assay kit [49] and is thus a commercially available 3D model for testing of dermatological products aimed at wound healing.

It is thus evident from the examples provided in this section that 3D wound healing models more closely mimic *in vivo* wound healing processes and thus may be a better choice for testing of pharmaceutical products aimed at wound healing. Wound healing models are also commercially available which may improve the use of *in vitro* models as a replacement for animal testing.

5. Skin cell cultures for psoriasis

As an inflammatory skin condition that involves the immune system, psoriasis has the characteristic appearance of silver scales that arise from increased proliferation of the keratinocytes in the epidermal layer [68, 69]. Psoriasis is typically treated with topical steroid medications or with vitamin D analogue medications such as calcipotriol as well as with moisturizing agents [68]. The treatment of psoriasis also includes systemic medications that suppress the immune system such as methotrexate and cyclosporine as well as biologic drugs [68, 70].

Psoriasis has been linked to various mental health illnesses such as anxiety and depression which are thought to result from having a chronic visible skin condition [71], and therefore the need for developing new pharmaceutical products and testing of these products is evident. As psoriasis is primarily treated with topical medications [70], having *in vitro* cell cultures or models for testing the safety and efficacy of these topical medications is essential.

Skin cell cultures that are 2D for dermatological conditions such as psoriasis and other autoimmune disorders exist [39]. 2D cell cultures of psoriasis initially consisted of psoriatic keratinocytes; however, this proved to be an ineffective model for psoriasis as the cells were not able to grow and lost their psoriatic genes with time [72]. As a result, alternative methods such as adding cytokines to normal human keratinocytes to induce psoriatic features were developed [73]. As stated previously, however, testing of medications on 2D cell culture models does not always translate to *in vivo* responses [7], and thus, 3D models to better mimic *in vivo* responses for testing of topical medications for psoriasis have been developed.

Barker et al. [74] introduce an *in vitro* model of psoriatic human skin, whereby a monolayer cell culture of psoriatic keratinocytes on top of collagen and fibroblasts from the dermis was created, onto which an epidermal layer was formed approximately 3 weeks later. Similarly, a 3D skin substitute using psoriatic skin cells was developed by Jean et al. [75] with either both psoriatic keratinocytes and fibroblasts or only one type of psoriatic cells (either keratinocytes or fibroblasts were psoriatic). Skin biopsies were obtained from patients with plaque psoriasis, and keratinocytes were extracted and seeded on to mouse fibroblasts and incubated at a temperature of 37°C [75]. For skin substitutes, ascorbic acid was used to culture the fibroblasts which formed dermal sheets that were altered to create a dermal layer onto which keratinocytes were then seeded to form an epidermal layer [75]. This method is the self-assembly method as the fibroblasts release their own extracellular matrix to maintain their growth [75]. A serum-free 3D psoriatic skin cell culture was similarly developed by Duque-Fernandez et al. [66]. This method is similar to the one presented by Jean et al. [75] and may be due to the fact that some authors are the same in both articles. The model by Duque-Fernandez et al. [76] was even used for testing percutaneous permeation of benzoic acid, caffeine, and hydrocortisone, using a Franz-diffusion cell method. This was compared to healthy skin substitutes and revealed that the psoriatic skin substitute had a greater permeability response to the three compounds [76].

For safety and efficacy testing of medications for psoriasis, commercially available 3D *in vitro* models also exist. Two examples of such models are the Psoriasis Skin Model by Creative Bioarray [77] and the model by MatTek Corporation [78]. According to Creative Bioarray [77], their skin model more closely mimics *in vivo* psoriatic skin vs. typical cell cultures as it allows for cell differentiation and can be used for testing of pharmaceutical products. The psoriasis model by MatTek Corporation can also be used for testing of dermatological products as it a 3D model composed of psoriatic fibroblasts while using normal human epidermal keratinocytes and is capable of mimicking *in vivo* psoriatic responses such as basal cell proliferation [78]. These examples were presented for a better understanding of the methods used in developing psoriatic human skin equivalents as well as the commercially available models for testing the safety and efficacy of dermatological products used for psoriasis.

6. 2D vs. 3D cell cultures: morphology and other characteristics

Cell cultures that are 2D and 3D clearly differ with respect to their cell morphology as well as other characteristics. Cell cultures that grow in controlled flat environments, such as a Petri dish, are 2D cell cultures [6], and as such have cells that are flat in morphology [7]. Cell cultures that are 3D involve cells that are combined and shaped into a 3D spheroids using surrounding milieu or specialized conditions [7]. Cell viability also differs between 2D and 3D cell cultures. For example, 2D breast cancer cell cultures showed greater cell viability than 3D breast cancer cells [79]. Drug efficacy may also differ in 2D and 3D cell cultures [79].

7. 2D vs. 3D cell cultures: advantages and disadvantages

Monolayer 2D cell cultures do not always provide an accurate representation of *in vivo* processes [7]. The use of a single cell type [80], such as the use of keratinocytes only in skin cell cultures, [8] or the lack of interactions in 2D cell cultures of cell-to-cell or cell-to-extracellular matrix may be possible causes [5, 80]. However, advantages to using a 2D cell culture include ease of use, cost, and abundant scientific literature surrounding its use [81, 82].

3D cell cultures, on the other hand, provide a better imitation of *in vivo* processes as they allow for better understanding of the interactions that take place between cell-to-cell and cell-to-extracellular matrix [7]. Cell cultures that are 3D have also revealed to possess gene expression abilities, thus mimicking *in vivo* processes [83]. However, regardless of their improvements vs. the 2D cell cultures, typically 3D cell cultures still lack in their ability to provide key functions such as cellular waste removal [7]. Microfabricated cell cultures or on-chip cell cultures are advantageous in this instance as they provide microfluidic channels that enable the flow of nutrients and waste removal as well as the ability to mimic vasculature [51, 52]. Thus microfabricated 3D cell cultures are arguably the *in vitro* method of cell cultures that most closely mimic *in vivo* processes [50].

With respect to skin cell cultures, those that are 3D have the advantages of containing a stratum corneum and thus the ability of testing pharmaceutical products on the stratum corneum

	Advantages	Disadvantages
2D cell cultures	• Ease of use, cost and abundant scientific literature surrounding use [81, 82]	• Do not always provide an accurate representation of *in vivo* processes [7] • Use of a single cell type, [80] for example with skin cell cultures, the use of keratinocytes only [8] • No cell-to-cell or cell-to-extracellular matrix interactions [5, 80]
3D cell cultures	• Better imitation of *in vivo* processes [7] • Better understanding of the interactions that take place between cell-to-cell and cell-to-extracellular matrix [7] • Possess gene expression abilities, thus mimicking *in vivo* processes [83] • Microfabricated cell cultures or on-chip cell cultures provide microfluidic channels that enable the flow of nutrients and waste removal as well as the ability to mimic vasculature [51, 52] • 3D skin cultures have a stratum corneum and thus the ability of testing pharmaceutical products on the stratum corneum as a skin barrier [40] • Longer skin cell culture use between 10 and 30 days [40]; more recent human skin equivalent (HSE) cultures can even be used for up to 20 weeks [41] • 3D bioprinting provides greater accuracy in placement of cells and extracellular matrix, potential of imbedding vasculature in the skin construct and have great plasticity [54] • 3D bioprinted skin models provide more uniform models vs. manually developed skin models [55]	• Typically lack in their ability to provide key functions such as cellular waste removal [7] • Cost associated with developing cultures [40, 54]

Table 1. Summary of the advantages and disadvantages of 2D vs. 3D cell cultures.

as a skin barrier [40]. Other advantages to the use of 3D skin cell cultures include longer cell culture use between 10 and 30 days [40]. More recent human skin equivalent (HSE) cultures can even be used for up to 20 weeks [41]. Skin substitutes that are derived from only the epidermal layer and primarily keratinocytes are limited in their use for testing of products related to particular types of skin conditions that involve the immune system, including testing of products for wound healing [47]. Full thickness models consisting of both the epidermal and dermal layers are thus beneficial [8]. 3D bioprinting of skin constructs are also advantageous as they provide a greater accuracy in placement of cells and extracellular matrices as well as having the potential of imbedding vasculature in the skin construct as bioprinting of vasculature is also possible [54]. Skin constructs made through bioprinting are also considered to have high plasticity [54]. Skin bioprinting may also be used for developing 3D models for drug testing, such as diseased skin models, and are considered to provide more uniform models vs. manually developed skin models [55]. The primary disadvantage to the use of 3D skin cell cultures, however, is the cost associated with developing these cultures [40, 54]. See **Table 1** for a summary of the information presented in this section.

8. Conclusion

Testing the efficacy and safety of dermatological products and medications using *in vitro* methods such as cell cultures can be considered as a replacement to animal testing. Cell cultures currently exist in both 2D and 3D forms. The very first cell cultures developed were of human skin which allowed for a better understanding of the physiological functions of the skin [5]. *In vitro* skin cell cultures developed initially were primarily 2D with keratinocytes as the primary cell types used [8, 9]. However, it became evident that an *in vitro* model consisting of both keratinocytes and fibroblasts is required to better mimic the physiological functions of the human skin, especially with relation to the wound healing properties of the skin [5]. For this reason, 3D cell cultures that allowed greater cell-to-cell and cell-to-extracellular matrix interactions were developed [7].

It is evident from the information presented in this chapter that 3D skin cell cultures can more closely mimic *in vivo* processes vs. 2D skin cell cultures for both wound healing and psoriasis. In addition, full thickness 3D skin equivalents that consist of both an epidermal and dermal layer are a better representation of the human skin [8]. Microfabricated cell systems, however, provide arguably an even better model for mimicking *in vivo* skin processes including vasculature and thus allowing for the testing of absorption of pharmaceutical products [50–52]. Additionally, 3D bioprinting provides greater accuracy in placement of cells and extracellular matrices along with the potential of imbedding vasculature in the skin construct as well as having great plasticity [54]. 3D bioprinted skin models also provide more uniform models vs. manually developed skin models [55].

Conflict of interest

No conflicts of interest.

Author details

Arezou Teimouri, Pollen Yeung and Remigius Agu*

*Address all correspondence to: remigius.agu@dal.ca

Dalhousie University, Halifax, Canada

References

[1] ThermoFisher Scientific. Introduction to Cell Culture. Available from: https://www.thermofisher.com/ca/en/home/references/gibco-cell-culture-basics/introduction-to-cell-culture.html [Accessed: 25-04-2018]

[2] Carter M, Shieh JC. Chapter 13—Cell culture techniques. In: Carter M, Shieh JC, editors. Guide to Research Techniques in Neuroscience. New York: Academic Press; 2010. pp. 281-296

[3] Philippeos C, Hughes RD, Dhawan A, Mitry RR. Introduction to cell culture. Methods in Molecular Biology. 2012;**806**:1

[4] MicroscopeMaster. Cell Culture—Basics, Techniques and Media. Available from: https://www.microscopemaster.com/cell-culture.html [Accessed: 25-04-2018]

[5] Coulomb B, Dubertret L. Skin cell culture and wound healing. Wound Repair and Regeneration. 2002;**10**(2):109-112

[6] Duval K, Grover H, Han LH, Mou Y, Pegoraro AF, Fredberg J, et al. Modeling physiological events in 2D vs. 3D cell culture. Physiology (Bethesda). 2017;**32**(4):266-277

[7] Edmondson R, Broglie J, Adcock A, Yang L. Three-dimensional cell culture systems and their applications in drug discovery and cell-based biosensors. Assay and Drug Development Technologies. 2014;**12**(4):207-218

[8] MacNeil S. Progress and opportunities for tissue-engineered skin. Nature. 2007;**445**(7130):874-880

[9] Rheinwald JG, Green H. Epidermal growth factor and the multiplication of cultured human epidermal keratinocytes. Nature. 1977;**265**(5593):421

[10] Nicholas MN, Jeschke MG, Amini-Nik S. Methodologies in creating skin substitutes. Cellular and Molecular Life Sciences. 2016;**73**(18):3453-3472

[11] Breslin S, O'Driscoll L. Three-dimensional cell culture: The missing link in drug discovery. Drug Discovery Today. 2013;**18**(5):240-249

[12] Davis J. Animal Cell Culture: Essential Methods. Chichester, West Sussex: Wiley; 2011

[13] Celis JE. Cell Biology, Four-Volume Set: A Laboratory Handbook. Amsterdam: Academic Press; 2006

[14] Casey G. Physiology of the skin. Nursing Standard (through 2013). 2002;**16**(34):47-51; quiz 53, 55

[15] Gaboriau HP, Murakami CS. Skin anatomy and flap physiology. Otolaryngologic Clinics of North America. 2001;**34**(3):555-569

[16] Rheinwald JG, Green H. Serial cultivation of strains of human epidermal keratinocytes: The formation of keratinizing colonies from single cells. Cell. 1975;**6**:331-344

[17] PromoCell. Fibroblasts. 2015. Available from: https://www.promocell.com/products/human-primary-cells/fibroblasts/# [Accessed: 10-05-2018]

[18] Molecular Devices. At the Cell Counter: IH3T3 Cells. Available from: https://www.molecular devices.com/cell-counter-nih3t3-cells [Accessed: 10-05-2018]

[19] Seo M, Kang TJ, Lee CH, Lee A, Noh M. HaCaT keratinocytes and primary epidermal keratinocytes have different transcriptional profiles of cornified envelope-associated genes to T helper cell cytokines. Biomolecules & Therapeutics. 2012;**20**(2):171-176

[20] Tung J. Welcome to the 3rd Dimension. 2017. Available from: https://sciencellonline.com/blog/welcome-to-the-3rd-dimension/ [Accessed: 26-04-2018]

[21] ThermoFisher Scientific. 3D Cell Culture Product Selection Guide for Organoids and Spheroids.2018.Availablefrom:http://assets.thermofisher.com/TFS-Assets/BID/brochures/3D-Organoid-Spheroid-Product-Selection-Guide-Global.pdf [Accessed: 20-06-2018]

[22] MilliporeSigma. 3D-Cell culture: An Overview of Advanced Tools and Techniques. Available from: https://www.sigmaaldrich.com/technical-documents/articles/biology/3d-cell-culture-technology.html [Accessed: 26-04-2018]

[23] Pampaloni F, Reynaud EG, Stelzer EHK. The third dimension bridges the gap between cell culture and live tissue. Nature Reviews Molecular Cell Biology. 2007;**8**(10):839-845

[24] Lin R, Chang H. Recent advances in three-dimensional multicellular spheroid culture for biomedical research. Biotechnology Journal. 2008;**3**(9):1285-1285

[25] Fennema E, Rivron N, Rouwkema J, van Blitterswijk C, de Boer J. Spheroid culture as a tool for creating 3D complex tissues. Trends in Biotechnology. 2013;**31**(2):108-115

[26] Moroni L, de Wijn JR, van Blitterswijk CA. Integrating novel technologies to fabricate smart scaffolds. Journal of Biomaterials Science. Polymer Edition. 2008;**19**(5):543-572

[27] Ravi M, Paramesh V, Kaviya SR, Anuradha E, Solomon FDP. 3D cell culture systems: Advantages and applications. Journal of Cellular Physiology. 2015;**230**(1):16-26

[28] Guruswamy M. The extracellular matrix. Young Scientists Journal. 2014;**16**:35

[29] Brown BN, Badylak SF. Extracellular matrix as an inductive scaffold for functional tissue reconstruction. Translational Research. 2014;**163**(4):268-285

[30] El-Ali J, Sorger PK, Jensen KF. Cells on chips. Nature. 2006;**442**(7101):403

[31] Whitesides GM. The origins and the future of microfluidics. Nature. 2006;**442**(7101):368-373

[32] Wu M, Huang S, Lee G. Microfluidic cell culture systems for drug research. Lab on a Chip. 2010;**10**(8):939

[33] Murphy SV, Atala A. 3D bioprinting of tissues and organs. Nature Biotechnology. 2014;**32**(8): 773-785

[34] Vijayavenkataraman S, Lu WF, Fuh JYH. 3D bioprinting of skin: a state-of-the-art review on modelling, materials, and processes. Biofabrication. 2016;**8**(3):032001

[35] Nakamura M, Kobayashi A, Takagi F, Watanabe A, Hiruma Y, Ohuchi K, et al. Biocompatible inkjet printing technique for designed seeding of individual living cells. Tissue Engineering. 2005;**11**(11-12):1658

[36] Xu T, Jin J, Gregory C, Hickman JJ, Boland T. Inkjet printing of viable mammalian cells. Biomaterials. 2005;**26**(1):93-99

[37] Wüst S, Müller R, Hofmann S. Controlled positioning of cells in biomaterials-approaches towards 3D tissue printing. Journal of Functional Biomaterials. 2011;**2**(3):119-154

[38] Guifang G, Ying H, Schilling AF, Karen H, Xiaofeng C. Organ bioprinting: Are we there yet? Advanced Healthcare Materials. 2018;**7**(1):1701018

[39] Bergers LIJC, Reijnders CMA, van den Broek LJ, Spiekstra SW, de Gruijl TD, Weijers EM, et al. Immune-competent human skin disease models. Drug Discovery Today. 2016;**21**(9):1479-1488

[40] Roguet R. Use of skin cell cultures for *in vitro* assessment of corrosion and cutaneous irritancy. Cell Biology and Toxicology. 1999;**15**(1):63-75

[41] El Ghalbzouri A, Commandeur S, Rietveld MH, Mulder AA, Willemze R. Replacement of animal-derived collagen matrix by human fibroblast-derived dermal matrix for human skin equivalent products. Biomaterials. 2009;**30**(1):71-78

[42] Hoffman RM. Histocultures and their use. In: Encyclopedia of Life Sciences (ELS). Chichester: John Wiley & Sons Ltd; October 2010. DOI: 10.1002/9780470015902.a0002573.pub2

[43] Ali N, Hosseini M, Vainio S, Taïeb A, Cario-André M, Rezvani HR. Skin equivalents: Skin from reconstructions as models to study skin development and diseases. The British Journal of Dermatology. 2015;**173**(2):391-403

[44] Li LN, Margolis LB, Hoffman RM. Skin toxicity determined *in vitro* by three-dimensional, native-state histoculture. In: Proceedings of the National Academy of Sciences of the United States of America. 1991;**88**(5):1908-1912

[45] Zhang Z, Michniak-Kohn B. Tissue engineered human skin equivalents. Pharmaceutics. 2012;**4**(1):26-41

[46] Xie Y, Rizzi S, Dawson R, Lynam E, Richards S, Leavesley D, et al. Development of a three-dimensional human skin equivalent wound model for investigating novel wound healing therapies. Tissue Engineering Part C: Methods. 2010;**16**(5):1111-1123

[47] Akagi T, Nagura M, Hiura A, Kojima H, Akashi M. Construction of three-dimensional dermo-epidermal skin equivalents using cell coating technology and their utilization as alternative skin for permeation studies and skin irritation tests. Tissue Engineering. Part A. 2017;**23**(11-12):481

[48] Matsusaki M, Fujimoto K, Shirakata Y, Hirakawa S, Hashimoto K, Akashi M. Development of full-thickness human skin equivalents with blood and lymph-like capillary networks by cell coating technology. Journal of Biomedical Materials Research Part A. 2015;**103**(10):3386-3396

[49] MatTek Corporation. EpidermFT™. Available from: https://www.mattek.com/products/epidermft/ [Accessed: 10-05-2018]

[50] Lee S, Jin S, Kim Y, Sung G, Chung J, Sung J. Construction of 3D multicellular microfluidic chip for an *in vitro* skin model. Biomedical Microdevices. 2017;**19**(2):1-14

[51] Mori N, Morimoto Y, Takeuchi S. Skin integrated with perfusable vascular channels on a chip. Biomaterials. 2017;**116**:48-56

[52] Song HJ, Lim HY, Chun W, Choi KC, Lee T, Sung JH, et al. Development of 3D skin-equivalent in a pump-less microfluidic chip. Journal of Industrial and Engineering Chemistry. 2018;**60**:355-359

[53] Duval C, Chagnoleau C, Pouradier F, Sextius P, Condom E, Bernerd F. Human skin model containing melanocytes: Essential role of keratinocyte growth factor for constitutive pigmentation—functional response to alpha-melanocyte stimulating hormone and forskolin. Tissue Engineering Part C: Methods. 2012;**18**(12):947-957

[54] He P, Zhao J, Zhang J, Li B, Gou Z, Gou M, et al. Bioprinting of skin constructs for wound healing. Burns & Trauma. 2018;**6**(1):5. http://doi.org/10.1186/s41038-017-0104-x

[55] Vultur A, Schanstra T, Herlyn M. The promise of 3D skin and melanoma cell bioprinting. Melanoma Research. 2016;**26**(2):205-206

[56] Lindley LE, Stojadinovic O, Pastar I, Tomic-Canic M. Biology and biomarkers for wound healing. Plastic and Reconstructive Surgery. 2016;**138**(3 Suppl):18S-28S

[57] Chin GA, Dieglemann RF, Schultz GS. 3 Cellular and molecular regulation of wound healing. In: Falabella A, Kirsner R, editors. Wound Healing. Baton Rouge: Chapman and Hall/CRC; 2005. pp. 17-37

[58] Adams SB, Sabesan VJ, Easley ME. Wound healing agents. Critical Care Nursing Clinics of North America. 2012;**24**(2):255-260

[59] Wang P, Huang B, Horng H, Yeh C, Chen Y. Wound healing. Journal of the Chinese Medical Association. 2018;**81**(2):94-101

[60] Stamm A, Reimers K, Strauß S, Vogt P, Scheper T, Pepelanova I. *In vitro* wound healing assays—state of the art. BioNanoMaterials. 2016;**17**(1-2):79-87

[61] Jonkman JEN, Cathcart JA, Xu F, Bartolini ME, Amon JE, Stevens KM, et al. An introduction to the wound healing assay using live-cell microscopy. Cell Adhesion & Migration. 2014;**8**(5):440-451

[62] Keese C, Wegener J, Walker S, Giaever I. Electrical wound-healing assay for cells *in vitro*. Proceedings of the National Academy of Sciences of the United States of America. 2004;**101**(6):1554-1559

[63] Liang C-C, Ann YP, Guan J-L. *In vitro* scratch assay: A convenient and inexpensive method for analysis of cell migration *in vitro*. Nature Protocols. 2007;**2**(2):329

[64] Belete H, Godin L, Stroetz R, Hubmayr R. Experimental models to study cell wounding and repair. Cellular Physiology and Biochemistry. 2010;**25**(1):71-80

[65] Klicks J, von Molitor E, Ertongur-Fauth T, Rudolf R, Hafner M. *In vitro* skin three-dimensional models and their applications. Journal of Cellular Biotechnology. 2017;**3**(1): 21-39

[66] Herman IM, Leung A. Creation of human skin equivalents for the *in vitro* study of angiogenesis in wound healing. In: Murray C, Martin S, editors. Angiogenesis Protocols: Second Edition. Totowa, NJ: Humana Press; 2009. pp. 241-248

[67] Sidgwick G, McGeorge D, Bayat A. Functional testing of topical skin formulations using an optimised ex vivo skin organ culture model. Archives of Dermatological Research. 2016;**308**(5):297-308

[68] Clarke P. Psoriasis. Australian Family Physician. 2011;**40**(7):468-473

[69] Menter A. Psoriasis. London: Manson Pub; 2011

[70] Brassinne DL, Failla V, Nikkels A. Psoriasis: State of the art 2013 part I: Clinical, historical, epidemiological and genetic aspects, Co-morbidities and pathogenesis. Acta Clinica Belgica. 2013;**68**(6):427-432

[71] Ferreira B, Abreu JLPDC, Reis JPGD, Figueiredo AMDC. Psoriasis and associated psychiatric disorders: A systematic review on etiopathogenesis and clinical correlation. The Journal of Clinical and Aesthetic Dermatology. 2016;**9**(6):36-43

[72] Jeong I, Lee HJ. Psoriasis skin models as promising tools in psoriasis research. Biomedical Journal of Scientific & Technical Research. 2018;**2**(3):1-4

[73] Bracke S, Desmet E, Guerrero-Aspizua S, Tjabringa SG, Schalkwijk J, Van Gele M, et al. Identifying targets for topical RNAi therapeutics in psoriasis: Assessment of a new *in vitro* psoriasis model. Archives of Dermatological Research. 2013;**305**(6):501-512

[74] Barker CL, McHale MT, Gillies AK, Waller J, Pearce DM, Osborne J, et al. The development and characterization of an *in vitro* model of psoriasis. Journal of Investigative Dermatology. 2004;**123**(5):892-901

[75] Jean J, Lapointe M, Soucy J, Pouliot R, et al. Journal of Dermatological Science. 2009; **53**(1):19-25

[76] Duque-Fernandez A, Gauthier L, Simard M, Jean J, Gendreau I, Morin A, et al. A 3D-psoriatic skin model for dermatological testing: The impact of culture conditions. Biochemistry and Biophysics Reports. 2016;**8**:268-276

[77] Creative Bioarray. Psoriasis Skin Model. Available from: https://www.creative-bioarray.com/psoriasis-skin-model.htm [Accessed: 18-05-2018]

[78] MatTek Corporation. Psoriasis. 2018. Available from: https://www.mattek.com/products/psoriasis/ [Accessed: 18-05-2018]

[79] Breslin S, O'Driscoll L. The relevance of using 3D cell cultures, in addition to 2D monolayer cultures, when evaluating breast cancer drug sensitivity and resistance. Oncotarget. 2016;**7**(29):45745

[80] Haycock JW. 3D cell culture: A review of current approaches and techniques. Methods in Molecular Biology. 2011;**695**:1

[81] Gargotti M, Lopez-Gonzalez U, Byrne H, Casey A. Comparative studies of cellular viability levels on 2D and 3D *in vitro* culture matrices. Cytotechnology. 2018;**70**(1):261-273

[82] Mimetas. 2D Versus 3D Cell Cultures. Available from: https://mimetas.com/article/2d-versus-3d-cell-cultures [Accessed: 27-04-2018]

[83] Abbott A. Cell culture: Biology's new dimension. Nature. 2003;**424**(6951):870

3

Authenticating Hybrid Cell Lines

Raymond W. Nims, Amanda Capes-Davis,
Christopher Korch and Yvonne A. Reid

Abstract

Hybrid (both intra-species and inter-species) cell lines arise through intentional or non-intentional fusion of somatic cells having different origins. Hybrid cell lines can pose a problem for authentication testing to confirm cell line identity, since the results obtained may not conform to the results expected for the two parental cell types. Thus, depending on the identity testing methodology, a hybrid cell may display characteristics of one of the parental cell type or of both. In some instances, the hybrid cell line may display characteristics that are different from those displayed by either parental cell type; these differences may not necessarily indicate cellular cross-contamination. Testing should be performed as soon as possible after an intended fusion has occurred, so that a baseline reference profile is available for later comparison. In this article, we describe the various approaches that have been used for identifying hybrid cell lines and the results that might be expected when using various technologies for this purpose.

Keywords: cell fusion, hybrid cell line, authentication, immunostaining, isoenzyme analysis, karyotyping, STR profiling, SNP profiling

1. Introduction

Fusion of cells occurs normally in vivo, such as of muscle cells, bone cells, macrophages, during fertilization of germ cells, and in placenta formation [1]. Somatic cell hybrid cell lines (or more simply hybrid cell lines) are cell lines that arise through intentional or nonintentional fusion of somatic cells having different origins [2]. Intra-species and inter-species (cross-species) cell fusions have been described since the 1950s [3] and can occur either spontaneously or can be mediated by human oncogenic viruses (such as Sendai virus, Epstein-Barr virus, human papilloma viruses, hepatitis B and C viruses, human T-cell lymphotropic virus type 1

(HTLV-1), herpesviruses-8/Kaposi sarcoma herpesvirus (HHV-8/KSHV)) [4, 5], polyethylene glycol [6], or electrical pulses (electrofusion) [7–12], resulting in viable syncytial cells (giant cells or polykaryotes) with hybrid genotypes, namely heterokaryons. Mouse-human hetero-hybridoma technology has advanced significantly with the use of electrofusion technology [13]. Currently, electrically induced cell fusion is also being used to develop cancer cells with increased immunogenicity by fusion with dendritic cells for development of anti-tumor vaccines [14, 15].

Cell fusion can be important in the establishment and evolution of cell lines (e.g., [16]) and can lead to cancer progression and metastasis via genetic instability [17–20]. Hybrid cell lines also can arise spontaneously. Numerous examples have been documented [21], including a case where a patient-derived xenograft model underwent spontaneous fusion with normal mouse stromal cells, forming a hybrid cell that was more tumorigenic than the parental lines [22]. Spontaneous cell-cell fusion can act as a mechanism for DNA exchange between malignant and non-malignant cells and for horizontal transmission of malignancy [21, 23]. Spontaneous cell-cell fusion can be challenging to detect. Some cases are only detected incidentally—for example, when unexpected chromosomes are detected during cytogenetic analysis [24].

Intentionally created hybrid cell lines have been used for a variety of purposes, including monoclonal antibody production by mouse × mouse and mouse × human hybridomas [25], gene mapping studies [26], studies of gene expression [27], study of cancer initiation, progression, and metastasis [21, 23, 24, 28, 29], evaluation of drug resistance mechanisms [30]; as well as in the field of virology [31, 32]. Perhaps the most commonly employed inter-species hybrid cell lines, currently, are mouse × human somatic cell hybrids. Examples of intra-species cell hybrids might include mouse × mouse (inter-strain) hybrid cells [33] or hybridomas created by fusion of mouse splenic cells and mouse myeloma cells [34].

Hybrid cell lines can be challenging to authenticate and to confirm that they are valid research models. This paper reviews historical and more recent technologies that have played a role in the authentication of inter-species and intra-species hybrid cell lines. As part of cell line authentication, the identity of a cell line is expected to be established to the species level, or if possible, to the individual donor level. There are a variety of approaches that may be used for this purpose. Over the years, these have included isoenzyme, cytogenetic, and immunological analyses, and more recently, a variety of molecular methods such as restriction length fragment polymorphism (RFLP), nested PCR analysis of mitochondrial genes, short tandem repeat (STR) profiling, single nucleotide polymorphism (SNP) profiling, sequence-based human leukocyte antigen (HLA) typing, and next generation sequencing. Each of these approaches may also be applicable to the authentication of intra-species or inter-species hybrid cell lines, although the results to be expected and, therefore, the interpretation of such results in arriving at the identification of the cell line may differ from those for non-hybrid cell lines. Inter-species hybrid cell lines have a propensity to lose chromosomes during continued passage of the cultures [18]. Such loss occurs especially in the case of hybrids of human and rodent cells, such as human × mouse or human × rat hybrids. In these cases, the human chromosomes tend to be lost with continued passage of the cultures. The results expected to be obtained with several of the authentication techniques mentioned below tend, therefore, to evolve over time as the hybrid cells are cultured [35–38]. This is especially true in the case of karyotyping and total DNA content, but also may impact isoenzyme analysis and molecular-based methods.

2. Methodologies for authenticating hybrid cell lines

2.1. Isoenzyme analysis

Isoenzyme analysis was one of the first methods to be used (as early as 1970s) for determining the species-level identity of cell lines. This method is still being used [39], despite the fact that reagents for performance of the method are not commercially available. There is a considerable amount of historical data, in the public domain, for non-hybrid and hybrid cell lines, therefore, discussion of these results has relevance in deciphering hybrid cell line identities.

In isoenzyme analysis, the gel electrophoresis banding patterns and relative migration distances of intracellular enzyme isoforms are used to confirm the expected animal species of origin for test cells. Normalized migration distances obtained for the set of enzymes evaluated are compared to a set of tabular values for various animal species, and through a process of elimination, the most likely animal species of origin for the test cell is determined. Although the results of isoenzyme analysis historically have been used to confirm species-level (intraspecies) identity of a cell line, the method also can be used to demonstrate the existence of an inter-species cell mixture [40] or to authenticate inter-species hybrid cell lines [27, 31–32, 36–38, 41, 42].

When evaluating inter-species cell mixtures using isoenzyme analysis, bands migrating as expected for each of the parental species comprising the mixture are observed, provided that a sufficient percentage of cells of both species are present in the mixture [31, 38–41]. In the case of inter-species hybrid cell lines, however, a variety of possible outcomes may be obtained when authenticating using isoenzyme analysis. These outcomes might include, for instance, bands for certain enzymes that migrate as expected for both parental species or for only one of the two parental species, or bands that migrate differently than expected for either parental species (**Figure 1**).

As is evident from **Figure 1**, interpretation of an isoenzyme analysis electropherogram for a hybrid cell is not as straightforward as it is for a cell mixture. The chromosomes contributed to the hybrid cell by the two parental cells determine the outcome of the isoenzyme analysis results for any given enzyme, as the genes encoding the enzymes evaluated in this method are scattered among the various chromosomes of the various animal species [37]. In fact, isoenzyme analysis was performed commonly in early gene mapping studies because linkage between genes encoding an isoenzyme and a gene of interest could be used to assign the chromosomal location for the gene of interest. Due to uncertainty of the assortment of parental chromosomes (and encoded enzyme genes) into a hybrid cell, it is not possible to predict in advance the phenotype and, therefore, the electrophoretic characteristics of enzymes being evaluated using isoenzyme analysis. This is depicted well by the results of authentication of a series of human × bovine hybrid cell lines (**Table 1**) by van Olphen and Mittal [32].

Authenticating an intentionally created hybrid cell line using isoenzyme analysis, therefore, entails evaluation of the hybrid as soon as possible after fusion of the parental cells. The migration patterns displayed by the enzymes evaluated are then considered to be the reference pattern to be expected for the hybrid cell during subsequent authentication assays. This is similar to the case for DNA fingerprinting. When reviewing historical data of cell line

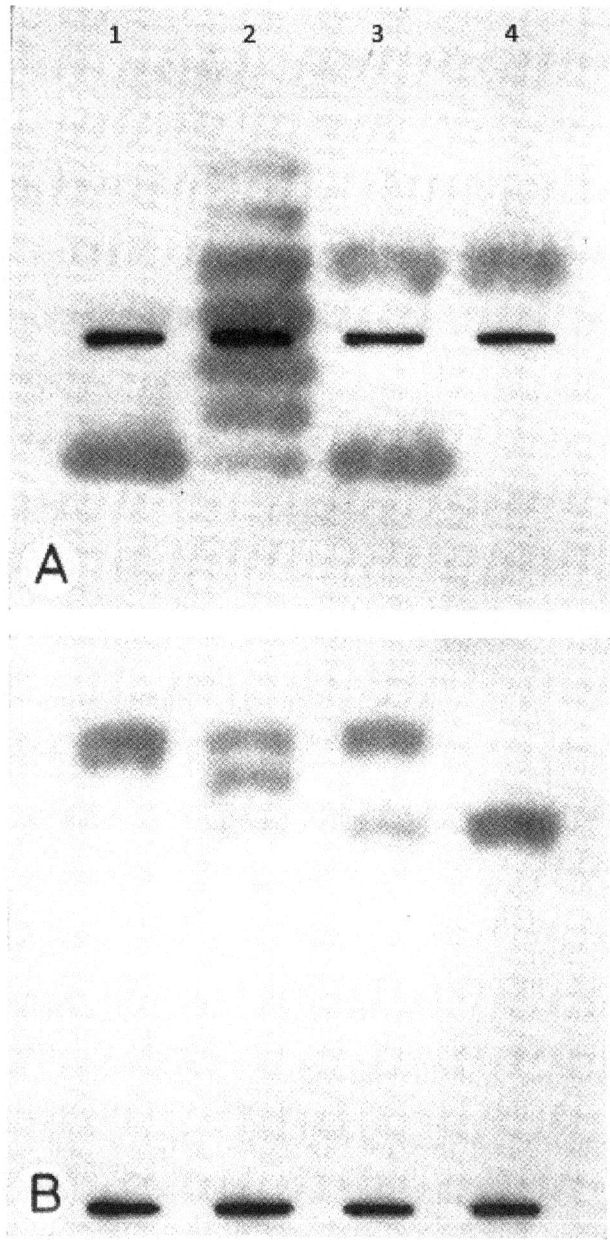

Figure 1. Isoenzyme analysis of (A) lactate dehydrogenase; and (B) 6-phosphogluconate dehydrogenase in parental cells, a cell mixture, and a hybrid cell. Lane 1: parental rat-SV40 cell; lane 2: hybrid H3 (rat-SV40 × mouse 3T3 TK^-); lane 3: mixture of rat-SV40 and mouse 3T3 TK^-; lane 4: parental mouse 3T3 TK^- (from [31]). The black lines in each lane indicate the origins (the slots in the wells into which the protein is loaded).

authentication by isoenzyme analysis, allowance must be made for loss of parental chromosomes during extended culture, as this may result in loss of electrophoretic bands associated with certain enzyme isoforms over time. Reagents for performing isoenzyme analysis are no longer commercially available, so other methods to be described below are now more commonly being used for hybrid cell authentication.

2.2. Immunostaining for surface antigens

The species-level identification of cells through use of antisera directed against species-specific cell surface markers has also been applied to the authentication of inter-species cell hybrids [32].

Method	Parental 293-Puro	Parental MDBK-Neo	Hybrid cell line		
			BHH2C	BHH3	BHH8
Immunofluorescent staining for surface antigen	Human	Bovine	Human	Bovine	Bovine
Karyotyping					
Number of human chromosomes	62	0	71	27	19
Number of bovine chromosomes	0	60	4	48	47
Total number of chromosomes[a]	62	60	97	113	103
Isoenzyme analysis					
Glucose-6-phosphate dehydrogenase	Human	Bovine	Human	Bovine +[b]	Bovine +
Malate dehydrogenase	Human	Bovine	Human	Bovine +	Bovine +
Lactate dehydrogenase	Human	Bovine	Human +[c]	Bovine +	Bovine +
Nucleoside phosphorylase	Human	Bovine	Human	Bovine	Bovine
Flow cytometry					
Total DNA content	171.1 ± 5.3	166.0 ± 4.9[d]	254.7 ± 6.8	297.7 ± 10.3	288.6 ± 7.5

[a]Total chromosomes includes human, bovine, and unidentified chromosomes.
[b]Bands expected for bovine were observed, along with extra bands.
[c]Bands expected for human were observed, along with extra bands.
[d]Mean ± standard deviation, units are relative DNA content.

Table 1. Authentication results for three hybrid (human × bovine) cell lines (data from [32]).

In the case of an inter-species cell mixture, immunostaining reagents directed against conserved surface antigens of each parental species would each be expected to demonstrate reactivity. In the case of inter-species hybrids, the result that might be obtained is not so easily predicted in advance. For instance, as shown in **Table 1**, the surface antigens that are actually detected in a set of hybrids may be derived from only one or the other of the parental species (**Figure 2**).

It is also possible, depending upon the species-specific antisera employed and the chromosomal make-up of the hybrid cell, for an inter-species hybrid cell to display surface staining for antigens of both parental species. For instance, Kano et al. [43] reported that all human × mouse hybrids evaluated in their study displayed mouse surface antigens, while most but not all also displayed human surface antigens. Surface antigens characteristic of both parental species were displayed by all human × hamster and human × rat hybrid cells evaluated. Gallagher et al. [44] reported similar results in their analysis of the surface antigens in human [HeLa] × mouse hybrid [3T3.4E] cells. Surface antigens characteristic of both parental cells were displayed by the hybrid cells.

It is also possible that by the staining of interspecies hybrid cells, one may detect a surface antigen that is not expressed by either parental cell. For instance, van Someren et al. [41]

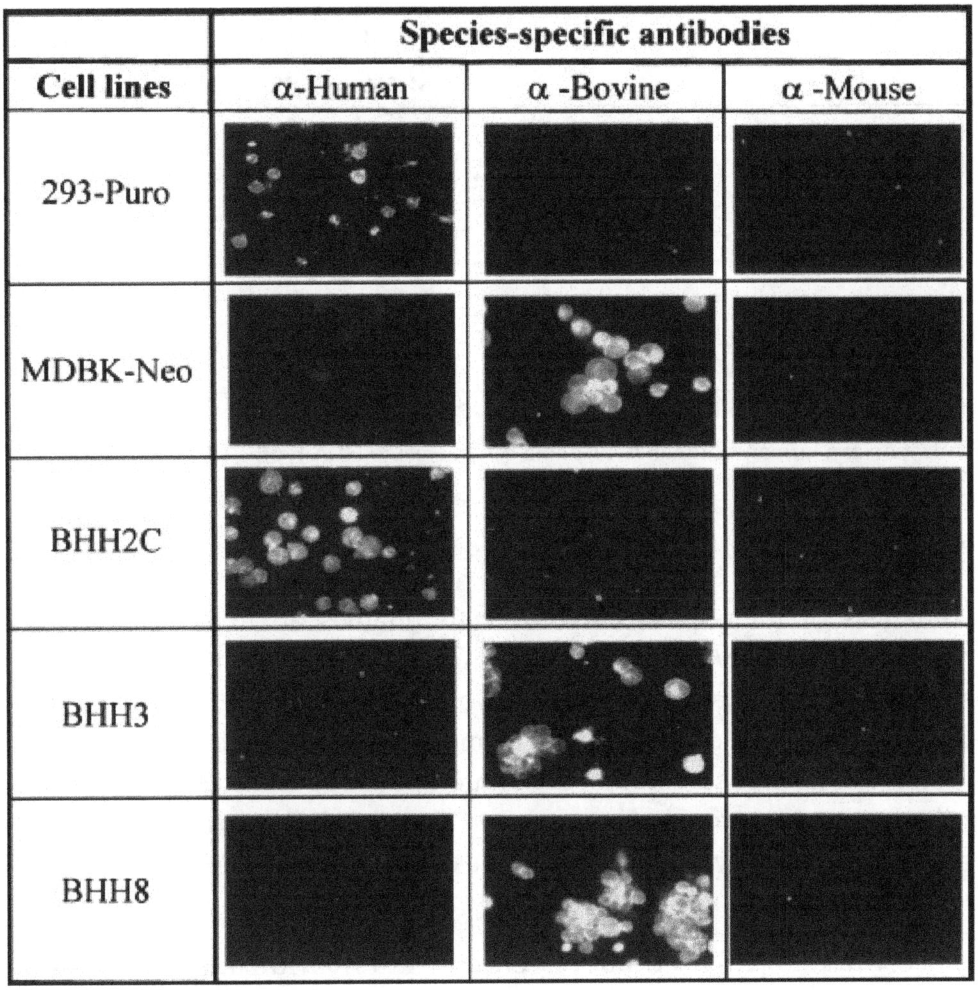

Figure 2. Use of immunostaining against species-specific surface antigens to characterize parental human [293-Puro] and bovine [MDBK-Neo] cells, and three human × bovine hybrid cell lines [BHH2C, BHH3, and BHH8]. Antisera are designed α-human (anti-human), α-bovine (anti-bovine), and α-mouse (anti-mouse). The latter was used as a negative control reagent (from [32]).

examined a large number of human × Chinese hamster hybrid cells using human leukocyte antigen (HLA) typing antisera. The parental human cells exhibited reactivity against HLA typing serum 3 only. Most of the hybrids evaluated retained reactivity against this typing serum, and a subset of these displayed reactivity against one or more additional typing sera (i.e., sera 1, 2, 9, and 10) for which the human parental cell was negative.

Immunostaining for HLA antigens may be used for detecting hybrids of parental human cells with differing HLA types, therefore conferring limited utility of this approach for detecting intra-species (human × human) cell hybrids [44, 45]. The results of HLA typing of a human × human hybridoma cell line [46] are displayed in **Table 2**.

As with isoenzyme analysis, immunostaining for surface antigens should be performed as soon as possible once a hybrid cell is created. The results may not in all cases remain the same throughout management of the cell culture over time. For instance, a loss of one or more of the parental surface antigen reactivities may coincide with loss of chromosomal material, and perhaps function, with time in culture.

Parameter	Parental cell lines		Hybrid cell line
	GM15006TGOB	EC_{EBV}	GMEC-101
Chromosomes (mean ± standard deviation)	38 ± 3	38 ± 2	84 ± 7
HLA type	A2	A1,2	A1,2
	B18	B5,17	B17
	C7	C6,7	C7
	Bw6	Bw4	Bw4,6
	DR5	DR6	DR5,6

Table 2. Authentication of a human × human hybridoma cell line using HLA typing (data from [46]).

2.3. Karyotypic (cytogenetic) analysis

Karyotypic analysis of inter-species hybrid cells enables an investigator to visualize the rearrangement and addition/deletion of chromosomes that are typically (but not always) observed in such hybrids as a result of the fusion of the two parental cells (**Tables 1** and **2**). The analysis may amount to determination of modal chromosome number and/or the range of chromosomal numbers observed in a set of metaphase spreads [38, 45]. The chromosomes comprising the karyotype may also be analyzed for morphology (telocentricity, acrocentricity, and banding pattern analysis, see **Scheme 1**), enabling assignment of chromosomes to one parental species or the other [26, 32, 36, 46, 47].

Jacobsen and co-workers [22] used fluorescence in situ hybridization (FISH) to demonstrate inter-species cell hybrids between human breast cancer and mouse stromal cells in patient-derived xenografts. The authors labeled human and mouse Cot-1 DNA (enriched in repetitive DNA sequences) with different fluorophores and used these as FISH probes. They were able to highlight the origin of individual nuclei in formalin-fixed, paraffin-embedded tissue

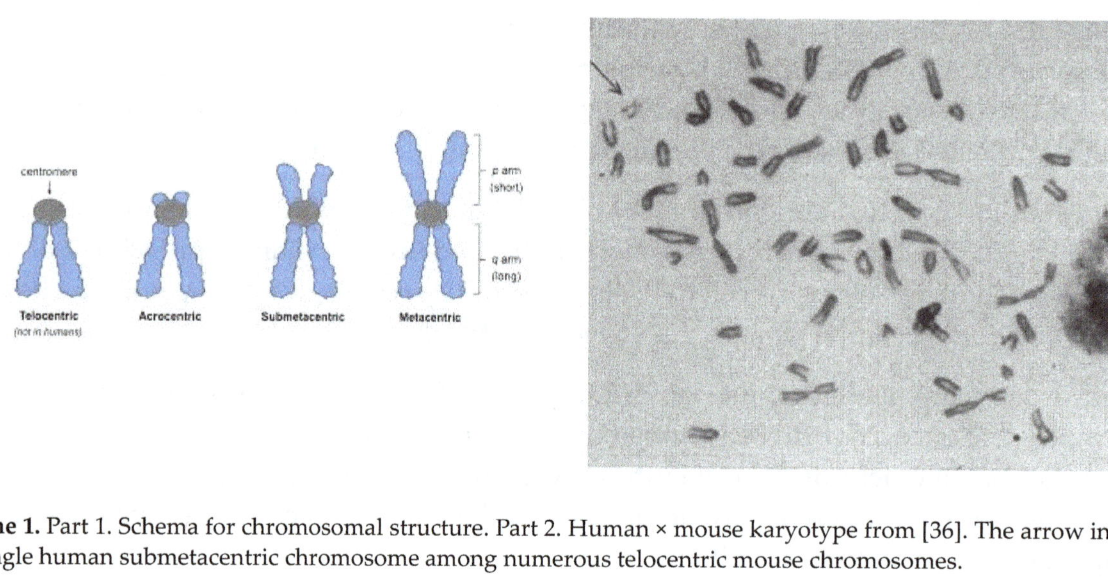

Scheme 1. Part 1. Schema for chromosomal structure. Part 2. Human × mouse karyotype from [36]. The arrow indicates the single human submetacentric chromosome among numerous telocentric mouse chromosomes.

sections, and the origin of individual chromosomes in metaphase spreads, and were able to detect hybrid chromosomes consisting of both human and mouse DNA.

At the time of creation, hybrid cells contain a complement of chromosomes, a portion of which are attributable to one of either parental cells, while some may be of unknown origin (**Table 1**). Inter-species chromosomal rearrangements may also occur in somatic cell hybrids [26]. In many, but not all cases, loss of chromosomes attributed to one or the other parental cell is experienced as the hybrid cells are cultured [26, 31, 32, 35, 36, 46].

2.4. DNA content

Measurement of total or nuclear DNA content is used in characterization of somatic cell hybrids, but is not necessarily intended as an identity test for authenticating such cells. The excess number of chromosomes present in certain cell hybrids relative to the parental cells (discussed above) is also reflected in an increase in DNA content in the hybrid cells. This increase may be detected using microspectrophotometric analysis for nuclear DNA [38] or flow cytometric analysis of propidium iodide-stained cells for total DNA [32]. For instance, while authenticating a series of human × bovine hybrid cell lines, van Olphen and Mittal [32] found that the total DNA content of the hybrid cells was 51–77% greater than the average value for the parental cells (**Table 1**).

Levels of nuclear DNA in bovine × mouse hybrid cells corresponded to the relative increases in chromosome count for the hybrids [38]. A hybrid with mean chromosomal count of 53 (vs. parental values of 44 and 44) was found to have nuclear DNA content similar to the parental mouse cell, while a hybrid with mean chromosome count of 89 displayed a bimodal nuclear DNA content, with one peak similar to that of the parental mouse cell and another peak at around twice the parental cell peak value [38].

Both total DNA and chromosome count for a hybrid cell may evolve with continued passage of a culture, due to the propensity for loss of chromosomes derived from one or both parental cells, as mentioned above.

2.5. Nucleic acid sequence-based methods

The ability to detect intra-species and inter-species cell hybrids has been greatly facilitated by the development of nucleic acid sequencing methods, such as DNA barcoding (PCR- or sequence-based approaches targeting mitochondrial genes), STR analysis, and next generation sequencing.

2.5.1. DNA barcoding

Ono et al. [48] used a nested PCR targeting the cytochrome b gene of 7 animal species (human, mouse, rat, rabbit, cat, cow, and pig) to authenticate two inter-species hybrid cell lines. These included the 4G12 hybridoma cell line (human B lymphocyte × mouse myeloma) and the N18-RE-105 hybridoma cell line (mouse glioma × rat neural retina). Cytoplasmic isoenzymes from the two hybridoma cell lines were found to display human- or rat-specific migration patterns in isoenzyme analysis, but yielded a result expected for mouse in the nested PCR (**Figure 3**). The authors concluded that the preferential retention of mouse mitochondria in

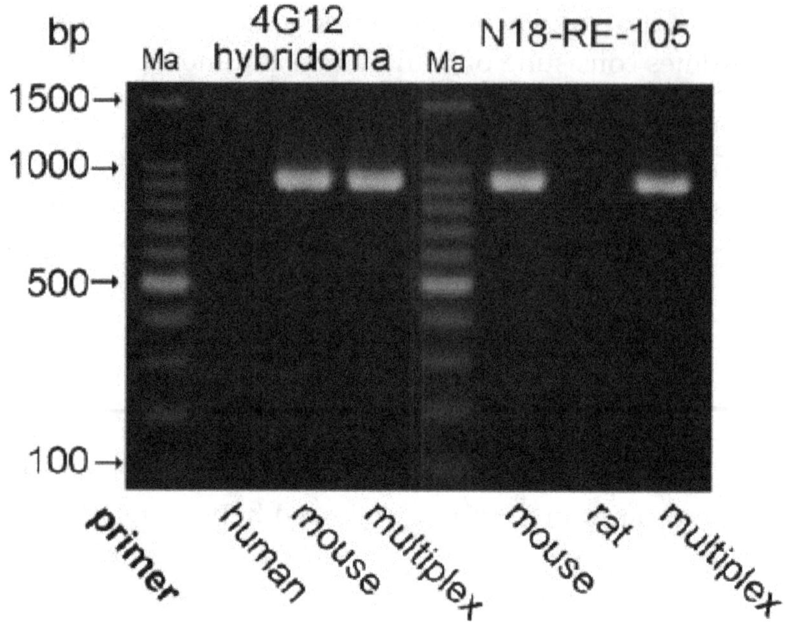

Figure 3. Use of nested PCR to evaluate a human × mouse hybrid cell line [4G12] and a mouse × rat hybrid cell line [N18-RE-105]. Multiplex group 1 (targeting human, mouse, rat, rabbit, cat, cow, and pig) or the corresponding species-specific primer pairs were used. Note that only the mouse DNA is detected in these two hybrid cells (from [48]).

hybrid cells in which one of the parental cells was mouse, as observed also by Attardi and Attardi [47], diminishes the utility of the nested PCR method for identifying inter-species cell hybrids.

2.5.2. STR profiling

Chan et al. [49] used STR profiling to identify presumed intra-species human hybrid cell lines comprised of HeLa × EBV-negative NPC (Epstein-Barr virus-negative nasopharyngeal carcinoma) cells. Four EBV-negative NPC cell lines (CNE-1, CNE-2, HNE-1, and HNE-2) were found to have STR profiles similar to each other, and also shared at least one allele with HeLa across 16 STR loci, as well as additional alleles at several of the STR loci that were attributed to an unknown EBV-negative NPC cell. High-throughput RNA sequencing by Strong et al. [50] resulted in similar conclusions and raised similar concerns for three other EBV-negative NPC cell lines (HONE-1, AdAH, and NPC-KT).

Because these particular EBV-negative NPC cell lines have additional alleles, they do not satisfy the usual match criteria for human cell line authentication [51]. However, it is important to consider all available evidence when deciding if cross-contamination has occurred. The data from Chan et al. [49] showed that CNE-1, CNE-2, HNE-1, and HNE-2 carry an allelic variant (D13S317 13.3) that is characteristic of HeLa derivatives [51]. Strong et al. [50] showed that CNE-1, CNE-2, HONE-1, AdAH, and NPC-KT carry human papillomavirus 18 (HPV18), which is an unexpected finding for NPC cell lines, and display viral and cellular genomic rearrangements that are consistent with HeLa. Looking at all the evidence, it is reasonable to conclude that these EBV-negative NPC cell lines represent not simply cross-contamination with HeLa, but rather somatic cell hybridization with HeLa.

The mechanism responsible for somatic cell hybridization in these seven EBV-negative NPC cell lines is not known. Cell-cell fusion may have occurred between HeLa and an unknown NPC cell line through exposure to Sendai virus. NPC-KT, one of the cell lines investigated by Strong et al. [50], was established by using Sendai virus to fuse AdAH and primary NPC cells [52]. NPC-KT also carried EBV, which can cause cell–cell fusion in monolayer cultures [5, 53]. If the originating laboratory unknowingly performed this work on a culture that was cross-contaminated with HeLa, it may have resulted in a HeLa fusion cell line, which may have subsequently cross-contaminated other cell lines used by the NPC research community. HeLa cells also contain the gene for HPV18 viral protein E5, which is fusogenic [54]. The E5 protein of HPV16 is a fusogenic membrane protein and if expressed in two cells, the cells can fuse [55, 56]. So if HeLa and another cell line expressed the HPV18 analogue protein E5, this could have also induced fusion.

Yoshino et al. [33] used an STR profiling approach to authenticate a series of mouse cell lines, including an inter-strain hybrid (Balb/c mouse × C3H/He mouse) cell line. In this approach, F1 hybrid cells derived from the two parental mouse strains displayed different alleles at each locus, corresponding to the alleles contributed by the two parental strains. Loss of heterozygosity occurring during extended culture was thought to result in loss of one of the parental alleles at the D5 Mit201.1 locus of the dinucleotide STR marker (**Table 3**).

Almeida et al. [34] described results obtained during authentication of an intra-strain (Balb/c mouse) hybridoma (P3X63Ag8.653 × Balb/c mouse splenic cell). In this case, not unexpectedly, identical alleles were detected for eight of nine STR loci evaluated. At the mouse STR 9-2 locus, the parental P3X63Ag8.653 cell was heterozygous, displaying alleles with 15 and 16 repeats, while the hybrid cell contained only the allele with 15 repeats.

Authentication of hybridoma cell lines is difficult, because of the inbred rodent populations that are used for hybridoma generation. Koren et al. [57] proposed a unique solution using degenerate primers to amplify and sequence the variable regions of the monoclonal antibodies produced by their hybridoma cell lines. Because these regions are highly diverse, they can potentially be used to uniquely identify the hybridoma cell line from which a monoclonal antibody is generated. Koren et al. [57] used this approach to resolve a misidentified cell line in their own laboratory, but the method would be useful for any laboratory working with hybridoma cell lines.

Strain/cell line	Size of amplicons for alleles at STR loci					
	D1 Mit159.1	D2 Mit395.1	D4 Mit170.1	D5 Mit201.1	D13 Mit256.1	D17 Mit51.1
Parental Balb/c	141.8	135.5	242.5	94.9	88.3	155.1
Parental C3H/He	185.1	123.8	236.4	92.5	78.4	140.1
Hybrid UV.CC3.11.1	142.0, 185.2	135.5, 123.9	242.6, 236.4	94.8	88.4, 78.5	155.2, 140.2

Table 3. Authentication of an inter-strain hybrid (mouse) cell line by STR profiling (data from [33]).

2.5.3. Next generation sequencing

Inter-species and intra-species cell fusion may be detectable by next generation sequencing techniques because of the extensive amount of DNA sequencing and the unbiased selection of the DNA segments (i.e., not using species-specific primers for PCR or selection of DNA fragments). A difficulty may occur in genomic regions that are highly conserved between species for which only small sequence differences exist (e.g., a single or a few bases). In such cases, it may be difficult to determine whether the observed difference is a single nucleotide variation (SNV) between two species or between two samples from the same species. Also, SNPs/SNVs are generally transitional sequence changes (i.e., either purine to purine or pyrimidine to pyrimidine) and may not provide sufficient information to determine whether a sample contains cells from two different species or from two different individuals of the same species. To overcome this difficulty, one must sequence DNA segments that are highly variable and unique to different individuals. These might include human, mouse [34], or rat [58, 59] STR and human [60, 61] or mouse SNP arrays [62, 63]. Multiple genomic regions must be evaluated for those cases in which a few or even a single chromosome from one species is retained by the hybrid cell, as is the case with many hybridoma cell lines.

3. Discussion

Hybrid cell lines represent a special problem for the various approaches that have been utilized for authentication up to now. Firstly, the endpoints that are used in cell authentication assays are ultimately, if not directly, dependent upon the genetic make-up of the cell. Intra-species and inter-species hybrid cells are difficult to test because they contain an assortment of genetic material conferred from the two parental cell types during the fusion process. Thus, specific isoforms of enzymes, the presence or absence of surface antigens, chromosome count, and total DNA content are each subject to the assortment of genetic material that is present in the hybrid cell line following the fusion process. This difficulty also applies to the molecular-based methods that are so useful for determining the authenticity of cells. Thus, one cannot predict, in advance, the results that will be obtained during authentication of a hybrid cell using one of these analytical techniques.

Secondly, not all of the genetic material in the hybrid cell is stable, as it is not uncommon for one or more chromosomes to be lost from hybrid cells on continued passage of the culture. This means that the authentication profile of a hybrid cell may evolve over time in culture.

Due to these considerations, a hybrid cell should be evaluated as soon as possible after fusion to get a baseline (reference) profile. Any drift or change in subsequent profiles may imply changes within the chromosome number or expression profiles. The profile of authentication resulting from one or more of these methods can then be used as a sort of fingerprint or reference against which subsequent authentication can be compared (as in authentication of a master cell bank or a working cell bank). Evolution of such a reference pattern may occur as chromosomes are lost, sometimes quite soon following fusion, from a hybrid during continued culture. This evolution typically involves the loss of characteristics of one of the parental cells (be it parental chromosome(s), species-specific isoenzyme bands or hybrid bands, loss of heterozygosity in SNP or STR profiles, loss of surface staining characteristic of one of the parental cells, etc.). On the other hand, gains in chromosome counts, addition of new isoenzyme bands, appearance of new alleles in SNP or STR profiles, or increases in DNA content with time in culture would not be expected, and such would be considered a red flag during authentication. For instance, such a result might indicate the presence of a cross-contaminating cell type.

4. Executive summary

- Fusion of two different cell types to create an inter-species or intra-species hybrid results in an unpredictable assortment of genetic material derived from one or the other parental cell into the hybrid.
- The outcome of the fusion process in terms of genetic content contributed by the two parental cells will impact the results of the methodologies typically used for determining cell line authenticity.
- Isoenzyme analysis may indicate the presence of electrophoretic bands migrating as expected for one parental cell, the other parental cell, or bands migrating differently than expected for either parental cell.
- Hybrid cells may retain surface antigens characteristic of one parental cell or the other, or of both.
- Chromosome number (total, and number derived from one parental cell or the other) and total DNA content will vary from hybrid to hybrid.
- Human chromosomes found in inter-species hybrid cells tend to be unstable, and are often lost over time in culture.
- Regardless of the authentication method to be used, it is recommended that a baseline evaluation be performed as soon as possible after the fusion process used to create the hybrid cell, and that the result be used as a reference against which future authentication results may be compared.

5. Future perspectives

Hybrid cell lines require new methods to ensure that cell-cell fusion is detected and such cultures can be authenticated to demonstrate their validity as research models. Detection of spontaneous cell-cell fusion is particularly important. Somatic cell hybridization may arise when using feeder layers in vitro or when working with patient-derived xenograft models in vivo. In many cases, cell–cell fusion is associated with increased tumorigenicity or the development of malignant behavior in adjacent cell populations [21, 22]. This has the potential to alter the behavior of patient-derived xenograft models, which may be used as the final step before a novel drug proceeds to clinical evaluation [64].

Current authentication methods are not always effective to detect hybrid cell lines. The advantages and limitations of the available methods are displayed in **Table 4**. Common to all methods is the fact that the chromosomal contributions from each parental cell are not possible to predict in advance. This means that chromosomal makeup and number and corresponding genetic information (e.g., surface antigens, HLA types, enzyme isoforms, alleles at STR loci) will be unique to each fusion cell. In hybrid cells, STR profiles become more

Method	Principle	Advantages	Disadvantages
Isoenzyme analysis	Electrophoretic mobilities of cytosolic enzymes in the fusion cell correspond to one or both parental cell isoforms	Rapid; inexpensive; visual endpoint; useful for inter-species hybrids	Reagents no longer commercially available; results must be compared to a reference result; certain isoforms may be lost with continued passage of the hybrid cells
Immunostaining	Surface antigens of one or both parental cells may be retained	Rapid, visual endpoint; useful for inter-species hybrids; HLA immunostaining enables detection of intra-species (human) fusion cells	Requires species- or HLA type-specific immunostaining reagents and fluorescent microscopy; results must be compared to a reference result; certain surface antigens may be lost with continued passage of the hybrid cells
Karyotyping (cytogenetic analysis)	Chromosomes contributed by two parental cells may be directly observed in the fusion cell	Visual endpoint; useful for inter-species hybrids; can directly determine chromosomal makeup derived from the parental cells; useful for inter-species hybrids	Preparing and interpreting karyotypes takes expertise; human chromosomes in inter-species hybrids are often lost with passage in culture
DNA content	Fusion cells may contain more total DNA content than either parental cell	Total DNA content is simple to measure; DNA content is useful in detecting intra- as well as inter-species hybrid cells	Total DNA content in hybrid cells is often not stable with passage in culture as chromosomes may be lost
DNA barcoding	Mitochondrial DNA sequences are conserved within species	Rapid; DNA barcoding is used for species of origin confirmation of cells	Nested PCR must be created with species-specific primers for parental cells; preferential retention of mouse mitochondria in inter-species bybrid cells diminishes utility of this method
STR profiling	Multiple (8 or more) STR loci provide identity to the donor level	Rapid; STR profiling can enable authentication of human, dog, or mouse cells to the individual donor level; useful for authenticating mouse hybridoma cells	Not typically useful for inter-species hybrid cells; loss of heterozygosity during extended passage in culture may result in loss of STR alleles
Next generation sequencing	Agnostic sequencing of variable DNA sequences such as STR or SNP arrays	Detects inter- and intra-species hybrid cells	Multiple genomic regions must be evaluated in hybrids containing few or single chromosomes from one of the parental cells

Table 4. Limitations and advantages of methods for authenticating hybrid cells.

complex and difficult to interpret, while mitochondrial-based methods may not be effective for species detection if mitochondria from one species are retained preferentially (as is true for the mouse). However, many methods can be optimized to allow for detection of cell-cell hybrids. For example, SNP genotyping is increasingly used for cell line authentication and has been used as a test method for patient-derived xenograft models [64]. SNP panels could be modified to include species-specific marker sets, or human and mouse SNP panels could be run in parallel to confirm species and strain identifications and search for additional markers. Although this type of comprehensive assessment is not usually performed, it could be incorporated into testing pipelines if laboratories are aware of the spontaneous (unintentional) cell fusion issue and look specifically for such markers of cell fusion. A role has been suggested for cell-cell fusion in cancer, stem cell plasticity, and trans-differentiation [17–20, 54, 59, 65–67]. A better set of tools is needed to explore hybrid cell lines and the role of somatic cell hybridization in health and disease.

Abbreviations

DNA	deoxyribonucleic acid
EBV	Epstein-Barr virus
FISH	fluorescence in situ hybridization
HHV-8/KSHV	herpesviruses-8/Kaposi sarcoma herpesvirus
HLA	human leukocyte antigen
HPV	human papillomavirus
HTLV-1	human T-cell lymphotropic virus type 1
NPC	nasopharyngeal carcinoma
PCR	polymerase chain reaction
RFLP	restriction length fragment polymorphism
RNA	ribonucleic acid
SNP	single nucleotide polymorphism
SNV	single nucleotide variation
STR	short tandem repeat
SV40	simian virus 40

Author details

Raymond W. Nims[1]*, Amanda Capes-Davis[2], Christopher Korch[3] and Yvonne A. Reid[4]

*Address all correspondence to: rnims@rmcpharma.com

1 RMC Pharmaceutical Solutions, Inc., Longmont, CO, USA

2 CellBank Australia, Children's Medical Research Institute, The University of Sydney, Westmead, New South Wales, Australia

3 Division of Medical Oncology, University of Colorado Denver, Aurora, CO, USA

4 ATCC, Manassas, VA, USA

References

[1] Cell Fusion. Encyclopedia of Genetics, Genomics, Proteomics and Informatics. Dordrecht: Springer; 2008. DOI: 10.1007/978-1-4020-6754-9_2621

[2] Somatic Cell Hybrids. Encyclopedia of Genetics, Genomics, Proteomics and Informatics. Dordrecht: Springer; 2008. DOI: 10.1007/978-1-4020-6754-9_15854

[3] Ephrussi B, Weiss MC. Interspecific hybridization of somatic cells. Proceedings of the National Academy of Science, USA. 1965;**53**:1040-1042

[4] Gao P, Zheng J. Oncogenic virus-mediated cell fusion: New insights into initiation and progression of oncogenic viruses–related cancers. Cancer Letters. 2011;**303**:1-8

[5] Hussain T, Mulherkar R. Lymphoblastoid cell lines: A continuous in vitro source of cells to study carcinogen sensitivity and DNA repair. International Journal of Molecular and Cellular Medicine. 2012;**1**:75-87

[6] Pedrazzoli F, Chrysantzas I, Dezzani L, Rosti V, Vincitorio M, Sitar G. Cell fusion in tumor progression: The isolation of cell fusion products by physical methods. Cancer Cell International. 2011;**11**:32. https://doi.org/10.1186/1475-2867-11-32

[7] Scheurich P, Zimmermann U, Mischel M, Lamprecht I. Membrane fusion and deformation of red blood cells by electric fields. Zeitschrift für Naturforschung C. 1980;**35**:1081-1085

[8] Pilwat G, Richter H-P, Zimmermann U. Giant culture cells by electric field-induced fusion. FEBS Letters. 1981;**133**:169-174

[9] Zimmermann U. Electric field-mediated fusion and related electrical phenomena. Biochimica et Biophysica Acta. 1982;**694**:227-277

[10] Zimmermann U, Vienken J. Electric field-induced cell-to-cell fusion. Journal of Membrane Biology. 1982;**67**:165-182

[11] Zimmermann U, Vienken J, Pilwat G, Arnold WM. Electro-fusion of cells: Principles and potential for the future. Ciba Foundation Symposia. 1984;**103**:60-85

[12] Zimmermann U, Urnovitz HB. Principles of electrofusion and electropermeabilization. Methods in Enzymology. 1987;**151**:194-221

[13] Trontelj K, Rebersek M, Kanduser M, Serbec VC, Sprohar M, Miklavcic D. Optimization of bulk cell electrofusion in vitro for production of human-mouse heterohybridoma cells. Bioelectrochemistry. 2008;**74**:124-129

[14] Hock BD, Roberts G, McKenzie JL, Gokhale P, Salm N, McLellan AD, et al. Exposure to the electrofusion process can increase the immunogenicity of human cells. Cancer Immunology and Immunotherapy. 2005;**54**:880-890

[15] Schoeman RM, van den Beld WTE, Kemna EWM, Wolbers F, Eijkel JCT, van den Berg A. Electrofusion of single cells in picoliter droplets. Scientific Reports. 2018;**8**:3714. DOI: 10.1038/s41598-018-21993-8

[16] Willkomm L, Bloch W. State of the art in cell-cell fusion. Methods in Molecular Biology. 2015;**1313**:1-19

[17] Bastida-Ruiz D, Van Hoesen K, Cohen M. The dark side of cell fusion. International Journal of Molecular Science. 2016;**17**(pii):E638. DOI: 10.3390/ijms17050638

[18] Zhou X, Merchak K, Lee W, Grande JP, Cascalho M, Platt JL. Cell fusion connects oncogenesis with tumor evolution. American Journal of Pathology. 2015;**185**:2049-2060

[19] Duelli D, Lazebnik Y. Cell fusion: A hidden enemy? Cancer Cell. 2003;**3**:445-448

[20] Duelli D, Lazebnik Y. Cell-to-cell fusion as a link between viruses and cancer. Nature Reviews in Cancer. 2007;**7**:968-976

[21] Goldenberg DM. Horizontal transmission of malignancy by cell-cell fusion. Expert Opinions in Biological Therapy. 2012;**12**:S133-S139

[22] Jacobsen BM, Harrell JC, Jedlicka P, Borges VF, Varella-Garcia M, Horwitz KB. Spontaneous fusion with, and transformation of mouse stroma by, malignant human breast cancer epithelium. Cancer Research. 2006;**66**:8274-8279

[23] Searles SC, Santosa EK, Bui JD. Cell-cell fusion as a mechanism of DNA exchange in cancer. Oncotarget. 2017;**9**:6156-6173

[24] Camps J, Salaverria I, Garcia MJ, Prat E, Beá S, Pole JC, et al. Genomic imbalances and patterns of karyotypic variability in mantle-cell lymphoma cell lines. Leukemia Research. 2006;**30**:923-934

[25] Köhler G, Milstein C. Derivation of specific antibody-producing tissue culture and tumor lines by cell fusion. European Journal of Immunology. 1976;**6**:511-519

[26] Raimondi E, Mezzelani A, Castiglioni B, Moralli D, Rognoni G, De Carli L. Characterization of hamster-bovine somatic cell hybrids by in situ hybridization and chromosome banding. Hereditas. 1993;**118**:191-194

[27] Melmed S, Fagin JA. Isolation and characterization of rat-mouse somatic cell hybrids secreting growth hormone and prolactin. Experimental Cell Research. 1986;**162**:475-485

[28] Gauck D, Keil S, Niggemann B, Zänker KS, Dittmar T. Hybrid clone cells derived from human breast epithelial cells and human breast cancer cells exhibit properties of cancer stem/initiating cells. BMC Cancer. 2017;**17**:515. DOI: 10.1186/s12885-017-3509-9

[29] Ramakrishnan M, Mathur SR, Mukhopadhyay A. Fusion-derived epithelial cancer cells express hematopoietic markers and contribute to stem cell and migratory phenotype in ovarian carcinoma. Cancer Research. 2013;**73**:5360-5370

[30] Giavazzi R, Scholar E, Hart IR. Isolation and preliminary characterization of an adriamicyn-resistant murine fibrosarcoma cell line. Cancer Research. 1983;**43**:2216-2222

[31] van der Noordaa J, van Haagen A, Walboomers JMM, van Someren H. Properties of somatic cell hybrids between mouse cells and simian virus 40-transformed rat cells. Journal of Virology. 1972;**10**:67-72

[32] van Olphen AL, Mittal SK. Development and characterization of bovine × human hybrid cell lines that efficiently support the replication of both wild-type bovine and human adenoviruses and those with E1 deleted. Journal of Virology. 2002;**76**:5882-5892

[33] Yoshino K, Saijo K, Noro C, Nakamura Y. Development of a simple method to determine the mouse strain from which cultured cells originated. Interdisciplinary Biology Central. 2010;**2**:1-9. DOI: 10.4051/ibc.2010.2.4.0014

[34] Almeida JL, Hill CR, Cole KD. Mouse cell line authentication. Cytotechnology. 2014; **66**:133-147

[35] Weiss MC, Green H. Human-mouse hybrid cell lines containing partial complements of human chromosomes and functioning human genes. Proceedings of the National Academy of Science, USA. 1967;**58**:1104-1111

[36] Migeon BR, Miller CS. Human-mouse somatic cell hybrids with single human chromosome (group E): Link with thymidine kinase activity. Science. 1968;**162**:1005-1006

[37] Nichols EA, Ruddle FH. A review of enzyme polymorphisms, linkage and electrophoretic conditions for mouse and somatic cell hybrids in starch gels. Journal of Histochemistry and Cytochemistry. 1973;**21**:1066-1081

[38] Chinchar VG, Floyd AD, Chinchar GD, Taylor MW. Characterization of hybrids between bovine (MDBK) and mouse (L-cells) cell lines. Biochemical Genetics. 1979;**17**:133-148

[39] Machado JZ, de Miranda ACS, Cruz AS. G6PD and MDH isoenzyme electrophoresis for cell lines authentication. Virus Reviews and Research. 2016;**21**:1-4

[40] Nims RW. Historical and current methods for detecting inter- and intra-species cell mixtures and thereby assuring cell line purity. In: Tsongalis GJ, editor. Advances in Molecular Pathology. Vol. 1. 2018. In press

[41] van Someren H, Westerveld A, Hagemeijer A, Mees JR, Meera Khan PM. Human antigen and enzyme markers in man-Chinese hamster somatic cell hybrids: Evidence for

synteny between the *HL-A*, *PGM*$_3$, *ME*$_1$, and *IPO-B* loci. Proceedings of the National Academy of Science, USA. 1974;**71**:962-965

[42] Croce CM, Talavera A, Basilico C, Miller OJ. Suppression of production of mouse 28S ribosomal RNA in mouse-human hybrids segregating mouse chromosomes. Proceedings of the National Academy of Science, USA. 1977;**74**:694-697

[43] Kano K, Baranska W, Knowles BB, Koprowski H, Milgrom F. Surface antigens on interspecies hybrid cells. Journal of Immunology. 1969;**103**:1050-1060

[44] Gallagher T, Chardonnet Y, Viac J, Patet J, Lefèvre A. Antigenic immunostaining patterns in somatic hybrids of human HeLa cells and mouse fibroblasts 3T3.4E propagated in conventional medium and delipidized serum. Virchows Archiv Part B Cell Pathology Including Molecular Pathology. 1989;**57**:339-347

[45] Kozbor D, Burioni R, Ar-Rushdi A, Zmijewski C, Croce CM. Expression of members of the immunoglobin gene family in somatic cell hybrids between human B and T cells. Proceedings of the National Academy of Science, USA. 1987;**84**:4969-4973

[46] Hulette CM, Effros RB, Dillard LC, Walford RL. Production of a human monoclonal antibody to HLA by human-human hybridoma technology. American Journal of Pathology. 1985;**121**:10-14

[47] Attardi B, Attardi G. Fate of mitochondrial DNA in human-mouse somatic cell hybrids. Proceedings of the National Academy of Science, USA. 1972;**69**:129-133

[48] Ono K, Satoh M, Yoshida T, Ozawa Y, Kohara A, Takeuchi M, et al. Species identification of animal cells by nested PCR targeted to mitochondrial DNA. In Vitro Cellular and Developmental Biology—Animal. 2007;**43**:168-175

[49] Chan SY-Y, Choy KW, Tsao S-W, Tao Q, Tang T, Chung GT-Y, et al. Authentication of nasopharyngeal carcinoma tumor lines. International Journal of Cancer. 2008;**122**:2169-2171

[50] Strong MJ, Badoo M, Nanbo A, Xu M, Puetter A, Lin Z. Comprehensive high-throughput RNA sequencing analysis reveals contamination of multiple nasopharyngeal carcinoma cell lines with HeLa cell genomes. Journal of Virology. 2014;**88**:10696-10704

[51] ANSI/ATCC ASN-0002-2011. Authentication of human cell lines: Standardization of STR profiling. In: https://webstore.ansi.org/RecordDetail.aspx?sku=ANSI%2FATCC+ASN-0002-2011

[52] Takimoto T, Furukawa M, Hatano M, Umeda R. Epstein-Barr virus nuclear antigen-positive nasopharyngeal hybrid cells. The Annals of Otology, Rhinology, and Laryngology. 1984;**93**:166-169

[53] Bayliss GJ, Wolf H. Epstein-Barr virus-induced cell fusion. Nature. 1980;**287**:164-165

[54] Gao P, Zheng J. High-risk HPV E5-induced cell fusion: A critical initiating event in the early stage of HPV-associated cervical cancer. Virology Journal. 2010;**7**:238. DOI: 10.1186/1743-422X-7-238

[55] Hu L, Plafker K, Vorozhko V, Zuna RE, Hanigan MH, Gorbsky GJ, et al. Human papillomavirus 16 E5 induces bi-nucleated cell formation by cell-cell fusion. Virology. 2008;**384**:125-134

[56] Hu L, Ceresa BP. Characterization of the plasma membrane localization and orientation of HPV16 E5 for cell-cell fusion. Virology. 2009;**393**:135-143

[57] Koren S, Kosmac M, Colja Venturini A, Montanic S, Curin Serbec V. Antibody variable-region sequencing as a method for hybridoma cell-line authentication. Applied Microbiology and Biotechnology. 2008;**78**:1071-1078

[58] Bryda EC, Riley LK. Multiplex microsatellite marker panels for genetic monitoring of common rat strains. Journal of the American Association for Laboratory Animal Science. 2008;**47**:37-41

[59] Shore SK, Bacheler LT, de Riel JK, Barrows LR, Lynch M. Cloning and characterization of a rat-specific repetitive DNA sequence. Gene. 1986;**45**:87-93

[60] Huang J, Wei W, Zhang J, Liu G, Bignell GR, Stratton MR, et al. Whole genome DNA copy number changes identified by high density oligonucleotide arrays. Human Genomics. 2004;**1**:287-299

[61] Sebat J, Lakshmi B, Troge J, Alexander J, Young J, Lundin P, et al. Large-scale copy number polymorphism in the human genome. Science. 2004;**305**:525-528

[62] Didion JP, Buus RJ, Naghashfar Z, Threadgill DW, Morse HC 3rd, de Villena FP. SNP array profiling of mouse cell lines identifies their strains of origin and reveals cross-contamination and widespread aneuploidy. BMC Genomics. 2014;**15**:847. DOI: 10.1186/1471-2164-15-847

[63] Morgan AP, Fu C-P, Kao C-Y, Welsh CE, Didion JP, Yadgary L, et al. The mouse universal genotyping array: From substrains to subspecies. G3: Genes, Genomes, Genetics. 2015;**6**:263-279

[64] El-Hoss J, Jing D, Evans K, Toscan C, Xie J, Lee H, et al. A single nucleotide polymorphism genotyping platform for the authentication of patient derived xenografts. Oncotarget. 2016;**737**:60475-60490

[65] Eisenberg LM, Eisenberg CA. Stem cell plasticity, cell fusion, and transdifferentiation. Birth Defects Research, Part C Embryo Today: Reviews. 2003;**69**:209-218

[66] Holland H, Livrea M, Ahnert P, Koschny R, Kirsten H, Meixensberger J, et al. Intracranial hemangiopericytoma: Case study with cytogenetics and genome wide SNP-A analysis. Pathology, Research and Practice. 2011;**207**:310-316

[67] Liu C, Lin J, Li L, Zhang Y, Chen W, Cao Z, et al. HPV16 early gene E5 specifically reduces miRNA-196a in cervical cancer cells. Scientific Reports. 2015;**5**:7653. DOI: 10.1038/srep07653

4

Two-Dimensional (2D) and Three-Dimensional (3D) Cell Culturing in Drug Discovery

Jitcy Saji Joseph, Sibusiso Tebogo Malindisa and Monde Ntwasa

Abstract

Cell culture is an indispensable *in vitro* tool used to improve our perception and understanding of cell biology, the development of tissue engineering, tissue morphology, mechanisms of diseases and drug action. Efficient cell culturing techniques both *in vitro* and *in vivo* allow researchers to design and develop new drugs in preclinical studies. Two-dimensional (2D) cell cultures have been used since 1900s and are still a dominant method in many biological studies. However, 2D cell cultures poorly imitate the conditions *in vivo*. Recently three-dimensional (3D) cell cultures have received remarkable attention in studies such as drug discovery and development. Optimization of cell culture conditions is very critical in ensuring powerful experimental reproducibility, which may help to find new therapies for cancer and other diseases. In this chapter, we discuss the 2D and 3D cell culture technologies and their role in drug discovery.

Keywords: 2D cell culture, 3D cell culture, drug discovery, cell-based assays

1. Introduction

The discovery and development of new drugs is a very lengthy and costly process. The cost of developing a new drug and bringing it to the market is between $800 million and $2 billion, and can take up to 15 years. In part, termination of the development process is due to failure at late preclinical stages of development at great expenditure [1]. The drug discovery and development process for new drugs consists of four phases; drug discovery, preclinical development, clinical development and regulatory approval. Most drugs fail at phase II and phase III clinical stages due to poor efficacy and safety issues [2]. The high attenuation rates in drug discovery suggest that the main reasons for drug failure are inappropriate preclinical testing methods and *in vitro* models, which do not sufficiently produce information needed for prediction of drug efficacy and safety issues [3]. Hence, one of the main areas expected to improve the success rate of drug development process could be the use of new technologies in preclinical testing and *in vitro* models, in order to get better accurate data.

Cell-based assays are crucial in the drug discovery and development process. Mammalian cell culture provides a defined platform for investigating cell and tissue physiology and pathophysiology outside of the organism. For over a century, traditional 2D cell culture was used in drug discovery. In 2D cell culture, cells are grown on flat dishes optimized for cell attachment and growth (**Figure 1**). Nowadays, 2D cell culture models are still used to test cellular drug responses to drug candidates. Although 2D cell culture is generally accepted and has increased understanding of drug mechanisms of action, there are limitations associated with it. The main limitation is that the cells grown as a monolayer on flat petri plates or flasks. This is a stiff platform, offering unnatural growth kinetics and cell attachments. Therefore, natural microenvironments of the cells are not fully represented [4]. Recently, significant work by researchers produced improvements in the form of better *in vitro* cell culture models that resemble *in vivo* conditions. Three-dimensional cell cultures are such products and better mimic tissue physiology in multicellular organisms (**Figure 1**) [5].

While traditional monolayer cultures still are predominant in cellular assays used for high-throughput screening (HTS), 3D cell cultures techniques for applications in drug discovery are making rapid progress [6, 7]. In this chapter, we provide an overview of 2D and 3D cell culture techniques, and their role in the discovery of new drugs.

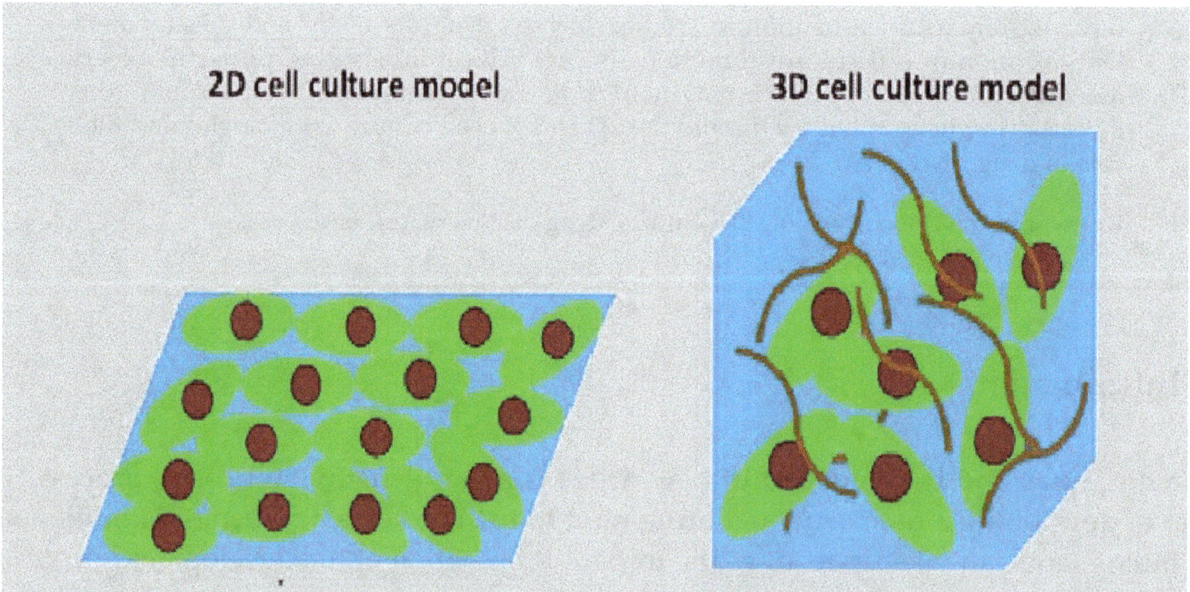

Figure 1. Simplified sketch of 2D and 3D cell culture.

2. Cell culture system

Cell culture involves the dispersal of cells in an artificial environment that is composed of an appropriate surface, nutrient supply, and optimal conditions of humidity, temperature and gaseous atmosphere [6]. Usually cells are grown for days or weeks in a sterile 37°C humidified incubator with 5% CO_2 until a sufficient number of cells are reached. This system allows the study of cellular response to different environmental cues such as physiological stimulants or agonists/antagonists, potential drugs or pathogens.

2.1. Two-dimensional (2D) cell culture system

Two-dimensional culture conditions vary widely for each cell type. Appropriate cell culture medium suitable for the growth of particular cells has to be used. Various laboratories use different recipes of cell culture media prepared in the laboratory or commercially produced. The commercially produced cell culture medium is obtained sterile and ready to use in liquid or powder form and is usually dissolved in sterile water. Most laboratories obtain commercial components, which are mixed in the lab to make a complete culture medium for optimal cell growth. In addition, the culture media are usually supplemented with antibiotics and/or fungicides to inhibit contamination (**Table 1**).

Many continuous mammalian cell lines can be maintained on a relatively simple medium such as MEM supplemented with serum and antibiotics. However, most laboratories use DMEM as mammalian cells can be easily grown in DMEM supplemented with serum as well as antibiotics. When working with specialized cell types, a specialized cell culture medium may be required to maintain the growth of cells such as RPMI-1640 medium that is mostly used to grow cells in suspension such as HL-60 (promyelocytic leukemia) with varying serum amounts.

2.1.1. Sub-culturing cells

As cells reach confluency, they must be sub-cultured or passaged. The first step in sub-culturing adherent cells is to detach them from the cell culture plate or flask. This is done by subjecting them to trypsin-EDTA or by physically scraping them off the plate using a sterile cell scraper. One must take care because some mechanical and chemical methods have the

	Adherent cells		**Non-adherent cell lines**
	Cancer cell lines	**Non-cancerous cell lines**	
Cell culture medium	89% DMEM or MEM with high glucose, L-glutamine + 10% FBS + 1% penicillin/streptomycin	89% DMEM or MEM with medium/low glucose, L-glutamine + 10% FBS + 1% penicillin/streptomycin	89% RPMI-1640 + 10% FBS + 1% penicillin/streptomycin

DMEM, Dulbecco's Modified Eagles Medium; MEM, minimum essential medium; RPMI, Roswell Park Memorial Institute; FBS, fetal bovine serum.

Table 1. Common 2D cell culture media recipes.

potential to damage the cellular structure and possibly kill cells. Once detached, pre-warmed medium is added to stop the activity of trypsin-EDTA or to dilute the cell suspension. Varying amounts of the cell suspension are then transferred into fresh culture vessels and the appropriated amount of pre-warmed medium added and further incubated in 37°C incubator with humidified atmosphere of 5% CO_2.

2.1.2. Two-dimensional cell cultures in drug discovery and development

Many types of *in vitro* assays are performed in Drug Discovery and Development Research (DDDR), however, use of cell cultures receives extensive use. For example, determination of drug absorption, distribution, metabolism, excretion and toxicity (ADMETox) or drug pharmacokinetics is initially assessed in *in vitro* experiments involving cell cultures. Various cell lines in 2D cultures are used to determine different aspects of ADMETox. For instance, the Human colon carcinoma cells (Caco-2) are commonly used to determine absorption of drug candidates. Cultured Caco-2 cells form tight junctions in a monolayer and mimic intestinal epithelium. Additionally, Caco-2 cells express proteins that are involved in drug transport making them a good model for testing drug absorption [8]. Another cell line commonly used to test absorption is the Madin-Darby canine kidney (MDCK-MDR1) cell line, which mimics efflux activity of P-glycoprotein and allows faster performance of transport assays [9]. Hepatic metabolism plays a critical role in the removal of xenobiotics. Hepatocytes are usually the best model to study drug metabolism [10]. Although immortalized hepatocyte cell lines such as HepG2 and HepaRG are used to test drug metabolism and excretion, freshly isolated hepatocytes are the best model as they exhibit complete expression of metabolic enzymes [10, 11].

Although 2D cell cultures are used widely in DDDR and play a big role in preclinical drug testing, data generated from their use often do not translate to what occurs *in vivo*. Nowadays, 3D cell cultures and co-cultures receive more attention as they exhibit protein expression patterns and intracellular junctions that are similar to *in vivo* states compared to classic monolayer cultures.

3. Three-dimensional cell culture system

Three-dimensional cell culture was developed to improve the structure of cells and physiological equivalence of *in vitro* experiments performed. It refers to the culture of living cells inside micro assembled devices with a 3D structure mimicking tissue and organ specific microarchitecture [12]. In 3D cell culturing, growth of cells in their 3D physical shape allows better cell-to-cell contact and intercellular signaling networks [13]. The 3D environment also facilitates developmental processes allowing cells to differentiate into more complex structures [14].

3.1. Three-dimensional cell culture techniques

Three-dimensional cell culture techniques are classified as Scaffold-based or non-scaffold-based techniques. Researchers are required to select the most appropriate model for their cell-based assay.

3.1.1. *Scaffold-based cell culture*

Scaffold-based culture technologies give physical support to basic mechanical structures to extra-cellular matrix (ECM)-like matrices, on which cells can aggregate, proliferate and migrate [15]. In scaffold-based techniques, cells are implanted into the matrix and the chemical and physical properties of the scaffold material mold the characteristics of cell. The ultimate aim of a scaffold is to produce characteristics for the native cell function within the ECM. The 3D scaffold is usually biocompatible and it characterizes the shape and function of the assimilated cell structure [16]. The design of scaffold is based on the tissue of interest and the bigger or complex the scaffold is; the more difficult or harder the extraction of cells for analysis becomes [17]. Regardless of the tissue type, there are important factors to consider when designing the scaffold as described in **Table 2**.

Scaffolds are manufactured from natural and synthetic materials by a plethora of fabrication techniques. The main natural materials used for scaffold synthesis are different components of the ECM including fibrin, collagen and hyaluronic acid [22–24]. In addition, natural derived materials such as silk and gelatin may also be used [25]. Synthetic materials used for scaffold synthesis include polymers, titanium, bioactive glasses and peptides [26–28]. Polymers have been widely used as biomaterials for the fabrication of scaffolds, due to their unique properties such as high porosity, small pore size, high surface to volume ratio, biodegradation and mechanical properties [29, 30]. Scaffolds are designed to support cell adhesion, cell-biomaterial interactions, adequate transport of gases and nutrients for cell growth and survival and to avoid toxicity [31]. The fabrication technique for scaffold synthesis depends on the size and surface properties of the material and recommended role of the scaffold. The relevant fabrication techniques for a particular target tissue must be identified to facilitate proper cell distribution and guide their growth into 3D space. The various techniques for scaffolds fabrication are given in **Table 3**.

Scaffold-based 3D culture can be broadly divided into two approaches—hydrogels and solid-state scaffolds.

Property	Purpose	References
Biocompatibility	Ability to provide normal cellular function	[18]
Bioactivity	Ability to activate fast tissue attachment to the implant surface	[18]
Biodegradability	Allow cells to produce their own ECM	[19]
Mechanical response	Scaffold should be strong enough to allow surgical handling during implantation and must have enough mechanical integrity for the completion of the remodeling process	[20]
Scaffold architecture	Porous interconnected structure provide cellular penetration and adequate diffusion of nutrients to cells and mean pore size should large enough to allow cells to migrate into the structure	[21]

Table 2. Scaffold requirements.

Scaffold fabrication techniques	Advantages	References
Solvent casting/particulate leaching	Easy method, pore size can be controlled, desired crystallinity, highly porous structure	[32]
Melt molding	Able to construct scaffolds of any shape by changing the mold geometry, free of organic solvents, controlled pore size and porosity	[33]
Gas foaming	Controlled porosity and pore size, free of strong organic solvents	[34, 35]
Fiber bonding	Large surface area for cell attachment, interconnected fiber structure and high porosity	[36]
Freeze drying	High porosity and interconnectivity, controlled pore size, leaching step not required, work at low temperature	[37, 38]
Electrospinning	Controlled over porosity and pore size, produces ultra-thin fibers with special orientation and large surface area	[39, 40]
Fiber mesh	Variable pore size, large surface area for cell attachment	[41, 42]
Porogen leaching	High porosity, controlled pore size and geometry, bigger pore size and increased pore interconnectivity	[43, 44]
Micro molding	It is biologically degradable, mechanical and physical complexity	[45]

Table 3. The different scaffold fabrication techniques and their advantages.

3.1.1.1. Hydrogel scaffolds

Hydrogels are water swollen polymeric materials formed by chemical reactions of monomers that generate main-chain free radicals that make cross-link junctions or by hydrogen bonding [46]. Hydrogels are one of the most used scaffolds because they mimic the ECM to a certain extent [17]. Hydrogels are highly hydrated hydrophilic polymer networks with pores and void space between the polymers [47]. The hydrophilic structure facilitates absorption and retention of large quantities of water. It is regarded as a powerful method when applied for biomedical purposes [48]. Because hydrogels have properties such as soft and rubbery consistence, low surface tension and high water content, they are more suitable substitutes for natural tissues [49]. Sources of hydrogels can be natural, synthetic or a mixture of both (hybrid) materials, offering a broad spectrum of chemical and mechanical properties. The natural materials used for hydrogels are collagen, gelatin, alginate, fibrin, hyaluronic acid, agarose, chitosan and laminin [50–53]. Natural hydrogels confer l adhesive properties, high cell viability, controlled proliferation and differentiation. Collagen is the most widely used natural polymer for hydrogel preparation and it is the main component of tissues such as ligament, bone, cartilage skin and tendon [54, 55].

Synthetic hydrogels can mimic biological properties of ECM and are ideal material to use for 3D scaffolds. They have well defined chemical, physical and mechanical properties to achieve stiffness and porosity [56]. The main synthetic materials used to formulate hydrogels are polyacrylic acid, polyethylene glycol (PEG), polyvinyl alcohol, polyglycolic acid (PGA) and poly (2-hydroxy ethyl methacrylate [57–60]. Synthetic hydrogels are the most used hydrogels because of their longer service life, high gel strength and water absorption capacity [61]. PEG and its derivatives are used mainly for synthetic hydrogels [62].

3.1.1.2. *Solid state scaffolds*

Culturing cells into a solid scaffold provides 3D space and helps generate natural 3D tissue-like structures. Solid scaffolds for 3D culture can be designed with different materials such as ceramics, metals, glass and polymers. Polymers are mainly used to construct solid scaffolds of different sizes, varying shapes, porosity, stiffness and permeability [63]. The main advantage of solid scaffolds is their ability to create organized positioning of cells *in vitro* in a controllable and reproducible manner [64]. The cell adhesion, growth and behavior in solid scaffold significantly depends on factors such as scale and topography of the internal structure, material used for its construction, the surface chemical properties, permeability and mechanical properties [65]. Solid scaffolds are commercially available, and are distributed sterile and ready to use. One of the main solid scaffolds is described below. An example is the porous scaffold. Porous scaffold creates a 3D microenvironment for cells to enter and maintain their natural 3D structure. It has a homogenous interconnected pore network, allowing cells to interact effectively to create tissue like structures and provides improved nutrient supply to the center of the device [64]. Sponge or foam porous scaffold have been especially used for bone regrowth and organ vascularization. Porous scaffold can be synthesized with specific porosity, pore size, crystallinity and surface area to volume ratio [66]. Synthetic biodegradable polymers such as polylactic-co-glycolic acid (PLGA), polyether ester (PEE), poly-L-lactic acid (PLLA) and PGA are the main materials used for porous scaffolding [67].

3.1.2. *Scaffold-free 3D cultures*

3.1.2.1. *Scaffold-free 3D spheroid cultures*

Scaffold-free-based 3D systems facilitate the development of multi-cellular aggregates, commonly known as spheroids, and can be generated from wide range of cell types [68]. Common examples of spheroids comprise tumor spheroids, embryonic bodies, mammospheres, neurospheres and hepatospheres. A cellular spheroid 3D model has a variety of properties such as (i) naturally mimicking/imitating various aspects of solid tissues; (ii) establishing geometry and ideal physiological cell-to-cell interactions; (iii) cells form their own ECM components and better cell-ECM interactions; (iv) excellent gradient for efficient diffusion growth factors as well as the (v) removal of metabolic waste [69]. The size of the spheroid can be based on the primary number cells seeded and it can increase in size where until they show oxygen and nutrient gradients similar to target tissue [70]. Spheroids are either self-assembling or are forced to grow as cell clusters [71]. Spheroids can be easily analyzed by imaging using light fluorescence, and confocal microscopy and that is an added advantage of spheroids compared to other 3D models. There are different approaches for facilitating spheroid cultures as described below.

Hanging drop method co-culture used to generate tissue-like cellular aggregates for molecular and biochemical analysis in a physiological suitable model. The hanging drop method was first developed in 1994 and became the basis of the non-scaffold method for the formation of multicellular spheroids. In hanging drop method, cells are cultured in a drop of media suspended on the lid of a cell culture dish, which is carefully inverted and placed on top of the dish containing media to maintain a humid atmosphere. Suspended cells then come together and form 3D spheroids at the apex of the droplet of media [72, 73]. This method has many advantages such as cost effectiveness, controlled spheroid size, and various cell types can be co-cultured and produced into spheroids [74, 75]. Moreover, it has been reported that 3D cell culture generated with hanging drop method have 100% reproducibility [69]. Due to limited volume of droplets generated with this technique, it is difficult to maintain spheroids and change the medium. Presently, there are many commercial devices for hanging drop culture (**Figure 2**).

The use of low adhesion plates helps to promote self-aggregation of cells into spheroids [76]. Low adhesion plates have been developed as the commercial product of the liquid overlay technique, which is a low cost highly reproducible culture method that easily promotes 3D aggregates or spheroids [77]. Low adhesion plates are spheroid microplates with round, V-shaped bottoms and very low attachment surfaces to generate self-aggregation and spheroid formation. Plates are designed with hydrophilic or hydrophobic coating, which reduces cell from attaching to the surface. The main advantage of low adhesion plates is the potential to produce one spheroid per well making it appropriate for medium-throughput screening,

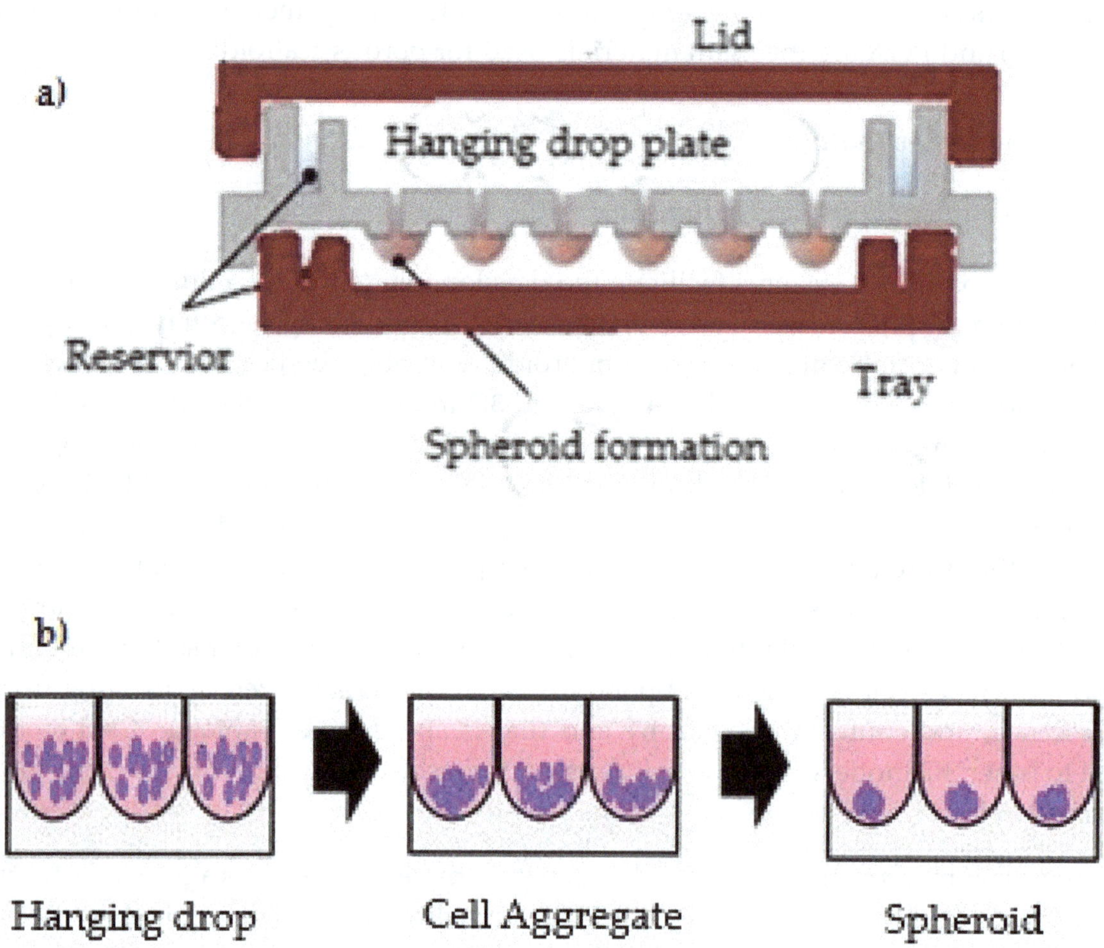

Figure 2. (a) A schematic of the hanging drop plate and (b) Schematic of spheroid formation techniques for hanging drop spheroids.

as well as creating defined geometry suitable for multicellular culture [78]. These plates have initial higher volume capacity than hanging droplets and there is no need to manipulate the spheroids.

Spheroids can also be cultured by using bioreactors under specific dynamic conditions [79]. The dynamic conditions are generated by stirring or rotating using spinner flask or NASA (National Aeronautics and Space Administration) rotating wall vessel, respectively [80]. The rotating wall vessel produces larger sized spheroids than spinner flask [81]. Bioreactors provide greater spheroid production control and reproducibility [82]. However, production of spheroids through this method requires expensive instruments and high quality cell culture medium.

3.1.2.2. Scaffold-free organoid cultures

Organoids are *in vitro* derived 3D cell aggregates that are capable of self-renewal, self-organization, and exhibit organ functionality [83]. Organoids are produced either from stem cells or primary tissues by providing suitable physical (support for cell attachment and survival) and biochemical (modulate signaling pathways) cues [84]. Organoids are classified into tissue organoids and stem cell organoids, based on how the organ buds are created [85]. Distinctive examples of tissue organoids culture are intestine, prostate, mammary and salivary glands. Stem cell organoids are created from either embryonic stem cells or primary stem cells (neonatal tissue) or induced pluripotent cells. Presently, different *in vitro* organoids have been set to simulate numerous tissues such as functional organoids for pancreas [86], liver [85], intestine [87], kidney [88], lung [89], retina [90], stomach [91] and thyroid [92]. Organoids mimic some of the structure and function of real organs [83]. Several approaches have been used to obtain organoids. The first approach is to culture cells as a monolayer on an ECM coated surface; organoids are then produced after the cells differentiate. The second is a mechanically supported cell culture to provide further differentiation of primary tissues. The third approach is to produce embryoid bodies through hang drop culture or on the low adhesion plates [93]. The main disadvantages of organoids are the lack of vasculature, lack of key cell types found *in vivo* and some organoids only replicate early stages of organ development [83].

3.2. Three-dimensional cell culture in drug discovery and development

Cell-based assays are the major tool used to evaluate the potency of a new compound in drug discovery. Three dimensional cell culture technologies have been used in different stages of drug discovery including diseases modeling, target identification and validation, screening, target selection, potency profiling and toxicity assessment. **Table 4** indicates the 3D models used in different stages of drug discovery. Three-dimensional culture models behave similarly to the cells *in vivo*, and are therefore used in the early stage of the drug discovery process, especially in cytotoxicity tests [94] such as MTT, Flow Cytometry and so on. The most effective cell-based assays with 3D cultures are cell viability, proliferation, signaling and migration [95]. It is now broadly accepted that cells act differently in 3D environments compared to 2D ones, especially when it comes to drug discovery—many prospective cancer therapeutics look favorable in the 2D cell culture dish, but fall painfully later on in clinical development.

Drug discovery stages	3D model	References
Disease modeling	Spheroids	[104, 105]
	Organoids	[106, 107]
Target identification	Spheroids	[108]
	Organoids	
Screening	Spheroids	[109–111]
Efficacy profiling	Spheroids	[112]
Toxicity profiling	Spheroids	[114, 115]
	Organoids	[113, 114]

Table 4. Three-dimensional culture techniques used in different stages of drug discovery.

Three-dimensional cell cultures promise to bridge the gap between traditional 2D cell culture and *in vivo* animal models. Studies have shown that cellular response to drug treatment in 3D cell culture are more similar to what occurs *in vivo* compared to 2D cell culture [96–98]. In addition, a number of studies show that cells cultured in 3D models are more resistant to anticancer drugs than those in 2D cultures [99, 100]. For example, the cell viability of ovarian cancer cells in 3D spheroid cell cultures after paclitaxel treatment was reduced by 40%, while the same treatment led to 80% reduced cell viability in 2D cell cultures [101]. The stronger drug resistance in 3D culture can be attributed to different factors including, phenotype and genotype changes [100], signals from cellular interactions between cells and ECM [102], activation of genes involved in cell survival and drug sensitivity due to limited diffusion through the spheroid [103].

Spheroid 3D cell cultures have been used for modeling the microenvironments, signaling, invasion and immune characteristics of cancer, also for studying cancer stem cells [104]. Studies have shown that cancer cell line spheroids have been used to analyze different characteristics of the cancer invasion process such as endothelial cell to tumor cell contact [116] and invasion of cells in a spheroid into the nearby 3D ECM structure [117]. Additionally, organoid cell cultures have been used to model number of diseases infectious diseases, neurodevelopmental and neuronal degeneration disorders [83]. For example, intestinal organoids were used to investigate genetically reconstituted tumorigenesis [118], gastrointestinal infection with rotavirus [119], *Cryptosporidium parvum* infection [106], and colon cancer stem cell biology [107]. A large number of genetic disorders that have not been possible to model in animals can be modeled using organoid 3D cultures. For example, intestinal organoids derived from patient biopsies have been used to understand onset and progression of genetic disorders [120, 121]. Organoid 3D culture model is also a powerful tool for modeling neurodevelopmental disorders such as microencephaly, caused by Zika virus infection at early stages of brain development. Moreover, brain organoid model of neural stem cells was used to understand implications of Zika virus infection during neurogenesis [122]. These are some examples of uses of 3D cell cultures as models to study disease.

Gene expression patterns seen in 3D systems are more similar to *in vivo* conditions compared to 2D cell culture systems [123]. For instance, analysis of gene expression in mesothelioma cell lines cultured in spheroids shows the basic cause of chemoresistance in malignant mesothelioma [108]. In addition, cancer cell lines grown in 2D and 3D models show different gene expression levels of various genes responsible for proliferation, chemo sensitivity, angiogenesis and invasion [63]. Ovarian cells grown in 3D system shown higher level of gene expression of the cell receptors integrins compared to 2D cell culture [99]. Moreover, 3D cell cultures are cost effective and time saving for drug screening because they decrease drug trail time whilst generating accurate representation of *in vivo* conditions [6]. Screening using cell-based assays has been the initial point for identifying the potential compounds in the early stage of drug discovery. Most 3D cell culture models, together with HTS and HCS (high-content screening) processes shows promise in identifying clinically relevant compounds.

Characteristics	2D cell culture	3D cell culture	References
Morphology	Cells grow on a flat surface and have flat or stretched shape	Cells grow naturally into 3D aggregates/ spheroids in a 3D environment and natural shape retained	[126]
Cell shape	Single layer	Multiple layers	[6]
Cell to cell contact	Limited cell to cell contact, only on edges	Physiologic cell to cell contact similar to *in vivo*	[127]
Distribution of medium	Cells receive an equal amount of nutrients and growth factors from the medium during growth.	Cells do not receive an equal medium during growth. The core cell receive less growth factors and nutrients from the medium and tend to be in a hypoxic state, which is very similar to *in vivo* tissues, especially in tumors	[115, 127]
Cell proliferation	Generally, cells proliferate at a fast rate than *in vivo*	Cells proliferate faster or slower depending on the type of cell or 3D system used	[128–130]
Protein/gene expression	Protein and gene expression profiles differ compared with *in vivo* models	Protein and gene expression profiles more similar to *in vivo* models	[131]
Cell differentiation	Moderately differentiated	Properly differentiated	[132]
Response to stimuli	Poor response to mechanical stimuli of cells	Good response to mechanical stimuli of cells	[133]
Viability	Sensitive to cytotoxin	Greater viability and less susceptible to external factors	[134]
Drug sensitivity	Cells are more sensitive to drugs and drug show high efficacy	Cells are more resistant to drugs and drug show low potency	[135]
Cell Stiffness	High stiffness	Low stiffness	[105]
Sub-culturing time	Allows cell to be grown in culture for up to 1 week	Allows cells to be grown in culture for almost 4 weeks	[136]

Table 5. Characteristics of 3D cell culture versus 2D cell culture.

Three-dimensional cell culture models have been shown to be more accurate in assessing drug screening, selection and efficacy than 2D models of the diseases [115, 124]. For instance, spheroids obtained from patients were used to identify an effective therapy for 120 patients with HER2-negative breast cancer of all stages. The results indicated that spheroid 3D culture models display present guideline treatment recommendation for breast cancer [113]. In addition, 3D cell culture models are very powerful in analyzing drug induced toxicity. Organ buds of heart, liver, brain and kidney can be used to identify drug toxicity [83]. For instance, liver cell spheroid 3D culture used for investigating drug induced liver injury, function and diseases. Spheroids generated from human primary hepatocyte found to be phenotypically stable and retained morphology and viability for almost 5 weeks, providing toxicity analysis of drug molecules [115]. Liver spheroids and organoids also have been used to understand the metabolism of drug molecules.

However, many challenges remain in 3D cell culture technologies in the drug discovery process. Three-dimensional culture are different in terms of size, morphology, complexity and protocol for assaying compared to 2D cell culture, which can lead to challenges in systematic assessment, culture and assay protocol standardization. It also has complexity of identifying specific phenotypes for drug screening [125]. Moreover, some 3D models have limited permeability, which can impact cell viability and functions thus making it difficult to have accurate automated system for HTS. A summary of the differences between 2D and 3D cell cultures is given in **Table 5**.

4. Conclusion

Two-dimensional and 3D cell culture models have been widely used for improving the productivity of pharmaceutical research and development. It is evident that 3D culture systems hold great potential as a tool for drug discovery compared to 2D cell culture. This is due to the improved cell-cell and cell-ECM interactions, cell populations and structures that similar to *in vivo*. However, there are still hurdles to overcome before 3D systems can be widely used in industry. More studies are needed to promise reproducibility, high throughput analysis and compatibility to demonstrate standardized and validated 3D culture models. In future, development of screening compatible 3D models would help to identify early physiological relevant efficacy and toxicity data in drug discovery.

List of abbreviations

ADMETox	absorption, distribution, metabolism, excretion and toxicity
CaCo-2	human colon carcinoma
CO_2	carbon dioxide
DDDR	drug discovery and development research

DMEM	Dulbecco's Modified Eagle Medium
ECM	extracellular matrix
EDTA	ethylenediaminetetraacetic acid
HCS	high-content screening
HEP-G2	liver hepatocellular carcinoma
HER-2	human epidermal growth factor receptor 2
HTS	high-throughput screening
MDCK-MDR1	Madin-Darby canine kidney cells
MEM	minimum essential medium
MTT	3-(4,5-dimethylthiazol-2-yl)-2,5-diphenyltetrazolium bromide
PEE	polyether ester
PEG	polyethylene glycol
PGA	polyglycolic acid
PLGA	polylactic-co-glycolic acid
PLLA	poly-L-lactic acid
RMPI	Roswell Park Memorial Institute medium
2D	two dimensional
3D	three dimensional

Author details

Jitcy Saji Joseph, Sibusiso Tebogo Malindisa and Monde Ntwasa*

*Address all correspondence to: ntwasmm@unisa.ac.za

Department of Life and Consumer Sciences, College of Agriculture and Environmental Sciences, University of South Africa, South Africa

References

[1] Kola I, Landis J. Can the pharmaceutical industry reduce attrition rates? Nature Reviews Drug Discovery. 2004;**3**:711-715

[2] Arrowsmith J, Miller P. Trial watch: Phase II and phase III attrition rates 2011-2012. Nature Reviews. Drug Discovery. 2013;**12**:569

[3] Kim JB. Three-dimensional tissue culture models in cancer biology. Seminars in Cancer Biology. 2005;**15**:365-377

[4] Cukierman E, Pankov R, Stevens DR, et al. Taking cell-matrix adhesions to the third dimension. Science. 2001;**294**:1708-1712

[5] Pampaloni F, Reynaud EG, Stelzer EH. The third dimension bridges the gap between cell culture and live tissue. Nature Reviews. Molecular Cell Biology. 2007;**8**:839-845

[6] Edmondson R, Broglie JJ, Adcock AF, Yang L. Three-dimensional cell culture systems and their applications in drug discovery and cell-based biosensors. Assay and Drug Development Technologies. 2014;**12**:207-218

[7] Montanez-Sauri SI, Beebe DJ, Sung KE. Microscale screening systems for 3D cellular microenvironments: Platforms, advances, and challenges. Cellular and Molecular Life Sciences. 2015;**72**:237-249

[8] Cai Y, Xu C, Chen P, Hu J, Hu R, Huang M, et al. Development, validation, and applications of a novel 7-day Caco-2 culture system. Journal of Pharmacological and Toxicological Methods. 2014;**70**(2):175-181

[9] Jin X, Loung LT, Reese N, Gaona H, Collazo-Velez V, Vuong C, et al. Comparison of MDCK-MDR1 and Caco-2 cell based permeability assays for anti-malarial drug screening and drug investigations. Journal of Pharmacological and Toxicological Methods. 2014;**70**(2):188-194

[10] Berry MN, Grivell AR, Grivell MB, Phillips JW. Isolated hepatocytes-past present and future. Cell Biology and Toxicology. 1997;**13**:223-233

[11] Schaeffner I, Petters J, Aurich H, Frohberg P, Christ B. A microtiterplate-based screening assay to assess diverse effects on cytochrome P450 enzyme activities in primary rat hepatocytes by various compounds. Assay and Drug Development Technologies. 2005;**3**:27-38

[12] Dongeum H, Geraldine AH, Donald EI. From three dimensional cell culture to organ on chips. Trends in Cell Biology. 2011;**21**(12):745-754

[13] Abbott. Cell culture biology's new dimensions. Nature. 2003;**424**:870-872

[14] Cukierman E, Pankov R, Yamada KM. Cell interactions with three-dimensional matrices. Current Opinion in Cell Biology. 2002;**14**:633-639

[15] Freed GV, Biron RJ, et al. Biodegradable polymer scaffolds for tissue engineering. Biotechnology. 1994;**12**(7):689-693

[16] Place E, George J, Williams C, Stevens M. Synthetic polymer scaffolds for tissue engineering. Chemical Society Reviews. 2009;**38**:1139-1151

[17] 3D cell culture 101: An introduction to 3D cell culture to tools and techniques 3 d biomatrix-white-paper-3d-cell-culture-101. Science. 2016

[18] O'Brien FJ. Biomaterials & scaffolds for tissue engineering. Materials Today. 2011;**14**:88-95

[19] Mikos AG, McIntire LV, Anderson JM, Babensee JE. Host response to tissue engineered devices. Advanced Drug Delivery Reviews. 1998;**33**(1-2):111-139

[20] Dietmar WH. Scaffolds in tissue engineering bone and cartilage. Biomaterials. 2000;**21**: 2529-2543

[21] Henry CK, Bruce KM, Clive DM, et al. Engineering thick tissues—The vascularisation problem. European Cells and Materials. 2007;**14**:1-19

[22] Baharvand H, Hashemi SM, Ashtian SK, et al. Differentiation of human embryonic stem cells into hepatocytes in 2D and 3D culture systems in vitro. The International Journal of Developmental Biology. 2006;**50**:645-652

[23] Willerth SM, Arendas KJ, Gottlieb DI, et al. Optimization of fibrin scaffolds for differentiation of murine embryonic stem cells into neural lineage cells. Biomaterials. 2006;**27**:5990-6003

[24] Gerecht S, Burdick JA, Ferreira LS, et al. Hyaluronic acid hydrogel for controlled self-renewal and differentiation of human embryonic stem cells. Proceedings of the National Academy of Sciences. 2007;**104**:11298-11303

[25] Awad HA, Wickham MQ, Leddy HA, et al. Chondrogenic differentiation of adipose-derived adult stem cells in agarose, alginate, and gelatin scaffolds. Biomaterials. 2004;**25**:3211-3222

[26] Gunatillake PA, Adhikari R. Biodegradable synthetic polymers for tissue engineering. European Cells & Materials. 2003;**5**:1-16

[27] Van den Dolder J, Spauwen PHM, Jansen JA. Evaluation of various seeding techniques for culturing osteogenic cells on titanium fiber mesh. Tissue Engineering. 2003;**9**:315-325

[28] Lu HH, El-Amin SF, Scott KD, et al. Three-dimensional, bioactive, biodegradable, polymer-bioactive glass composite scaffolds with improved mechanical properties support collagen synthesis and mineralization of human osteoblast-like cells in vitro. Journal of Biomedical Materials Research. Part A. 2003;**64A**:465-474

[29] Ji K, Ghosh XZ, et al. Electrospun three-dimensional hyaluronic acid nanofibrous scaffolds. Biomaterials. 2006;**27**:3782-3792

[30] Piskin E. Biodegradable polymers as biomaterials. Journal of Biomaterials Science Polymer Edition. 1994;**6**:775-795

[31] Langer R, Tirrell. Desigining a material and biological and medicine. Nature. 2004;**428**(6982): 487-792

[32] Xiang Z, Liao R, Kelly MS, Spector M. Collagen–GAG scaffolds grafted onto myocardial infarcts in a rat model: A delivery vehicle for mesenchymal stem cells. Journal of Tissue Engineering. 2006;**12**:2467-2478

[33] Thompson RC, Wake MC, Yaszemski, Mikos AG. Biodegradable polymer scaffolds to regenerate organs. Advances in Polymer Science. 1995;**122**:245-274

[34] Ikada Y. Scope of tissue engineering. In: Tissue Engineering: Fundamental and Applications. Elsevier/Science Direct. 2006. eBook ISBN: 9780080464008

[35] Mooney DJ, Baldwin DF, Suh NP, Vacanti JP, Langer R. Novel approach to fabricate porous sponges of poly(D,L-lactic-co-glycolic acid) without the use of organic solvents. Biomaterials. 1996;**17**(14):1417-1422

[36] Moroni L, Hamann D, Paoluzzi L, Pieper J, de Wijn JR, van Blitterswijk CA. Regenerating articular tissue by converging technologies. PLoS One. 2008;**3**(8):e3032

[37] Schoof H, Apel J, Heschel I, Rau G. Control of pore structure and size in freezedried collagen sponges. Journal of Biomedical Materials Research. 2001;**58**:352-357

[38] Mandal BB, Kundu SC. Cell proliferation and migration in silk fibroin 3D scaffolds. Biomaterials. 2009;**30**:2956-2965

[39] Liang D, Hsiao BS, Chu B. Functional electrospun nanofibrous scaffolds for biomedical applications. Advanced Drug Delivery Reviews. 2007;**59**:1392-1412

[40] Li WJ, Tuan RS. Fabrication and application of nanofibrous scaffolds in tissue engineering. Current Protocols in Cell Biology. 2009;**25**:Unit 25.2

[41] Chen G, Ushida T, Tateishi T. Development of biodegradable porous scaffolds for tissue engineering. Materials Science and Engineering C. 2002;**17**:63-69

[42] Martins AM, Pham QP, Malafaya PB, et al. The role of lipase and alphaamylase in the degradation of starch/poly(varepsilon-caprolactone) fiber meshes and the osteogenic differentiation of cultured marrow stromal cells. Tissue Engineering. Part A. 2009;**15**(2): 295-305

[43] Mikos AG, Sarakinos G, Leite SM, Vacanti JP, Langer R. Laminated threedimensional biodegradable foams for use in tissue engineering. Biomaterials. 1993;**14**:323-330

[44] Mano JF, Silva GA, Azevedo HS, et al. Natural origin biodegradable systems in tissue engineering and regenerative medicine: Present status and some moving trends. Journal of the Royal Society Interface. 2007;**4**:999-1030

[45] Fukuda J, Khademhosseini y, et al. Micromolding of photocrosslinkable chitosan hydrogel for spheroid microarray and co-culture. Biomaterials. 2006;**79**:522-532

[46] Yan C, Pochan DJ. Rheological properties of peptide-based hydrogels for biomedical and other applications. Chemical Society Reviews. 2010;**39**:3528-3540

[47] Daniele MA, Adams AA, Naciri J, North SH, Ligler FS. Interpenetrating networks based on gelatin methacrylamide and PEG formed using concurrent thiol click chemistries for hydrogel tissue engineering scaffolds. Biomaterials. 2014;**35**:1845-1856

[48] Wang T, Jiao Y, Chai Q, Yu X. Gold nanoparticles: Synthesis and biological applications. Nano LIFE. 2015;**5**:1542007

[49] Hamidi M, Azadi A, Rafiei P. Hydrogel nanoparticles in drug delivery. Advanced Drug Delivery Reviews. 2008;**60**:1638-1649

[50] Huh D, Hamilton GA, Ingber DE. From 3D cell culture to organs-on-chips. Trends in Cell Biology. 2011;**21**(12):745-754

[51] Ehrbar M, Djonov V, Schnell C, et al. Cell-demanded liberation of VEGF121 from fibrin implants induces local and controlled blood vessel growth. Circulation Research. 2004;**94**: 1124-1132

[52] Ahmed T, Dare E, Hincke M. Fibrin: A versatile scaffold for tissue engineering applications. Tissue Engineering. Part B, Reviews. 2008;**14**:199-215

[53] Allison D, Grande-Allen K. Review. Hyaluronan: A powerful tissue engineering tool. Tissue Engineering. 2006;**12**:2131-2140

[54] Glowacki J, Mizuno S. Collagen scaffolds for tissue engineering. Biopolymers. 2008;**89**: 338-344

[55] Pathak A, Kumar S. Biophysical regulation of tumor cell invasion: Moving beyond matrix stiffness. Integrative Biology. 2011;**3**:267-278

[56] Zhang YS, Khademhosseini A. Advances in engineering hydrogels. Science. 2017;**356**: eaaf3627

[57] Sawhney AS, Pathak CP, Hubbell JA. Bioerodible hydrogels based on photopolymerized poly(ethylene glycol)-co-poly(a-hydroxy acid) diacrylate macromers. Macromolecules. 1993;**26**(4):581-587

[58] Martens P, Anseth KS. Characterization of hydrogels formed from acrylate modified poly(vinyl alcohol) macromers. Polymer. 2000;**41**(21):7715-7722

[59] Chirila TV, Constable IJ, Crawford GJ, et al. Poly(2-hydroxyethel methacrylate) sponges as implant materials: In vivo and in vitro evaluation of cellular invasion. Biomaterials. 1993;**14**(1):26-38

[60] Raeber GP, Lutolf MP, Hubbell JA. Molecularly engineered PEG hydrogels: A novel model system for proteolytically mediated cell migration. Biophysical Journal. 2005;**89**: 1374-1388

[61] Hoffman AS. Hydrogels for biomedical applications. Advanced Drug Delivery Reviews. 2012;**64**:18-23

[62] Caiazzo M, Okawa Y, Ranga A, et al. Defined three-dimensional microenvironments boost induction of pluripotency. Nature Materials. 2016;**15**:344-352

[63] Gurski L, Petrelli N, Jia X, Farach-Carson M. Three-dimensional matrices for anti-cancer drug testing and development. Oncology Issues. 2010;**25**:20-25

[64] Knight E, Przyborski S. Advances in 3D cell culture technologies enabling tissue-like structures to be created in vitro. Journal of Anatomy. 2015;**227**(6):746-756

[65] Haycock JW. 3D cell culture: A review of current approaches and techniques. Methods in Molecular Biology. 2011;**695**:1-15

[66] Ourimechi EM, Vergnaud JM. Process of drug transfer with three different polymeric systems with transdermal drug delivery. Computational and Theoretical Polymer Science. 2000;**10**:391-401

[67] Hou Q, Grijipma DW, Feijen J. Porous polymeric structures of tissue engineering prepared by a coagulation, compression, moulding and salt leaching technique. 2003;**24**:1937-1947

[68] Sutherland RM. Cell and environment interactions in tumor microregions: The multicell spheroid model. Science. 1988;**240**:177-184

[69] Breslin S, O'Driscol L. Three-dimensional cell culture: The missing link in drug discovery. Drug Discovery Today. 2013;**18**:240-249

[70] Ekert JE, Johnson K, Strake B, et al. Three-dimensional lung tumor microenvironment modulates therapeutic compound responsiveness in vitro—Implication for drug development. PLoS One. 2014;**9**:e92248

[71] Yamanda KM, Cukierman E. Modeling of tissue morphogenesis and cancer in 3D. Cell. 2007;**130**:601-610

[72] Keller GM. In-vitro differentiation of embryonic stem-cells. Current Opinion in Cell Biology. 1995;**7**:862-869

[73] Rimann M, Graf-Hausner U. Synthetic 3D multicellular systems for drug development. Current Opinion in Biotechnology. 2012;**23**:803-809

[74] Pham P. Breast cancer stem cell culture and proliferation. In: Breast Cancer Stem Cells & Therapy Resistance. Cham: Springer International Publishing; 2015. pp. 41-55

[75] Hsiao AY, Tung YC, Qu X, Patel LR, Pieta KJ, Takayama S. 384 hanging drop arrays give excellent Z-factors and allow versatile formation of co-culture spheroids. Biotechnology and Bioengineering. 2012;**109**:1293-1304

[76] Vinci M, Gowan S, Boxall F, et al. Advances in establishment and analysis of three-dimensional tumor spheroid-based functional assays for target validation and drug evaluation. BMC Biology. 2012;**10**:29

[77] Carlsson J, Yuhas JM. Liquid-overlay culture of cellular spheroids. Recent results in cancer research. Cancer. 1984;**95**:1-23

[78] Thoma DS, Buranawat B, Hammerle CHF, Held U, Jung RE. Efficacy of soft tissue augmentation around dental implants and in partially edentulous areas: A systematic review. Journal of Clinical Periodontology. 2014;**41**:S77-S91

[79] Yu X, Chen X, Chai Q, Ayres N. Synthesis of polymer organogelators using hydrogen bonding as physical cross-links. Colloid & Polymer Science. 2016;**294**:59-68

[80] Shin CS, Han BKB, Park K, Panitch A. 3D cancer tumor models for evaluating chemotherapeutic efficacy. Biomaterials for Cancer Therapeutics. 2013:445-460

[81] Lelkes PI, Galvan DL, Hayman GT, Goodwin TJ, et al. Simulated microgravity conditions enhance differentiation of cultured PC 12 cells towards the neuroendocrine phenotype. In Vitro Cellular & Developmental Biology. 1998;**34**(4):316-325

[82] Ou KL, Hosseinkhani H. Development of 3D in vitro technology for medical applications. International Journal of Molecular Sciences. 2014;**15**:17938-17962

[83] Lancaster MA, Knoblauch JA. Organogenesis in a dish: Modeling development and disease using organoid technologies. Science. 2014;**345**:124-125

[84] Clevers H. Modeling development and diseases with oragnoids. Cell. 2016;**165**:1586-1597

[85] Huch M, Dorrell C, Boj SF, et al. In vitro expansion of single Lgr5+ liver stem cells induced by Wnt-driven regeneration. Nature. 2013;**494**:247-250

[86] Greggio C, De Franceschi F, Figueiredo-Larsen M, et al. Artificial three-dimensional niches deconstruct pancreas development in vitro. Development. 2013;**140**:4452-4462

[87] Spence JR, Mayhew CN, Rankin SA, et al. Directed differentiation of human pluripotent stem cells into intestinal tissue in vitro. Nature. 2011;**470**:105-109

[88] Takasato M, Er PX, Chiu HS, et al. Kidney organoids from human iPS cells contain multiple lineages and model human nephrogenesis. Nature. 2015;**526**:564-568

[89] Lee JH, Bhang DH, Beede A, et al. Lung stem cell differentiation in mice directed by endothelial cells via a BMP4-NFATc1-thrombospondin-1 axis. Cell. 2014;**156**:440-455

[90] Nakano T, Ando S, Takata, et al. Self-formation of optic cups and storable stratified neural retina from human ESCs. Cell Stem Cell. 2012;**10**:771-785

[91] Barker N, Huch M, Kujala P, et al. Lgr5(+ve) stem cells drive self-renewal in the stomach and build long-lived gastric units in vitro. Cell. 2010;**6**:25-36

[92] Antonica F, Kasprzyk DF, Opitz R, et al. Generation of functional thyroid from embryonic stem cells. Nature. 2012;**491**:66-71

[93] Turner DA, Baillie-Johnson P, Arias AZ. Organoids and the genetically encoded self-assembly of embryonic stem cells. BioEssays. 2015;**38**:181-191

[94] Karmen B, Marija T, Ivo Z, Dubravko J. Three-dimensional cell cultures as a new tool in drug discovery. Periodicum Biologorum. 2016;**118**:59-65

[95] Comley J. 3D cell culture: Easier said than done! Drug discovery world, summer. 2010;**11**: 25-41

[96] Lee J, Cuddihy MJ, Kotov NA. Three-dimensional cell culture matrices: State of the art. Tissue Engineering. Part B, Reviews. 2008;**14**:61-86

[97] Zietarska M, Maugeri CM, Filali-Mouhim A, et al. Molecular description of a 3D in vitro model for the study of epithelial ovarian cancer (EOC). Molecular Carcinogenesis. 2007;**46**:872-885

[98] Shield K, Ackland ML, Ahmed N, Rice GE. Multicellular spheroids in ovarian cancer metastases: Biology and pathology. Gynecologic Oncology. 2009;**113**:143-148

[99] Loessner D, Stok KS, Lutolf MP, Hutmacher DW, Clements JA, Rizzi SC. Bioengineered 3D platform to explore cell–ECM interactions and drug resistance of epithelial ovarian cancer cells. Biomaterials. 2010;**31**:8494-8506

[100] Karlsson H, Fryknäs M, Larsson R, et al. Loss of cancer drug activity in colon cancer HCT-116 cells during spheroid formation in a new 3-D spheroid cell culture system. Experimental Cell Research. 2012;**318**:1577-1585

[101] Nguyen TA, Yin TI, Reyes D, Urban GA. Microfluidic chip with integrated electrical cell-impedance sensing for monitoring single cancer cell migration in three-dimensional matrixes. Analytical Chemistry. 2013;**85**:11068-11076

[102] Walker DM, Boey G, McDonald LA. The pathology of oral cancer. Pathology. 2003;**35**: 376-383

[103] Trédan O, Galmarini CM, Patel K, Tannock IF. Drug resistance and the solid tumor microenvironment. Journal of the National Cancer Institute. 2007;**99**(19):1441-1454

[104] Weiswald LB, Bellet D, Dangles-Marie V. Spherical cancer models in tumor biology. Neoplasia. 2015;**17**:1-15

[105] Dieter SM, Ball CR, Hoffmann CM, et al. Distinct types of tumor-initiating cells form human colon cancer tumors and metastases. Cell Stem Cell. 2011;**9**:357-365

[106] Castellanos-Gonzalez A, Cabada MM, Nichols J, et al. Human primary intestinal epithelial cells as an improved in vitro model for *Cryptosporidium parvum* infection. Infection and Immunity. 2013;**81**:1996-2001

[107] Yeung TM, Gandhi SC, Wilding JL, et al. Cancer stem cells from colorectal cancer derived cell lines. Proceedings of the National Academy of Sciences of the United States of America. 2010;**107**:3722-3727

[108] Barbone D, Van Dam L, Follo C, et al. Analysis of gene expression in 3D spheroids highlights a survival role for ASS1 in mesothelioma. PLoS One. 2016;**11**:e0150044

[109] Drost J, van Jaarsveld RH, Ponsioen B, et al. Sequential cancer mutations in cultured human intestinal stem cells. Nature. 2015;**521**:43-47

[110] Senkowski W, Zhang X, Olofsson MH, et al. Three-dimensional cell culture-based screening identifies the anthelmintic drug nitazoxanide as a candidate for treatment of colorectal Cancer. Molecular Cancer Therapeutics. 2015;**14**:1504-1516

[111] Kenny HA, Lal-Nag M, White EA, et al. Quantitative high throughput screening using a primary human three-dimensional organotypic culture predicts In vivo efficacy. Nature Communications. 2015;**6**:6220

[112] Tong JG, Valdes YR, Barrett JW, et al. Evidence for differential viral oncolytic efficacy in an In vitro model of epithelial ovarian cancer metastasis. Molecular Therapy Oncolytics. 2015;**2**:15013

[113] Halfter K, Hoffmann O, Ditsch N, et al. Testing chemotherapy efficacy in HER2 negative breast cancer using patient-derived spheroids. Journal of Translational Medicine. 2016;**14**:112

[114] Gunness P, Mueller D, Shevchenko V, et al. 3D organotypic cultures of human HepaRG cells: A tool for in vitro toxicity studies. Toxicological Sciences. 2013;**133**:67-78

[115] Bell CC, Hendriks DF, Moro SM, et al. Characterization of primary human hepatocyte spheroids as a model system for drug-induced liver injury, liver function and disease. Scientific Reports. 2016;**6**:25187

[116] Ghosh S, Joshi MB, Ivanov D, et al. Use of multicellular tumor spheroids to dissect endothelial cell–tumor cell interactions: A role for T-cadherin in tumor angiogenesis. FEBS Letters. 2007;**581**:4523-4528

[117] Blacher S, Erpicum C, Lenoir R, et al. Cell invasion in the spheroid sprouting assay: A spatial organisation analysis adaptable to cell behaviour. PLoS One. 2004;**9**(5):e97019

[118] Onuma K, Ochiai M, Orihashi K, et al. Genetic reconstitution of tumorigenesis in primary intestinal cells. Proceedings of the National Academy of Sciences of the United States of America. 2013;**110**:11127-11132

[119] Finkbeinera SR, Zenga XL, Utamaa B, Atmara RB, Shroyerc NF, Mary K, et al. Stem cell-derived human intestinal organoids as an infection model for rotaviruses. MBio. 2012;**3**(4):e00159-e00112

[120] Dekkers JF, Wiegerinck CL, de Jonge HR, et al. A functional CFTR assay using primary cystic fibrosis intestinal organoids. Nature Medicine. 2013;**19**:939-945

[121] Bigorgne AE, Farin HF, Lemoine R, et al. TTC7A mutations disrupt intestinal epithelial apicobasal polarity. The Journal of Clinical Investigation. 2013;**124**:328-337

[122] Garcez PP, Loiola EC, Madeiro da Costa R, et al. Zika virus impairs growth in human neurospheres and brain organoids. Science. 2016;**352**:3

[123] Ghosh S, Spagnoli GC, Martin I, et al. Three dimentional culture of melanoma cells profoundly affects gene expression profile: A high density oligonucleotide array study. Journal of Cellular Physiology. 2005;**204**:52-531

[124] Hickman JA, Graeser R, de Hoogt R, et al. IMI PREDECT consortium. Three-dimensional models of cancer for pharmacology and cancer cell biology: Capturing tumor complexity in vitro/ex vivo. Biotechnology Journal. 2014;**9**:1115-1128

[125] Booij TH, Klop MJ, Yan K, et al. Development of a 3D tissue culture-based high-content screening platform that uses phenotypic profiling to discriminate selective inhibitors of receptor tyrosine kinases. Journal of Biomolecular Screening. 2016;**21**:912-922

[126] Huang H, Ding Y, Sun XS, Nguyen TA. Peptide hydrogelation and cell encapsulation for 3D culture of MCF-7 breast cancer cells. PLoS One. 2013;**8**:59482

[127] Li Z, Cui Z, et al. Three-dimensional perfused cell culture. Biotechnology Advances. 2014; **32**:243-254

[128] Chitcholtan K, Sykes P, Evans J. The resistance of intracellular mediators to doxorubicin and cisplatin are distinct in 3D and 2D endometrial cancer. Journal of Translational Medicine. 2012;**10**:1-16

[129] Fallica B, Mafia JS, Villa S, Makin G, Zaman M. Alteration of cellular behavior and response to PI3K pathway inhibition by culture in 3D collagen gels. PLoS One. 2012;**7**:48024

[130] Luca AC, Mersch S, Deenen R, et al. Impact of the 3D microenvironment on phenotype, gene expression, and EGFR inhibition of colorectal cancer cell lines. PLoS One. 2013;**8**:e59689

[131] Price KJ, Tsykin A, Giles KM, et al. Matrigel basement membrane matrix influences expression of microRNAs in cancer cell lines. Biochemical and Biophysical Research Communications. 2012;**427**:343-348

[132] Chitcholtan K, Asselin E, Parent S, Sykes PH, Evans JJ. Differences in growth properties of endometrial cancer in three dimensional (3D) culture and 2D cell monolayer. Experimental Cell Research. 2013;**319**:75-78

[133] Li Y, Huang G, Li M, et al. An approach to quantifying 3D responses of cells to extreme strain. Scientific Reports. 2016;**6**:19550

[134] Elkayam T, Amitay-Shaprut S, Dvir-Ginzberg M, Harel T, Cohen S. Enhancing the drug metabolism activities of C3A-a human hepatocyte cell line--by tissue engineering within alginate scaffolds. Tissue Engineering. 2006;**12**:1357-1368

[135] Bokhari M, Carnachan RJ, Cameron NR, Przyborsk SA. Culture of HepG2 liver cells on three dimensional polystyrene scaffolds enhances cell structure and function during toxicological challenge. Journal of Anatomy. 2007;**211**:567-576

[136] Baker BM, Chen CS. Deconstructing the third dimension—How 3D culture microenvironments alter cellular cues. Journal of Cell Science. 2012;**125**:3015-3024

5

Time-Lapse Microscopy

John L. Collins, Bart van Knippenberg, Kai Ding and Alexander V. Kofman

Abstract

Time-lapse microscopy is a powerful, versatile and constantly developing tool for real-time imaging of living cells. This review outlines the advances of time-lapse microscopy and refers to the most interesting reports, thus pointing at the fact that the modern biology and medicine are entering the thrilling and promising age of molecular cinematography.

Keywords: time-lapse, microscopy, real-time, imaging, cell

1. Introduction

Originally described as *time-lapse cinemicrography (microphotography)* [1], the modern *time-lapse microscopy* (TLM) emerged as a powerful and continuously improving tool for studying the cellular processes and cell-cell interactions with the applications ranging from fundamental aspects of molecular and cell biology to medical practice. The related *time-lapse photography* is more relevant to observing non-microscopic objects, such as plants and landscapes. TLM is the technique of capturing the sequence of microscopic images at regular intervals. TLM allows scientists to observe cellular dynamics and behavior of the population of living cells as well as of the single living cell within the population [2, 3]. Live cell imaging and the first non-sophisticated TLM techniques were pioneered at the very beginning of the twentieth century [4]. However, to be visible in the light microscope, the cells are to be subjected to fixation and staining, the processes that kill the cells. Introduction of phase-contrast microscopy in 1940s, development of fluorescent and multidimensional microscopy, flow cytometry and computational tools made live cell imaging a widespread approach and prompted scientists to consider TLM as an essential technique that carries an enormous promise for basic biological science and medicine. For this review, we focused on mammalian cell cultures, although TLM can also be efficiently employed to study prokaryotic cells and unicellular microorganisms. In the absence of up-to-date comprehensive review on TLM advances, our aim was to familiarize the readers with the current advances of TLM methodology and provide for the reference guide to the most interesting reports where TLM has been utilized both for biological research and clinical purposes.

2. Time-lapse microscopy: from making movies to bedside

2.1. Versatility of TLM

In this part, we will briefly review some selected publications, which highlight the rapid development of TLM as a versatile discovery tool within the broad scope of modern biology and medicine. Importance of TLM as a new method in biological research was highlighted by Burton [5]. The progress of tissue culture methods, phase-contrast microscopy (see below) and real-time imaging by TLM enabled scientists to overcome the major limitation of traditional microscopy; preparation of very thin transparent samples, which required tissue fixation and did not make it possible to investigate living cells, let alone, and biological processes over time in the same sample. Early reports demonstrated the feasibility of TLM for comparative studies of cultured cells [6–8] and for monitoring living blood and lymph cells [1], cell division [9, 10] and reaction of cells to varying contents of electrolytes in perfusion chambers [11]. TLM was helpful to decode the process of multinucleation in the developing skeletal muscles [12] and to describe the variable cytotoxic response toward allografts [13, 14].

TLM is a suitable tool to monitor *cell motility and migration,* including quantitative assessment of migration, such as the number of migrating cells and the distance [15–20]. In multicellular organisms, the directed and coordinated cell migration (chemotaxis) occurs during embryonic development, tissue regeneration and inflammatory response [21], while cancer cells migrate into surrounding tissues and the vasculature. To monitor chemotaxis, TLM can be used together with the Dunn chamber [21–23]; Boyden chamber [24, 25]; Bridge chamber [26]; LOCOMOTIS, the motility tracking system [27] and other types of chambers for cell visualization and TLM applications [28–30]. TLM was employed to study embryonic stem cells [31]; hematopoietic progenitor cells [20, 32–34]; mesenchymal stem cells [35]; activated lymphocytes forming lymphocytes colonies [36]; primordial germ cells, a migratory cell population that will eventually give rise to the gametes [37–39]; the migration route of progenitor cells in cell cultures obtained from live chicken embryos [40, 41]; microglial cells [42–44]; olfactory cells from schizophrenia patients [45]; neurons [46]; chemokines that drive migration of megakaryocytes from the proliferative osteoblastic niche within the bone marrow to the capillary-rich vascular niche, which is an essential step for platelet production [47]; migration of osteoclasts toward bone surfaces [48]; motility of cultured endothelial cells to study remodeling of their intercellular junctions [49]; generation of a complete polarized epithelial monolayer by the epithelial cells of mammary gland [17]; movement of cancer cells that were cultured under hypoxic conditions [50] and treated with salinomycin [24]; individual cell motility in fibroblastoid L929 cells [51]; human osteosarcoma MG-63 cells [52]; B35 neuroblastoma cells transiently expressing GFP and C6 glioma cells after staining with Hoechst 33258 [16] and motility of L5222 leukemia cells within the mesentery and migration of induced pluripotent cells during their early reprograming [53]. Of note, most studies are devoted to neural stem cells [18, 19, 54–63] due to growing clinical importance.

TLM allows investigators to visualize and characterize *cell-cell contacts* [46, 52, 64–71]. The most interesting reports are concerned with the contacts between the various types of stem/progenitor cells as well as the tumor-environment cell interactions: the importance of proper cell-cell contacts level for their correct positioning and cell polarity during organogenesis [39], glial-neuronal interactions [72–74], interactions between microglia and brain tumors [75], between astrocytes and neural progenitor cells [42], between mesenchymal stem cells and human myoblasts [76], dendritic cells [77], endothelial cells [78], cancer cells [79], extracellular matrix molecules [80], between erythroblastic islands in bone marrow [81], between neural progenitor cells [62, 82–84], between neural cells and hematopoietic stem cells that migrate to the central nervous system [85], hematopoietic stem cells and stromal cells [20], endothelial progenitor cells and cardiac myocytes [86], between induced pluripotent cells during the early reprogramming phase [53], vesicle traffic through intercellular bridges between prostate cancer cells [87] and synaptic contacts [88–94].

Cell division and *cell death* can be well investigated with TLM [50, 52, 95–97]. Division and growth of both labeled [96, 98–100] and non-labeled [101, 102] cells in culture [52, 95, 98, 103–105] and tissue slices [106], including monitoring of a single cell [95, 99, 107–112], can be observed and assessed with TLM. The fluorescent ubiquitination-based cell cycle indicator (FUCCI) system can effectively label individual G1, S/G2/M and G1/S-transition phase nuclei as red, green and yellow, respectively, to visualize the real-time cell cycle transitions in living mammalian cells [113–116]. Microinjection of complementary RNA to cyclin B1 was reported as a tool for TLM studying meiosis [117]. Real-time imaging was employed to monitor nuclear envelope breakdown, which is one of the major morphological changes during mitosis [118] and apoptosis [119]; nucleolar assembly after mitosis [120]; tracking of template DNA strands during mitosis [121, 122]; preferred mitotic orientation of daughter cells, which is important for their following self-organization and tissue formation [123, 124]; interkinetic nuclear migration toward the apical surface in epithelial cells [125, 126]; multinucleation of skeletal muscle cells [12]; asymmetric division of stem cells [127–129]; identification and characterization of cell division genes by combining RNA interference, time-lapse microscopy and computational image processing [130]; cytokinesis [131, 132]; cleavage furrow [133]; abscission by using TLM in combination with electron microscopy [134, 135] and mitotic synchronization in the cell population [136]. The observations related to *cellular senescence* and various forms of *cell death* include re-entry into the cell cycle [10, 124, 137–139] and variable frequency of divisions [140]; changes in mitotic and interphase duration [141–147]; short G1 phase, which is a distinctive feature of mouse embryonic stem cells [148]; delayed G2 phase [149]; *neosis*, the term used for karyokinesis via nuclear budding followed by asymmetric, intracellular cytokinesis [150]; secretion of exosomes with anti-apoptotic microRNAs [151]; apoptosis [119, 152–156]; phagocytosis of apoptotic cells [157]; necrosis [158]; autophagy [159]; mitotic catastrophe [52, 143] and phototoxicity [160].

TLM can also be used to study *intracellular dynamics* of subcellular organelles [161, 162], natural cellular proteins and reporters, introduced nanoparticles and even physiological effects of small inorganic molecules and gases. Time-lapse imaging was used to monitor and quantify movements and changes in mitochondria [163–165]; Golgi apparatus [166]; centrosomes and microtubules [167–170]; centromeres [171]; cellular membrane [172]; dendritic spines [173]; dynamics of interkinetic nuclear migration [174, 175]; intercellular uptake and distribution of nano-sized (less than 100 mkm) ceramic particles [176]; intracellular translocation of p65 and IkappaB-alpha proteins [177]; intracellular distribution of integrin beta1 and F-actin [178]; fluctuations in Notch signaling to maintain neural progenitors [179]; re-localization of PP1gamma, which is implicated in multiple cell cycle-related processes including regulation of chromosome segregation and cytokinesis [180]; movement of the replication origin region of the chromosome during the cell cycle in *Bacillus subtilis* [181]; dynamics of 53BP1 protein in DNA-damage response [182]; measuring gene dynamics with luciferase as a reporter [183]; colocalization of MAP kinases in mitochondria [184]; clustering of acetylcholine receptor on myotubes [185]; multiple chromosomal populations of topoisomerase II [186]; focal points for chromosome condensation and decondensation [187]; intracellular calcium dynamics [188, 189] and single-cell time-lapse imaging of intracellular O_2 [190].

Although TLM is mostly used with cultured mammalian cells and live cells in tissues, the significant number of reports indicates that TLM could be employed to observe and study prokaryotic cells and other unicellular and multicellular organisms as well as viruses. Here, we mention only few examples, such as time-lapse imaging of growth, cell-cell contacts and formation of spherical granules in *E. coli* [191–194]; time-lapse visualization of bacterial colony morphologies in the special bacterial chamber MOCHA [195]; screening and assessing effects of antibiotics, such as antibiotics-bacteria interactions [196–199] and studying yeasts [200–202] and viruses [203–207]. The smaller microorganisms, analogously to intracellular structures, usually require higher magnification and more sophisticated microscopic equipment.

2.2. TLM technical approaches

TLM monitoring of mammalian cells usually requires the inverted microscope, which is fully or partially enclosed by a cell incubator (environmental chamber), a partly sealed transparent box that maintains the temperature, humidity and even partial gas (carbon dioxide) pressure, protects cultured cells from the light and allows the investigator to manipulate with the microscope in order to choose the field of view and adjust other imaging parameters [208–210]. The TLM chambers and devices underwent significant improvements over the time, from the simple glass tissue chambers and manual capturing sequences of images to the automated high-resolution microscopes and sophisticated computerized equipment for long-term TLM observations [154, 162, 211–219]. The up-to-date portable live cell culture monitor (CytoSMART Technologies, Eindhoven, The Netherlands) works within the regular CO_2 incubator. The culture flask (T-flask, Petri dish, wells or any other transparent vessel) is positioned onto the lens of the device; the field of view is chosen by the investigator, and the cell growth and migration can be monitored and analyzed in the real-time mode by accessing the cloud [52].

The *phase-contrast* method of imaging is based on the ability of materials with a different refractive index to delay the passage of the light through the sample by different amounts, so that they appear darker or brighter. This is the most common TLM technique that is used since 1950s [1, 6, 7, 11] for studying different types of cells and microorganisms both alone and in combination with electron microscopy [220–222]. The so-called differential interference contrast (DIC) microscopy (Nomarski microscopy) also produces high-contrast images of transparent non-stained biological objects, and it has been broadly used for TLM [223–226]. Fluorescent TLM dating back in 1950s TLM [9, 227] can be used nowadays with fluorescent proteins-reporters [207, 228–231], fluorescent nanoparticles [232, 233] and membrane dyes [160, 234, 235]. As the further proof of TLM flexibility, we present some reports where TLM is combined with other advanced microscopy techniques: multiplexed or multifield (recording of many fields simultaneously) TLM [236, 237], confocal TLM [156, 171, 207, 238–242], multi-photon TLM [58, 243–245], the so-called four-dimensional imaging (three-dimensional images over time) [242, 246], time-lapse bioluminescence analysis [247], Forster resonance energy transfer (FRET) microscopy [248], time-lapse optical coherence tomography [249–251], *in toto* imaging to image and track every single cell movement and division during the development of organs and tissues [241] and other innovative approaches [50, 252]. TLM can be used to monitor not only cultured cells (cell population and single cell [109] but also living cells in tissue slices up to a depth of 60 micrometers in brain slices, in regions where cell bodies remain largely uninjured by the tissue preparation and are visible in their natural environment [229, 253]. For real-time observation of corneal cells in a living mouse, a novel microscope system was designed, which consists of an upright fluorescence microscope for visualization of corneal cells, a mouse-holding unit for immobilization of the animal and the eye and a set of gimbals which permit observation of a wide area of corneal surface without refocusing [254].

TLM would not be possible without an *automated image analysis*, which is used to extract meaningful data from the bulk of images. Automated cell tracking faces problems associated with high cell density; cell mobility; cell division; multiple cell parameters such as object size, position or texture; cell lysis or overlap of cells [255]. A variety of algorithms, including *segmentation* (the process of partitioning a digital image into multiple sets of pixels or segments) algorithms, have been developed, and they are constantly improving. For most datasets, a *preprocessing* step is needed before information can be extracted. Irregular illumination and shading effects can be removed by using a *background subtraction method*. Other commonly used techniques include *contrast enhancement* and *noise filtering* [256]. In some cases, *registration* is needed to align subsequent image frames and compensate for unwanted movements. Global movements can be caused by movement of the specimen or imaging equipment, but local deformations in the specimen might also have to be corrected for. This is especially the case when considering TLM of living animals, which is heavily affected by breathing and heartbeat [257]. At higher magnifications, when studying intracellular dynamics, cell migration itself might also be considered an unwanted movement that has to be corrected [258]. *Object detection* is a set of techniques to separate objects of interest from the background. The objects of interest can be cells or intracellular particles [130, 259]. Basic segmentation techniques can be sufficient to detect individual cells, although more advanced techniques are still being developed to cope with increasingly complex data [260, 261]. Finally, several *analysis* techniques are available to quantify the different types of cell behavior over time, for example, *trajectory analysis* for assessing trajectory length and directional persistence [262]. By now, various algorithms are designed for quantifying and tracking cell migration [3] and single cell motility [261, 263]; cell proliferation [264]; cell cycle and cell lineage analysis [107]; changes in mitotic and interphase duration [141]; cell-cell contacts [52]; studying specific cells and tissues [265] and specific intracellular processes such as transcription [99] or morphogenesis [266]; colocalization of cells and intracellular markers [184]; tracking cellular organelles [258]; highlighting the certain cell type within tissues or mixed cell cultures [267]; clustered, overlapping or dying cells [268]; *in toto* imaging of developing organisms, tissues and organs [241] and assessing development and selection of embryos for *in vitro* fertilization [269, 270].

2.3. TLM for assisted reproductive technology and its promise for clinical medicine

TLM is emerging as a promising clinical technique for selecting embryos for transplantation, although the discussion is still under way whether TLM may become an alternative to preimplantation genetic screening [271, 272]. The so-called morphokinetic analysis [273] by TLM is aimed to assess the number, development and quality (viability) of embryos by monitoring cleavage anomalies, multinucleation [274] or specific cell cycle kinetics [274, 275] and cleavage divisions [276], aneuploidy [277, 278], which is considered as a key causal factor of delays in embryonic development toward a blastocyte [278], and even chromosomal abnormalities [279]. Although more clinical research is required to finally prove that TLM can identify the best embryo for transfer and has an advantage over the conventional incubation of embryos [280], TLM is under consideration for patenting as a method for selecting embryos for implantation [281, 282]. TLM can also be used for sperm motility analysis [283].

One of the potential medical applications of TLM is the assessment of *ex vivo* engineered cells for cell therapy of degenerative and inherited disorders and other human pathologies like cancer [284–288]. TLM can also be used for diagnostics, for example, for detecting abnormalities in cell behavior in human dystrophic muscle cultures [289] or estimating tumor malignancy [290] in drug discovery [291], for testing gene therapeutic agents [292] and for evaluating side effects of antibiotics [293] and efficacy of chemotherapeutics [294, 295]. TLM is a valuable tool for understanding the pathogenesis of certain disorders, such as dysplastic erythroblast formation of erythroblasts from the patient with congenital dyserythropoietic anemia [296], thrombus formation [224], IgE-mediated mast cell degranulation and recovery [297], imaging of disease progression in deep brain areas using fluorescence microendoscopy [298], reprogramming in induced pluripotent cells [110] and other applications.

3. Conclusion

TLM is a powerful and versatile tool in modern biological research, with the immense potential for future clinical applications. One of the probably underexplored features of TLM is its promise to further characterize heterogeneity of cells within tissues [144], in particular, stem/progenitor cells and differentiating cells [299] as well as cancer cells [300]. Some of the above-mentioned methods are associated with unavoidable costs (expensive equipment, such as lenses, filters and sensors, and their damage due to high humidity within the incubator), non-natural impacts on living cells by the high excitation energy of lasers and bleaching/degradation of the fluorochromes over time, which influences quantification of long-running processes. However, the growing number of reports about new improvements and advances in TLM techniques and TLM-related applications that provide valuable information, which is not imageable by other techniques, makes it possible to conclude that the era of microcinematography in biomedical research has just begun.

Acknowledgements

We thank Mr. Joffry Maltha (CytoSMART Technologies) for assistance with preparation of this manuscript. The work was partially supported by the National Cancer Institute, USA, Award Number R37CA229417.

Conflict of interest

None declared.

Author details

John L. Collins[1], Bart van Knippenberg[2], Kai Ding[3] and Alexander V. Kofman[4*]

*Address all correspondence to: akofman@troy.edu

1 University of Tennessee at Martin, Martin, TN, USA

2 Cytosmart Technologies BV, Eindhoven, The Netherlands

3 Johns Hopkins University School of Medicine, Baltimore, MD, USA

4 Department of Biological & Environmental Sciences, Troy University, Troy, AL, USA

References

[1] Klausewitz W. Cytodiagnostic studies on living blood and lymph cells of some Amphibia by means of micro-time lapse film and phase contrast microscopy. Zeitschrift für Zellforschung und Mikroskopische Anatomie. 1953;**39**(1):1-35

[2] Baker M. Cellular imaging: Taking a long, hard look. Nature. 2010;**466**(7310):1137-1140

[3] Svensson CM, Medyukhina A, Belyaev I, Al-Zaben N, Figge MT. Untangling cell tracks: Quantifying cell migration by time lapse image data analysis. Cytometry Part A: The Journal of the International Society for Analytical Cytology. 2018;**93**(3):357-370

[4] Landecker H. Seeing things: From microcinematography to live cell imaging. Nature Methods. 2009;**6**(10):707-709

[5] Burton AL. Time-lapse phase-contrast cinephotomicrography: A new method in biological research. Canadian Medical Association Journal. 1962;**87**:20-26

[6] Borysko E, Sapranauskas P. A new technique for comparative phase-contrast and electron microscope studies of cells grown in tissue culture, with an evaluation of the technique by means of time-lapse cinemicrographs. Bulletin of the Johns Hopkins Hospital. 1954;**95**(2):68-79

[7] Kramis NJ. Time-lapse, phase contrast cine photomicrography of tissue culture cells. Journal of the Biological Photographic Association. 1956;**24**(1):27-29

[8] Rose GG. Time-lapse cinemicrography of cells in tissue culture. Bulletin of the Johns Hopkins Hospital. 1965;**116**:33-68

[9] Montgomery PO, Bonner WA. Ultra-violet time lapse motion picture observations of mitosis in newt cells. Experimental Cell Research. 1959;**17**(3):378-384

[10] Froese G. Division delay in HeLa cells in Chinese hamster cells. A time-lapse study. International Journal of Radiation Biology and Related Studies in Physics, Chemistry, and Medicine. 1966;**10**(4):353-367

[11] Overman RR, Pomerat CM. Electrolytes and plasma expanders. I. Reaction of human cells in perfusion chambers with phase contrast, time-lapse cine records. Zeitschrift für Zellforschung und Mikroskopische Anatomie. 1956;**45**(1):2-17

[12] Capers CR. Multinucleation of skeletal muscle in vitro. The Journal of Biophysical and Biochemical Cytology. 1960;**7**:559-566

[13] Terasaki PI, Cannon JA, Longmire WP Jr, Chamberlain CC. Antibody response to homografts: V. Cytotoxic effects upon lymphocytes as measured by time-lapse cinematography. Annals of the New York Academy of Sciences. 1960;**87**:258-265

[14] Sharp JA, Smiddy FG. Time-lapse cinemicrography of lymphoid tissue cultured in normal and in uraemic serum. Nature. 1962;**193**:191-192

[15] Jain P, Worthylake RA, Alahari SK. Quantitative analysis of random migration of cells using time-lapse video microscopy. Journal of Visualized Experiments: JoVE. 2012;**63**: e3585

[16] Dai L, Alt W, Schilling K, Retzlik J, Gieselmann V, Magin TM, et al. A fast and robust quantitative time-lapse assay for cell migration. Experimental Cell Research. 2005;**311**(2): 272-280

[17] Wick N, Thurner S, Paiha K, Sedivy R, Vietor I, Huber LA. Quantitative measurement of cell migration using time-lapse videomicroscopy and non-linear system analysis. Histochemistry and Cell Biology. 2003;**119**(1):15-20

[18] Arocena M, Zhao M, Collinson JM, Song B. A time-lapse and quantitative modelling analysis of neural stem cell motion in the absence of directional cues and in electric fields. Journal of Neuroscience Research. 2010;**88**(15):3267-3274

[19] Hossain WA, D'Sa C, Morest DK. Interactive roles of fibroblast growth factor 2 and neurotrophin 3 in the sequence of migration, process outgrowth, and axonal differentiation of mouse cochlear ganglion cells. Journal of Neuroscience Research. 2008;**86**(11):2376-2391

[20] Wagner W, Saffrich R, Wirkner U, Eckstein V, Blake J, Ansorge A, et al. Hematopoietic progenitor cells and cellular microenvironment: Behavioral and molecular changes upon interaction. Stem Cells. 2005;**23**(8):1180-1191

[21] Wells CM, Ridley AJ. Analysis of cell migration using the Dunn chemotaxis chamber and time-lapse microscopy. Methods in Molecular Biology. 2005;**294**:31-41

[22] Muinonen-Martin AJ, Veltman DM, Kalna G, Insall RH. An improved chamber for direct visualisation of chemotaxis. PLoS One. 2010;**5**(12):e15309

[23] Vasaturo A, Caserta S, Russo I, Preziosi V, Ciacci C, Guido S. A novel chemotaxis assay in 3-D collagen gels by time-lapse microscopy. PLoS One. 2012;**7**(12):e52251

[24] Kopp F, Hermawan A, Oak PS, Herrmann A, Wagner E, Roidl A. Salinomycin treatment reduces metastatic tumor burden by hampering cancer cell migration. Molecular Cancer. 2014;**13**:16

[25] Boyden S. The chemotactic effect of mixtures of antibody and antigen on polymorphonuclear leucocytes. The Journal of Experimental Medicine. 1962;**115**:453-466

[26] Zigmond SH. Ability of polymorphonuclear leukocytes to orient in gradients of chemotactic factors. The Journal of Cell Biology. 1977;**75**(2 Pt 1):606-616

[27] Lynch AE, Triajianto J, Routledge E. Low-cost motility tracking system (LOCOMOTIS) for time-lapse microscopy applications and cell visualisation. PLoS One. 2014;**9**(8):e103547

[28] Buhler H, Adamietz R, Abeln T, Diaz-Carballo D, Nguemgo-Kouam P, Hero T, et al. Automated multichamber time-lapse videography for long-term in vivo observation of migrating cells. In Vivo. 2017;**31**(3):329-334

[29] Mathieu E, Paul CD, Stahl R, Vanmeerbeeck G, Reumers V, Liu C, et al. Time-lapse lens-free imaging of cell migration in diverse physical microenvironments. Lab on a Chip. 2016;**16**(17):3304-3316

[30] Funahashi J, Nakamura H. Time-lapse imaging system with shell-less culture chamber. Development, Growth & Differentiation. 2014;**56**(4):305-309

[31] Coll JL, Ben-Ze'ev A, Ezzell RM, Rodriguez Fernandez JL, Baribault H, Oshima RG, et al. Targeted disruption of vinculin genes in F9 and embryonic stem cells changes cell morphology, adhesion, and locomotion. Proceedings of the National Academy of Sciences of the United States of America. 1995;**92**(20):9161-9165

[32] Denkers IA, Dragowska W, Jaggi B, Palcic B, Lansdorp PM. Time lapse video recordings of highly purified human hematopoietic progenitor cells in culture. Stem Cells. 1993; **11**(3):243-248

[33] Wuchter P, Leinweber C, Saffrich R, Hanke M, Eckstein V, Ho AD, et al. Plerixafor induces the rapid and transient release of stromal cell-derived factor-1 alpha from

human mesenchymal stromal cells and influences the migration behavior of human hematopoietic progenitor cells. Cell and Tissue Research. 2014;**355**(2):315-326

[34] Fonseca AV, Freund D, Bornhauser M, Corbeil D. Polarization and migration of hematopoietic stem and progenitor cells rely on the RhoA/ROCK I pathway and an active reorganization of the microtubule network. Journal of Biological Chemistry. 2010;**285**(41): 31661-31671

[35] Lee DH, Park BJ, Lee MS, Lee JW, Kim JK, Yang HC, et al. Chemotactic migration of human mesenchymal stem cells and MC3T3-E1 osteoblast-like cells induced by COS-7 cell line expressing rhBMP-7. Tissue Engineering. 2006;**12**(6):1577-1586

[36] Donnenberg AD, Cameron J. Cinemicroscopic studies of lymphocyte behavior in semisolid medium: The polyclonal origin of lymphocyte colonies. Experimental Hematology. 1985;**13**(1):29-35

[37] Dudley B, Molyneaux K. In vivo germ line stem cell migration: A mouse model. Methods in Molecular Biology. 2011;**750**:117-129

[38] Srihawong T, Kuwana T, Siripattarapravat K, Tirawattanawanich C. Chicken primordial germ cell motility in response to stem cell factor sensing. The International Journal of Developmental Biology. 2015;**59**(10-12):453-460

[39] Paksa A, Bandemer J, Hoeckendorf B, Razin N, Tarbashevich K, Minina S, et al. Repulsive cues combined with physical barriers and cell-cell adhesion determine progenitor cell positioning during organogenesis. Nature Communications. 2016;**7**:11288

[40] Song J, Yue Q, Munsterberg A. Time-lapse imaging of chick cardiac precursor cells. Methods in Molecular Biology. 2011;**769**:359-372

[41] Masyuk M, Morosan-Puopolo G, Brand-Saberi B, Theiss C. Combination of in ovo electroporation and time-lapse imaging to study migrational events in chicken embryos. Developmental Dynamics: An Official Publication of the American Association of the Anatomists. 2014;**243**(5):690-698

[42] Kornyei Z, Szlavik V, Szabo B, Gocza E, Czirok A, Madarasz E. Humoral and contact interactions in astroglia/stem cell co-cultures in the course of glia-induced neurogenesis. Glia. 2005;**49**(3):430-444

[43] Carbonell WS, Murase S, Horwitz AF, Mandell JW. Migration of perilesional microglia after focal brain injury and modulation by CC chemokine receptor 5: An in situ time-lapse confocal imaging study. The Journal of Neuroscience: The Official Journal of the Society for Neuroscience. 2005;**25**(30):7040-7047

[44] Braun H, Buhnemann C, Neumann J, Reymann KG. Preparation of a tissue-like cortical primary culture from embryonic rats using Matrigel and serum free start V medium. Journal of Neuroscience Methods. 2006;**157**(1):32-38

[45] Fan Y, Abrahamsen G, Mills R, Calderon CC, Tee JY, Leyton L, et al. Focal adhesion dynamics are altered in schizophrenia. Biological Psychiatry. 2013;**74**(6):418-426

[46] Famulski JK, Trivedi N, Howell D, Yang Y, Tong Y, Gilbertson R, et al. Siah regulation of Pard3A controls neuronal cell adhesion during germinal zone exit. Science. 2010; **330**(6012):1834-1838

[47] Mazharian A. Assessment of megakaryocyte migration and chemotaxis. Methods in Molecular Biology. 2012;**788**:275-288

[48] Nevius E, Pinho F, Dhodapkar M, Jin H, Nadrah K, Horowitz MC, et al. Oxysterols and EBI2 promote osteoclast precursor migration to bone surfaces and regulate bone mass homeostasis. The Journal of Experimental Medicine. 2015;**212**(11):1931-1946

[49] Guo R, Sakamoto H, Sugiura S, Ogawa M. Endothelial cell motility is compatible with junctional integrity. Journal of Cellular Physiology. 2007;**211**(2):327-335

[50] Kamlund S, Strand D, Janicke B, Alm K, Oredsson S. Influence of salinomycin treatment on division and movement of individual cancer cells cultured in normoxia or hypoxia evaluated with time-lapse digital holographic microscopy. Cell Cycle. 2017;**16**(21): 2128-2138

[51] Hartmann-Petersen R, Walmod PS, Berezin A, Berezin V, Bock E. Individual cell motility studied by time-lapse video recording: Influence of experimental conditions. Cytometry. 2000;**40**(4):260-270

[52] Dosch J, Hadley E, Wiese C, Soderberg M, Houwman T, Ding K, et al. Time-lapse microscopic observation of non-dividing cells in cultured human osteosarcoma MG-63 cell line. Cell Cycle. 2018;**17**(2):174-181

[53] Megyola CM, Gao Y, Teixeira AM, Cheng J, Heydari K, Cheng EC, et al. Dynamic migration and cell-cell interactions of early reprogramming revealed by high-resolution time-lapse imaging. Stem Cells. 2013;**31**(5):895-905

[54] Puche AC, Bovetti S. Studies of adult neural stem cell migration. Methods in Molecular Biology. 2011;**750**:227-240

[55] Ito H, Morishita R, Tabata H, Nagata K. Visualizing septin and cell dynamics in mammalian brain slices. Methods in Cell Biology. 2016;**136**:295-309

[56] Hughes EG, Kang SH, Fukaya M, Bergles DE. Oligodendrocyte progenitors balance growth with self-repulsion to achieve homeostasis in the adult brain. Nature Neuroscience. 2013;**16**(6):668-676

[57] Zhang RL, LeTourneau Y, Gregg SR, Wang Y, Toh Y, Robin AM, et al. Neuroblast division during migration toward the ischemic striatum: A study of dynamic migratory and proliferative characteristics of neuroblasts from the subventricular zone. The Journal of Neuroscience: The Official Journal of the Society for Neuroscience. 2007;**27**(12):3157-3162

[58] Zhao LR, Nam SC. Multiphoton microscope imaging: The behavior of neural progenitor cells in the rostral migratory stream. Neuroscience Letters. 2007;**425**(2):83-88

[59] Comte I, Kim Y, Young CC, van, der Harg JM, Hockberger P, Bolam PJ, et al. Galectin-3 maintains cell motility from the subventricular zone to the olfactory bulb. Journal of Cell Science. 2011;**124**(Pt 14):2438-2447

[60] Sonego M, Oberoi M, Stoddart J, Gajendra S, Hendricusdottir R, Oozeer F, et al. Drebrin regulates neuroblast migration in the postnatal mammalian brain. PLoS One. 2015;**10**(5):e0126478

[61] Sonego M, Gajendra S, Parsons M, Ma Y, Hobbs C, Zentar MP, et al. Fascin regulates the migration of subventricular zone-derived neuroblasts in the postnatal brain. The Journal of Neuroscience: The Official Journal of the Society for Neuroscience. 2013;**33**(30):12171-12185

[62] Kulesa P, Bronner-Fraser M, Fraser S. In ovo time-lapse analysis after dorsal neural tube ablation shows rerouting of chick hindbrain neural crest. Development. 2000;**127**(13): 2843-2852

[63] Noctor SC, Martinez-Cerdeno V, Ivic L, Kriegstein AR. Cortical neurons arise in symmetric and asymmetric division zones and migrate through specific phases. Nature Neuroscience. 2004;**7**(2):136-144

[64] Chen S, Bremer AW, Scheideler OJ, Na YS, Todhunter ME, Hsiao S, et al. Interrogating cellular fate decisions with high-throughput arrays of multiplexed cellular communities. Nature Communications. 2016;**7**:10309

[65] Merouane A, Rey-Villamizar N, Lu Y, Liadi I, Romain G, Lu J, et al. Automated profiling of individual cell-cell interactions from high-throughput time-lapse imaging microscopy in nanowell grids (TIMING). Bioinformatics. 2015;**31**(19):3189-3197

[66] Hirata Y, Li HW, Takahashi K, Ishii H, Sykes M, Fujisaki J. MHC class I expression by donor hematopoietic stem cells is required to prevent NK cell attack in allogeneic, but not syngeneic recipient mice. PLoS One. 2015;**10**(11):e0141785

[67] Ciccocioppo R, Cangemi GC, Kruzliak P, Gallia A, Betti E, Badulli C, et al. Ex vivo immunosuppressive effects of mesenchymal stem cells on Crohn's disease mucosal T cells are largely dependent on indoleamine 2,3-dioxygenase activity and cell-cell contact. Stem Cell Research & Therapy. 2015;**6**:137

[68] Tomura M, Mori YS, Watanabe R, Tanaka M, Miyawaki A, Kanagawa O. Time-lapse observation of cellular function with fluorescent probe reveals novel CTL-target cell interactions. International Immunology. 2009;**21**(10):1145-1150

[69] Jensen AM, Raff MC. Continuous observation of multipotential retinal progenitor cells in clonal density culture. Developmental Biology. 1997;**188**(2):267-279

[70] Shih YT, Wang MC, Peng HH, Chen TF, Chen L, Chang JY, et al. Modulation of chemotactic and pro-inflammatory activities of endothelial progenitor cells by hepatocellular carcinoma. Cellular Signalling. 2012;**24**(3):779-793

[71] Yamazaki K, Roberts RA, Spooncer E, Dexter TM, Allen TD. Cellular interactions between 3T3 cells and interleukin-3-dependent multipotent haemopoietic cells: A model system for stromal-cell-mediated haemopoiesis. Journal of Cellular Physiology. 1989;**139**(2):301-312

[72] Ioannidou K, Edgar JM, Barnett SC. Time-lapse imaging of glial-axonal interactions. Current Protocols in Neuroscience. 2015;**72**:2.23.1-2.23.14

[73] Ioannidou K, Anderson KI, Strachan D, Edgar JM, Barnett SC. Time-lapse imaging of the dynamics of CNS glial-axonal interactions in vitro and ex vivo. PLoS One. 2012;**7**(1): e30775

[74] Louhivuori LM, Jansson L, Turunen PM, Jantti MH, Nordstrom T, Louhivuori V, et al. Transient receptor potential channels and their role in modulating radial glial-neuronal interaction: A signaling pathway involving mGluR5. Stem Cells and Development. 2015;**24**(6):701-713

[75] Bayerl SH, Niesner R, Cseresnyes Z, Radbruch H, Pohlan J, Brandenburg S, et al. Time lapse in vivo microscopy reveals distinct dynamics of microglia-tumor environment interactions-a new role for the tumor perivascular space as highway for trafficking microglia. Glia. 2016;**64**(7):1210-1226

[76] Black AB, Dahlenburg H, Pepper K, Nacey C, Pontow S, Kuhn MA, et al. Human myoblast and mesenchymal stem cell interactions visualized by videomicroscopy. Human Gene Therapy Methods. 2015;**26**(6):193-196

[77] Silva AM, Oliveira MI, Sette L, Almeida CR, Oliveira MJ, Barbosa MA, et al. Resveratrol as a natural anti-tumor necrosis factor-alpha molecule: Implications to dendritic cells and their crosstalk with mesenchymal stromal cells. PLoS One. 2014;**9**(3):e91406

[78] Ern C, Krump-Konvalinkova V, Docheva D, Schindler S, Rossmann O, Bocker W, et al. Interactions of human endothelial and multipotent mesenchymal stem cells in cocultures. The Open Biomedical Engineering Journal. 2010;**4**:190-198

[79] Al-toub M, Vishnubalaji R, Hamam R, Kassem M, Aldahmash A, Alajez NM. CDH1 and IL1-beta expression dictates FAK and MAPKK-dependent cross-talk between cancer cells and human mesenchymal stem cells. Stem Cell Research & Therapy. 2015;**6**:135

[80] Su PJ, Tran QA, Fong JJ, Eliceiri KW, Ogle BM, Campagnola PJ. Mesenchymal stem cell interactions with 3D ECM modules fabricated via multiphoton excited photochemistry. Biomacromolecules. 2012;**13**(9):2917-2925

[81] Allen TD, Testa NG. Cellular interactions in erythroblastic islands in long-term bone marrow cultures, as studied by time-lapse video. Blood Cells. 1991;**17**(1):29-38. Discussion 9-43

[82] Baghbaderani BA, Behie LA, Mukhida K, Hong M, Mendez I. New bioengineering insights into human neural precursor cell expansion in culture. Biotechnology Progress. 2011;**27**(3):776-787

[83] Kulesa PM, Fraser SE. In ovo time-lapse analysis of chick hindbrain neural crest cell migration shows cell interactions during migration to the branchial arches. Development. 2000;**127**(6):1161-1172

[84] Fok-Seang J, Mathews GA, Ffrench-Constant C, Trotter J, Fawcett JW. Migration of oligodendrocyte precursors on astrocytes and meningeal cells. Developmental Biology. 1995;**171**(1): 1-15

[85] Gottschling S, Eckstein V, Saffrich R, Jonas A, Uhrig M, Krause U, et al. Primitive and committed human hematopoietic progenitor cells interact with primary murine neural cells and are induced to undergo self-renewing cell divisions. Experimental Hematology. 2007;**35**(12):1858-1871

[86] Koyanagi M, Brandes RP, Haendeler J, Zeiher AM, Dimmeler S. Cell-to-cell connection of endothelial progenitor cells with cardiac myocytes by nanotubes: A novel mechanism for cell fate changes? Circulation Research. 2005;**96**(10):1039-1041

[87] Vidulescu C, Clejan S, O'Connor KC. Vesicle traffic through intercellular bridges in DU 145 human prostate cancer cells. Journal of Cellular and Molecular Medicine. 2004; **8**(3):388-396

[88] Nishiyama N, Colonna J, Shen E, Carrillo J, Nishiyama H. Long-term in vivo time-lapse imaging of synapse development and plasticity in the cerebellum. Journal of Neurophysiology. 2014;**111**(1):208-216

[89] Kopel H, Schechtman E, Groysman M, Mizrahi A. Enhanced synaptic integration of adult-born neurons in the olfactory bulb of lactating mothers. The Journal of Neuroscience: The Official Journal of the Society for Neuroscience. 2012;**32**(22):7519-7527

[90] Wang L, Kisaalita WS. Administration of BDNF/ginsenosides combination enhanced synaptic development in human neural stem cells. Journal of Neuroscience Methods. 2011;**194**(2):274-282

[91] Li Q, Deng Z, Zhang Y, Zhou X, Nagerl UV, Wong ST. A global spatial similarity optimization scheme to track large numbers of dendritic spines in time-lapse confocal microscopy. IEEE Transactions on Medical Imaging. 2011;**30**(3):632-641

[92] Li J, Erisir A, Cline H. In vivo time-lapse imaging and serial section electron microscopy reveal developmental synaptic rearrangements. Neuron. 2011;**69**(2):273-286

[93] Yampolsky P, Pacifici PG, Lomb L, Giese G, Rudolf R, Roder IV, et al. Time lapse in vivo visualization of developmental stabilization of synaptic receptors at neuromuscular junctions. Journal of Biological Chemistry. 2010;**285**(45):34589-34596

[94] Walsh MK, Lichtman JW. In vivo time-lapse imaging of synaptic takeover associated with naturally occurring synapse elimination. Neuron. 2003;**37**(1):67-73

[95] Piltti KM, Cummings BJ, Carta K, Manughian-Peter A, Worne CL, Singh K, et al. Live-cell time-lapse imaging and single-cell tracking of in vitro cultured neural stem cells—Tools for analyzing dynamics of cell cycle, migration, and lineage selection. Methods. 2018;**133**:81-90

[96] Yang R, Wang M, Wang J, Huang X, Yang R, Gao WQ. Cell division mode change mediates the regulation of cerebellar granule neurogenesis controlled by the sonic hedgehog Signaling. Stem Cell Reports. 2015;**5**(5):816-828

[97] Ortega F, Berninger B, Costa MR. Primary culture and live imaging of adult neural stem cells and their progeny. Methods in Molecular Biology. 2013;**1052**:1-11

[98] Daynac M, Morizur L, Kortulewski T, Gauthier LR, Ruat M, Mouthon MA, et al. Cell sorting of neural stem and progenitor cells from the adult mouse subventricular zone and live-imaging of their cell cycle dynamics. Journal of Visualized Experiments: JoVE. 2015;**103**:53247

[99] Blanchoud S, Nicolas D, Zoller B, Tidin O, Naef F. CAST: An automated segmentation and tracking tool for the analysis of transcriptional kinetics from single-cell time-lapse recordings. Methods. 2015;**85**:3-11

[100] Pacini G, Marino A, Migliarini S, Brilli E, Pelosi B, Maddaloni G, et al. A Tph2GFP reporter stem cell line to model in vitro and in vivo serotonergic neuron development and function. ACS Chemical Neuroscience. 2017;**8**:1043-1052

[101] Suga M, Kii H, Niikura K, Kiyota Y, Furue MK. Development of a monitoring method for nonlabeled human pluripotent stem cell growth by time-lapse image analysis. Stem Cells Translational Medicine. 2015;**4**(7):720-730

[102] Haetscher N, Feuermann Y, Wingert S, Rehage M, Thalheimer FB, Weiser C, et al. STAT5-regulated microRNA-193b controls haematopoietic stem and progenitor cell expansion by modulating cytokine receptor signalling. Nature Communications. 2015;**6**:8928

[103] Azimi MS, Motherwell JM, Murfee WL. An ex vivo method for time-lapse imaging of cultured rat mesenteric microvascular networks. Journal of Visualized Experiments: JoVE. 2017;**120**. DOI: 10.3791/55183

[104] Mackay DR, Ullman KS, Rodesch CK. Time-lapse imaging of mitosis after siRNA transfection. Journal of Visualized Experiments: JoVE. 2010;**40**:1878

[105] Lin S, Fonteno S, Satish S, Bhanu B, Talbot P. Video bioinformatics analysis of human embryonic stem cell colony growth. Journal of Visualized Experiments: JoVE. 2010;**39**:1933

[106] Pilaz LJ, Silver DL. Live imaging of mitosis in the developing mouse embryonic cortex. Journal of Visualized Experiments: JoVE. 2014;**88**. DOI: 10.3791/51298

[107] Errington RJ, Chappell SC, Khan IA, Marquez N, Wiltshire M, Griesdoorn VD, et al. Time-lapse microscopy approaches to track cell cycle and lineage progression at the single-cell level. Current Protocols in Cytometry. 2013;**64**(1):12.4.1-12.4.13

[108] Scherf N, Franke K, Glauche I, Kurth I, Bornhauser M, Werner C, et al. On the symmetry of siblings: Automated single-cell tracking to quantify the behavior of hematopoietic stem cells in a biomimetic setup. Experimental Hematology. 2012;**40**(2):119-30 e9

[109] Ortega F, Costa MR, Simon-Ebert T, Schroeder T, Gotz M, Berninger B. Using an adherent cell culture of the mouse subependymal zone to study the behavior of adult neural stem cells on a single-cell level. Nature Protocols. 2011;**6**(12):1847-1859

[110] Smith ZD, Nachman I, Regev A, Meissner A. Dynamic single-cell imaging of direct reprogramming reveals an early specifying event. Nature Biotechnology. 2010;**28**(5):521-526

[111] Sillitoe K, Horton C, Spiller DG, White MR. Single-cell time-lapse imaging of the dynamic control of NF-kappaB signalling. Biochemical Society Transactions. 2007;**35**(Pt 2): 263-266

[112] Errington RJ, Marquez N, Chappell SC, Wiltshire M, Smith PJ. Time-lapse microscopy approaches to track cell cycle progression at the single-cell level. Current Protocols in Cytometry. 2005;**31**(1):12.4.1-12.4.11

[113] Hashimoto H, Yuasa S, Tabata H, Tohyama S, Hayashiji N, Hattori F, et al. Time-lapse imaging of cell cycle dynamics during development in living cardiomyocyte. Journal of Molecular and Cellular Cardiology. 2014;**72**:241-249

[114] Mort RL, Ford MJ, Sakaue-Sawano A, Lindstrom NO, Casadio A, Douglas AT, et al. Fucci2a: A bicistronic cell cycle reporter that allows Cre mediated tissue specific expression in mice. Cell Cycle. 2014;**13**(17):2681-2696

[115] Miwa S, Yano S, Kimura H, Yamamoto M, Toneri M, Murakami T, et al. Heterogeneous cell-cycle behavior in response to UVB irradiation by a population of single cancer cells visualized by time-lapse FUCCI imaging. Cell Cycle. 2015;**14**(12):1932-1937

[116] Jovic D, Sakaue-Sawano A, Abe T, Cho CS, Nagaoka M, Miyawaki A, et al. Direct observation of cell cycle progression in living mouse embryonic stem cells on an extracellular matrix of E-cadherin. Springerplus. 2013;**2**:585

[117] Holt JE, Lane SI, Jones KT. Time-lapse epifluorescence imaging of expressed cRNA to cyclin B1 for studying meiosis I in mouse oocytes. Methods in Molecular Biology. 2013;**957**:91-106

[118] Shankaran SS, Mackay DR, Ullman KS. A time-lapse imaging assay to study nuclear envelope breakdown. Methods in Molecular Biology. 2013;**931**:111-122

[119] Andrade R, Crisol L, Prado R, Boyano MD, Arluzea J, Arechaga J. Plasma membrane and nuclear envelope integrity during the blebbing stage of apoptosis: A time-lapse study. Biology of the Cell. 2009;**102**(1):25-35

[120] Hernandez-Verdun D, Louvet E, Muro E. Time-lapse, photoactivation, and photobleaching imaging of nucleolar assembly after mitosis. Methods in Molecular Biology. 2013;**1042**:337-350

[121] Drpic D, Barisic M, Pinheiro D, Maiato H. Selective tracking of template DNA strands after induction of mitosis with unreplicated genomes (MUGs) in drosophila S2 cells. Chromosome Research: An International Journal on the Molecular, Supramolecular and Evolutionary Aspects of Chromosome Biology. 2013;**21**(3):329-337

[122] Schultz N, Onfelt A. Video time-lapse study of mitosis in binucleate V79 cells: Chromosome segregation and cleavage. Mutagenesis. 1994;**9**(2):117-123

[123] Wong MN, Nguyen TP, Chen TH, Hsu JJ, Zeng X, Saw A, et al. Preferred mitotic orientation in pattern formation by vascular mesenchymal cells. American Journal of Physiology. Heart and Circulatory Physiology. 2012;**303**(12):H1411-H1417

[124] Siegel AL, Kuhlmann PK, Cornelison DD. Muscle satellite cell proliferation and association: New insights from myofiber time-lapse imaging. Skeletal Muscle. 2011;**1**(1):7

[125] Spear PC, Erickson CA. Interkinetic nuclear migration: A mysterious process in search of a function. Development, Growth & Differentiation. 2012;**54**(3):306-316

[126] Spear PC, Erickson CA. Apical movement during interkinetic nuclear migration is a two-step process. Developmental Biology. 2012;**370**(1):33-41

[127] Dong Z, Yang N, Yeo SY, Chitnis A, Guo S. Intralineage directional notch signaling regulates self-renewal and differentiation of asymmetrically dividing radial glia. Neuron. 2012;**74**(1):65-78

[128] Namba T, Mochizuki H, Suzuki R, Onodera M, Yamaguchi M, Namiki H, et al. Time-lapse imaging reveals symmetric neurogenic cell division of GFAP-expressing progenitors for expansion of postnatal dentate granule neurons. PLoS One. 2011;**6**(9):e25303

[129] Huang S, Law P, Francis K, Palsson BO, Ho AD. Symmetry of initial cell divisions among primitive hematopoietic progenitors is independent of ontogenic age and regulatory molecules. Blood. 1999;**94**(8):2595-2604

[130] Neumann B, Walter T, Heriche JK, Bulkescher J, Erfle H, Conrad C, et al. Phenotypic profiling of the human genome by time-lapse microscopy reveals cell division genes. Nature. 2010;**464**(7289):721-727

[131] Kosodo Y, Toida K, Dubreuil V, Alexandre P, Schenk J, Kiyokage E, et al. Cytokinesis of neuroepithelial cells can divide their basal process before anaphase. The EMBO Journal. 2008;**27**(23):3151-3163

[132] Krishan A. Cytochalasin-B: Time-lapse cinematographic studies on its effects on cytokinesis. The Journal of Cell Biology. 1972;**54**(3):657-664

[133] Kogo H, Fujimoto T. Concentration of caveolin-1 in the cleavage furrow as revealed by time-lapse analysis. Biochemical and Biophysical Research Communications. 2000; **268**(1):82-87

[134] Guizetti J, Mantler J, Muller-Reichert T, Gerlich DW. Correlative time-lapse imaging and electron microscopy to study abscission in HeLa cells. Methods in Cell Biology. 2010;**96**:591-601

[135] Sattler CA, Sawada N, Sattler GL, Pitot HC. Electron microscopic and time lapse studies of mitosis in cultured rat hepatocytes. Hepatology. 1988;**8**(6):1540-1549

[136] Cone CD Jr, Tongier M Jr. Mitotic synchronization of L-strain fibroblasts with 5-aminouracil as determined by time-lapse cinephotography. NASA TN D-5021. Technical Note United States National Aeronautics and Space Administration. 1969:1-27

[137] Stoll EA, Habibi BA, Mikheev AM, Lasiene J, Massey SC, Swanson KR, et al. Increased re-entry into cell cycle mitigates age-related neurogenic decline in the murine subventricular zone. Stem Cells. 2011;**29**(12):2005-2017

[138] Absher M, Ryan US. Comparison of pulmonary endothelial cell and fibroblast proliferation using time-lapse cinematographic analysis. Tissue & Cell. 1981;**13**(4):645-650

[139] Bedford JS, Mitchell JB. Mitotic accumulation of HeLa cells during continuous irradiation. Observations using time-lapse cinemicrography. Radiation Research. 1977;**70**(1): 173-186

[140] Collyn-d'Hooghe M, Hemon D, Gilet R, Curtis SB, Valleron AJ, Malaise EP. Comparative effects of 60Co gamma-rays and neon and helium ions on cycle duration and division probability of EMT 6 cells. A time-lapse cinematography study. International Journal of Radiation Biology and Related Studies in Physics, Chemistry, and Medicine. 1981;**39**(3):297-306

[141] Sigoillot FD, Huckins JF, Li F, Zhou X, Wong ST, King RW. A time-series method for automated measurement of changes in mitotic and interphase duration from time-lapse movies. PLoS One. 2011;**6**(9):e25511

[142] Dykstra B, Ramunas J, Kent D, McCaffrey L, Szumsky E, Kelly L, et al. High-resolution video monitoring of hematopoietic stem cells cultured in single-cell arrays identifies new features of self-renewal. Proceedings of the National Academy of Sciences of the United States of America. 2006;**103**(21):8185-8190

[143] Chu K, Teele N, Dewey MW, Albright N, Dewey WC. Computerized video time lapse study of cell cycle delay and arrest, mitotic catastrophe, apoptosis and clonogenic survival in irradiated 14-3-3sigma and CDKN1A (p21) knockout cell lines. Radiation Research. 2004;**162**(3):270-286

[144] Dover R, Potten CS. Heterogeneity and cell cycle analyses from time-lapse studies of human keratinocytes in vitro. Journal of Cell Science. 1988;**89**(Pt 3):359-364

[145] Zielke-Temme B, Hopwood L. Time-lapse cinemicrographic observations of heated G1-phase Chinese hamster ovary cells. I. Division probabilities and generation times. Radiation Research. 1982;**92**(2):320-331

[146] d'Hooghe MC, Hemon D, Valleron AJ, Malaise EP. Comparative effects of ionizing radiations on cycle time and mitotic duration. A time-lapse cinematography study. Radiation Research. 1980;**81**(3):384-392

[147] Colly-d'Hooghe M, Valleron AJ, Malaise EP. Time-lapse cinematography studies of cell cycle and mitosis duration. Experimental Cell Research. 1977;**106**(2):405-407

[148] Coronado D, Godet M, Bourillot PY, Tapponnier Y, Bernat A, Petit M, et al. A short G1 phase is an intrinsic determinant of naive embryonic stem cell pluripotency. Stem Cell Research. 2013;**10**(1):118-131

[149] Kinzel V, Bonheim G, Richards J. Phorbol ester-induced G2 delay in HeLa cells analyzed by time lapse photography. Cancer Research. 1988;**48**(7):1759-1762

[150] Sundaram M, Guernsey DL, Rajaraman MM, Rajaraman R. Neosis: A novel type of cell division in cancer. Cancer Biology & Therapy. 2004;**3**(2):207-218

[151] Yu B, Kim HW, Gong M, Wang J, Millard RW, Wang Y, et al. Exosomes secreted from GATA-4 overexpressing mesenchymal stem cells serve as a reservoir of anti-apoptotic microRNAs for cardioprotection. International Journal of Cardiology. 2015;**182**:349-360

[152] Nagy G, Pinter G, Kohut G, Adam AL, Trencsenyi G, Hornok L, et al. Time-lapse analysis of cell death in mammalian and fungal cells. DNA and Cell Biology. 2010;**29**(5):249-259

[153] Zahm JM, Baconnais S, Monier S, Bonnet N, Bessede G, Gambert P, et al. Chronology of cellular alterations during 7-ketocholesterol-induced cell death on A7R5 rat smooth muscle cells: Analysis by time lapse-video microscopy and conventional fluorescence microscopy. Cytometry Part A: The Journal of the International Society for Analytical Cytology. 2003;**52**(2):57-69

[154] Forrester HB, Vidair CA, Albright N, Ling CC, Dewey WC. Using computerized video time lapse for quantifying cell death of X-irradiated rat embryo cells transfected with c-myc or c-Ha-ras. Cancer Research. 1999;**59**(4):931-939

[155] Hurwitz C, Tolmach LJ. Time lapse cinemicrographic studies of x-irradiated HeLa S3 cells. I. Cell progression and cell disintegration. Biophysical Journal. 1969;**9**(4):607-633

[156] Hwang SY, Cho SH, Cho DY, Lee M, Choo J, Jung KH, et al. Time-lapse, single cell based confocal imaging analysis of caspase activation and phosphatidylserine flipping during cellular apoptosis. Biotechnic & Histochemistry: Official Publication of the Biological Stain Commission. 2011;**86**(3):181-187

[157] Saijo S, Nagata K, Masuda J, Matsumoto I, Kobayashi Y. Discrimination of early and late apoptotic cells by NBD-phosphatidylserine-labelling and time-lapse observation of phagocytosis of apoptotic cells by macrophages. Journal of Biochemistry. 2007;**141**(3): 301-307

[158] Wallberg F, Tenev T, Meier P. Time-lapse imaging of necrosis. Methods in Molecular Biology. 2013;**1004**:17-29

[159] Cai Q, Zakaria HM, Sheng ZH. Long time-lapse imaging reveals unique features of PARK2/parkin-mediated mitophagy in mature cortical neurons. Autophagy. 2012;**8**(6): 976-978

[160] Oh DJ, Lee GM, Francis K, Palsson BO. Phototoxicity of the fluorescent membrane dyes PKH2 and PKH26 on the human hematopoietic KG1a progenitor cell line. Cytometry. 1999;**36**(4):312-318

[161] Herman B, Albertini DF. A time-lapse video image intensification analysis of cytoplasmic organelle movements during endosome translocation. The Journal of Cell Biology. 1984;**98**(2):565-576

[162] Farnum CE, Turgai J, Wilsman NJ. Visualization of living terminal hypertrophic chondrocytes of growth plate cartilage in situ by differential interference contrast microscopy and time-lapse cinematography. Journal of Orthopaedic Research: Official Publication of the Orthopaedic Research Society. 1990;**8**(5):750-763

[163] Barasa A, Godina G, Buffa P, Pasquali-Ronchetti I. Biochemical lesions of respiratory enzymes and configurational changes of mitochondria in vivo. I. The effect of fluoroacetate: A study by phase-contrast microscopy and time-lapse cinemicrography. Zeitschrift für Zellforschung und Mikroskopische Anatomie. 1973;**138**(2):187-210

[164] Sison M, Chakrabortty S, Extermann J, Nahas A, James Marchand P, Lopez A, et al. 3D time-lapse imaging and quantification of mitochondrial dynamics. Scientific Reports. 2017;**7**:43275

[165] Gonzalez S, Fernando R, Berthelot J, Perrin-Tricaud C, Sarzi E, Chrast R, et al. In vivo time-lapse imaging of mitochondria in healthy and diseased peripheral myelin sheath. Mitochondrion. 2015;**23**:32-41

[166] Spacek J. Dynamics of the Golgi method: A time-lapse study of the early stages of impregnation in single sections. Journal of Neurocytology. 1989;**18**(1):27-38

[167] Pampalona J, Januschke J, Sampaio P, Gonzalez C. Time-lapse recording of centrosomes and other organelles in Drosophila neuroblasts. Methods in Cell Biology. 2015;**129**: 301-315

[168] Bang C, Cheng J. Dynamic interplay of spectrosome and centrosome organelles in asymmetric stem cell divisions. PLoS One. 2015;**10**(4):e0123294

[169] Wiese C, Mayers JR, Albee AJ. Analysis of centrosome function and microtubule dynamics by time-lapse microscopy in Xenopus egg extracts. Methods in Molecular Biology. 2009;**586**:89-113

[170] Braun A, Caesar NM, Dang K, Myers KA. High-resolution time-lapse imaging and automated analysis of microtubule dynamics in living human umbilical vein endothelial cells. Journal of Visualized Experiments: JoVE. 2016;**114**:54265

[171] Sullivan KF, Shelby RD. Using time-lapse confocal microscopy for analysis of centromere dynamics in human cells. Methods in Cell Biology. 1999;**58**:183-202

[172] Snapp EL, Lajoie P. Time-lapse imaging of membrane traffic in living cells. Cold Spring Harbor Protocols. 2011;**2011**(11):1362-1365

[173] Verkuyl JM, Matus A. Time-lapse imaging of dendritic spines in vitro. Nature Protocols. 2006;**1**(5):2399-2405

[174] Pearson RA, Luneborg NL, Becker DL, Mobbs P. Gap junctions modulate interkinetic nuclear movement in retinal progenitor cells. The Journal of Neuroscience: The Official Journal of the Society for Neuroscience. 2005;**25**(46):10803-10814

[175] Kosodo Y, Suetsugu T, Suda M, Mimori-Kiyosue Y, Toida K, Baba SA, et al. Regulation of interkinetic nuclear migration by cell cycle-coupled active and passive mechanisms in the developing brain. The EMBO Journal. 2011;**30**(9):1690-1704

[176] Abe S, Seitoku E, Iwadera N, Hamba Y, Yamagata S, Akasaka T, et al. Estimation of biocompatibility of nano-sized ceramic particles with osteoblasts, osteosarcomas and hepatocytes by static and time-lapse observation. Journal of Biomedical Nanotechnology. 2016;**12**(3):472-480

[177] Schwamborn R, Dussmann H, Konig HG, Prehn JHM. Time-lapse imaging of p65 and IkappaBalpha translocation kinetics following Ca(2+)-induced neuronal injury reveals biphasic translocation kinetics in surviving neurons. Molecular and Cellular Neurosciences. 2017;**80**:148-158

[178] Jokhadar SZ, Sustar V, Svetina S, Batista U. Time lapse monitoring of CaCo-2 cell shapes and shape dependence of the distribution of integrin beta1 and F-actin on their basal membrane. Cell Communication & Adhesion. 2009;**16**(1-3):1-13

[179] Shimojo H, Isomura A, Ohtsuka T, Kori H, Miyachi H, Kageyama R. Oscillatory control of Delta-like1 in cell interactions regulates dynamic gene expression and tissue morphogenesis. Genes & Development. 2016;**30**(1):102-116

[180] Trinkle-Mulcahy L, Andrews PD, Wickramasinghe S, Sleeman J, Prescott A, Lam YW, et al. Time-lapse imaging reveals dynamic relocalization of PP1gamma throughout the mammalian cell cycle. Molecular Biology of the Cell. 2003;**14**(1):107-117

[181] Webb CD, Graumann PL, Kahana JA, Teleman AA, Silver PA, Losick R. Use of time-lapse microscopy to visualize rapid movement of the replication origin region of the chromosome during the cell cycle in *Bacillus subtilis*. Molecular Microbiology. 1998;**28**(5):883-892

[182] Georgescu W, Osseiran A, Rojec M, Liu Y, Bombrun M, Tang J, et al. Characterizing the DNA damage response by cell tracking algorithms and cell features classification using high-content time-lapse analysis. PLoS One. 2015;**10**(6):e0129438

[183] Mazo-Vargas A, Park H, Aydin M, Buchler NE. Measuring fast gene dynamics in single cells with time-lapse luminescence microscopy. Molecular Biology of the Cell. 2014;**25**(22):3699-3708

[184] Villalta JI, Galli S, Iacaruso MF, Antico Arciuch VG, Poderoso JJ, Jares-Erijman EA, et al. New algorithm to determine true colocalization in combination with image restoration and time-lapse confocal microscopy to MAP kinases in mitochondria. PLoS One. 2011;**6**(4):e19031

[185] Wang MD, Axelrod D. Time-lapse total internal reflection fluorescence video of acetylcholine receptor cluster formation on myotubes. Developmental Dynamics: An Official Publication of the American Association of the Anatomists. 1994;**201**(1):29-40

[186] Swedlow JR, Sedat JW, Agard DA. Multiple chromosomal populations of topoisomerase II detected in vivo by time-lapse, three-dimensional wide-field microscopy. Cell. 1993;**73**(1):97-108

[187] Hiraoka Y, Minden JS, Swedlow JR, Sedat JW, Agard DA. Focal points for chromosome condensation and decondensation revealed by three-dimensional in vivo time-lapse microscopy. Nature. 1989;**342**(6247):293-296

[188] Owens DF, Kriegstein AR. Patterns of intracellular calcium fluctuation in precursor cells of the neocortical ventricular zone. The Journal of Neuroscience: The Official Journal of the Society for Neuroscience. 1998;**18**(14):5374-5388

[189] Stricker SA. Time-lapse confocal imaging of calcium dynamics in starfish embryos. Developmental Biology. 1995;**170**(2):496-518

[190] Dussmann H, Perez-Alvarez S, Anilkumar U, Papkovsky DB, Prehn JH. Single-cell time-lapse imaging of intracellular O2 in response to metabolic inhibition and mitochondrial cytochrome-c release. Cell Death & Disease. 2017;**8**(6):e2853

[191] Hoffman H, Frank ME. Time-lapse photomicrography of cell growth and division in *Escherichia coli*. Journal of Bacteriology. 1965;**89**:212-216

[192] Hoffman H, Frank ME. Time-lapse photomicrography of lashing, flexing, and snapping movements in *Escherichia coli* and *Corynebacterium* microcultures. Journal of Bacteriology. 1965;**90**(3):789-795

[193] Shapiro JA, Hsu C. *Escherichia coli* K-12 cell-cell interactions seen by time-lapse video. Journal of Bacteriology. 1989;**171**(11):5963-5974

[194] Hoffman H, Frank ME. Time-lapse photomicrography of the formation of a free spherical granule in an *Escherichia coli* cell end. Journal of Bacteriology. 1963;**86**:1075-1078

[195] Penil Cobo M, Libro S, Marechal N, D'Entremont D, Penil Cobo D, Berkmen M. Visualizing bacterial colony morphologies using time-lapse imaging chamber MOCHA. Journal of Bacteriology. 2018;**200**(2):e00413-17

[196] Ungphakorn W, Lagerback P, Nielsen EI, Tangden T. Automated time-lapse microscopy a novel method for screening of antibiotic combination effects against multidrug-resistant gram-negative bacteria. Clinical Microbiology and Infection: The Official Publication of the European Society of Clinical Microbiology and Infectious Diseases. 2018;**24**(7):778.e777-778.e714

[197] Durham NN, Noller EC, Burger MW, Best GK. Time-lapse cinematography of vancomycin--treated microbial cells. Canadian Journal of Microbiology. 1967;**13**(4):417-421

[198] Ungphakorn W, Malmberg C, Lagerback P, Cars O, Nielsen EI, Tangden T. Evaluation of automated time-lapse microscopy for assessment of in vitro activity of antibiotics. Journal of Microbiological Methods. 2017;**132**:69-75

[199] Louvet JN, Carrion C, Stalder T, Alrhmoun M, Casellas M, Potier O, et al. Vancomycin sorption on activated sludge gram(+) bacteria rather than on EPS; 3D confocal laser scanning microscopy time-lapse imaging. Water Research. 2017;**124**:290-297

[200] Schmidt GW, Frey O, Rudolf F. The CellClamper: A convenient microfluidic device for time-lapse imaging of yeast. Methods in Molecular Biology. 2018;**1672**:537-555

[201] Kumar A, Mendoza M. Time-lapse fluorescence microscopy of budding yeast cells. Methods in Molecular Biology. 2016;**1369**:1-8

[202] Kron SJ. Digital time-lapse microscopy of yeast cell growth. Methods in Enzymology. 2002;**351**:3-15

[203] Heymann JB, Cheng N, Newcomb WW, Trus BL, Brown JC, Steven AC. Dynamics of herpes simplex virus capsid maturation visualized by time-lapse cryo-electron microscopy. Nature Structural Biology. 2003;**10**(5):334-341

[204] Lemay P, Collyn-D'Hooghe M. Flow cytophotometric and time-lapse cinematographic study of human cells infected by adenovirus type 2 wild-type and two DNA-negative temperature-sensitive mutants. The Journal of General Virology. 1984;**65**(Pt 8):1419-1423

[205] Wright HT Jr, Kasten FH, McAllister RM. Human cytomegalovirus, observations of intracellular lesion development as revealed by phase contrast, time-lapse cinematography. Proceedings of the Society for Experimental Biology and Medicine Society for Experimental Biology and Medicine. 1968;**127**(4):1032-1036

[206] Smith KM, Brown RM Jr, Walne PL. Ultrastructural and time-lapse studies on the replication cycle of the blue-green algal virus LPP-1. Virology. 1967;**31**(2):329-337

[207] Duprex WP, Rima BK. Using green fluorescent protein to monitor measles virus cell-to-cell spread by time-lapse confocal microscopy. Methods in Molecular Biology. 2002;**183**: 297-307

[208] Riddle PN. Time-lapse cinemicroscopy. Methods in Molecular Biology. 1990;**5**:415-446

[209] Kulesa PM, Kasemeier-Kulesa JC. Construction of a heated incubation chamber around a microscope stage for time-lapse imaging. CSH Protocols. 2007;**2007**:pdb prot4792

[210] Tuttle C. Analysis of equipment and methods for time lapse motion photomicrography. Journal of Biological Photography. 1992;**60**(2):54-55

[211] Constable FL, Moffat MA. A glass tissue culture chamber for use in time-lapse cinematography. Journal of Clinical Pathology. 1958;**11**(5):455-457

[212] Fischler HA. Design of a phase contrast time-lapse cinemicrographic unit. Journal of the Biological Photographic Association. 1966;**34**(2):53-82

[213] Shand FL. An inexpensive high intensity lamp unit for phase-contrast time-lapse cinemicrography. Journal of the Royal Microscopical Society. 1967;**86**(4):437-440

[214] Dawson M, Johnstone AJ, Matthews JE. An electronic light source and modified Bolex H-16 camera for time-lapse cinemicrography. Journal of Microscopy. 1970;**91**(2):139-143

[215] Dawson M, Matthews JE. Apparatus for time-lapse cinemicrography. Journal of Microscopy. 1972;**96**(1):97-103

[216] Peters JH. Device for time lapse studies on living cells in vitro (author's transl). Microscopica Acta. 1979;**81**(3):217-225

[217] Vesely P, Maly M, Cumpelik J, Pluta M, Tuma V. Improved spatial and temporal resolution in an apparatus for time-lapse phase contrast cine micrography of cells in vitro. Journal of Microscopy. 1982;**125**(Pt 1):67-76

[218] Allen TD. Time lapse video microscopy using an animation control unit. Journal of Microscopy. 1987;**147**(Pt 2):129-135

[219] Endlich B, Radford IR, Forrester HB, Dewey WC. Computerized video time-lapse microscopy studies of ionizing radiation-induced rapid-interphase and mitosis-related apoptosis in lymphoid cells. Radiation Research. 2000;**153**(1):36-48

[220] Armstrong WD, Wilt JC, Pritchard ET. Vacuolation in the human amnion cell studies by time—Lapse photography and electron microscopy. American Journal of Obstetrics and Gynecology. 1968;**102**(7):932-948

[221] Choi BH, Cho KH, Lapham LW. Effects of methylmercury on human fetal neurons and astrocytes in vitro: A time-lapse cinematographic, phase and electron microscopic study. Environmental Research. 1981;**24**(1):61-74

[222] Balfour BM, Goscicka T, MacKenzie JL, Gautam A, Tate M, Clark J. Combined time-lapse cinematography and immuno-electron microscopy. The Anatomical Record. 1990; **226**(4):509-514

[223] LeSage AJ, Kron SJ. Design and implementation of algorithms for focus automation in digital imaging time-lapse microscopy. Cytometry. 2002;**49**(4):159-169

[224] Brieu N, Navab N, Serbanovic-Canic J, Ouwehand WH, Stemple DL, Cvejic A, et al. Image-based characterization of thrombus formation in time-lapse DIC microscopy. Medical Image Analysis. 2012;**16**(4):915-931

[225] Concha ML, Adams RJ. Oriented cell divisions and cellular morphogenesis in the zebrafish gastrula and neurula: A time-lapse analysis. Development. 1998;**125**(6):983-994

[226] Aletta JM, Greene LA. Growth cone configuration and advance: A time-lapse study using video-enhanced differential interference contrast microscopy. The Journal of Neuroscience: The Official Journal of the Society for Neuroscience. 1988;**8**(4):1425-1435

[227] Montgomery PO, Bonner WA, Roberts F. Ultra-violet flying spot time lapse motion picture observations of living cells. Texas Reports on Biology and Medicine. 1957;**15**(3): 386-395

[228] Malide D, Metais JY, Dunbar CE. In vivo clonal tracking of hematopoietic stem and progenitor cells marked by five fluorescent proteins using confocal and multiphoton microscopy. Journal of Visualized Experiments: JoVE. 2014;**90**:e51669

[229] Noctor SC. Time-lapse imaging of fluorescently labeled live cells in the embryonic mammalian forebrain. Cold Spring Harbor Protocols. 2011;**2011**(11):1350-1361

[230] Stadtfeld M, Varas F, Graf T. Fluorescent protein-cell labeling and its application in time-lapse analysis of hematopoietic differentiation. Methods in Molecular Medicine. 2005;**105**:395-412

[231] Ellenberg J, Lippincott-Schwartz J, Presley JF. Two-color green fluorescent protein time-lapse imaging. BioTechniques. 1998;**25**(5):838-42-8344-6

[232] Nakamura M, Miyamoto K, Hayashi K, Awaad A, Ochiai M, Ishimura K. Time-lapse fluorescence imaging and quantitative single cell and endosomal analysis of peritoneal macrophages using fluorescent organosilica nanoparticles. Nanomedicine: Nanotechnology, Biology and Medicine. 2013;**9**(2):274-283

[233] Cova L, Bigini P, Diana V, Sitia L, Ferrari R, Pesce RM, et al. Biocompatible fluorescent nanoparticles for in vivo stem cell tracking. Nanotechnology. 2013;**24**(24):245603

[234] Yun SW, Leong C, Bi X, Ha HH, Yu YH, Tan YL, et al. A fluorescent probe for imaging symmetric and asymmetric cell division in neurosphere formation. Chemical Communications. 2014;**50**(56):7492-7494

[235] Lee GM, Fong SS, Oh DJ, Francis K, Palsson BO. Characterization and efficacy of PKH26 as a probe to study the replication history of the human hematopoietic KG1a progenitor cell line. In Vitro Cellular & Developmental Biology. Animal. 2002;**38**(2):90-96

[236] Heye RR, Kiebler EW, Arnzen RJ, Tolmach LJ. Multiplexed time-lapse photomicrography of cultured cells. Journal of Microscopy. 1982;**125**(Pt 1):41-50

[237] Kallman RF, Blevins N, Coyne MA, Prionas SD. Novel instrumentation for multifield time-lapse cinemicrography. Computers and Biomedical Research, An International Journal. 1990;**23**(2):115-129

[238] Hamant O, Das P, Burian A. Time-lapse imaging of developing meristems using confocal laser scanning microscope. Methods in Molecular Biology. 2014;**1080**:111-119

[239] Boisset JC, Andrieu-Soler C, van Cappellen WA, Clapes T, Robin C. Ex vivo time-lapse confocal imaging of the mouse embryo aorta. Nature Protocols. 2011;**6**(11):1792-1805

[240] Nowotschin S, Ferrer-Vaquer A, Hadjantonakis AK. Imaging mouse development with confocal time-lapse microscopy. Methods in Enzymology. 2010;**476**:351-377

[241] Megason SG. In toto imaging of embryogenesis with confocal time-lapse microscopy. Methods in Molecular Biology. 2009;**546**:317-332

[242] Rupp PA, Kulesa PM. High-resolution, intravital 4D confocal time-lapse imaging in avian embryos using a Teflon culture chamber design. CSH Protocols. 2007;**2007**:pdb prot4790

[243] Carvalho L, Heisenberg CP. Imaging zebrafish embryos by two-photon excitation time-lapse microscopy. Methods in Molecular Biology. 2009;**546**:273-287

[244] Rebollo E, Gonzalez C. Time-lapse imaging of embryonic neural stem cell division in drosophila by two-photon microscopy. Current Protocols in Stem Cell Biology. 2010; **13**(1):1H.2.1-1H.2.9

[245] Pakan JM, McDermott KW. A method to investigate radial glia cell behavior using two-photon time-lapse microscopy in an ex vivo model of spinal cord development. Frontiers in Neuroanatomy. 2014;**8**:22

[246] Gerlich D, Ellenberg J. 4D imaging to assay complex dynamics in live specimens. Nature Cell Biology. 2003;(Suppl):S14-S19

[247] Dierickx P, Vermunt MW, Muraro MJ, Creyghton MP, Doevendans PA, van Oudenaarden A, et al. Circadian networks in human embryonic stem cell-derived cardiomyocytes. EMBO Reports. 2017;**18**(7):1199-1212

[248] Kuo HL, Ho PC, Huang SS, Chang NS. Chasing the signaling run by tri-molecular time-lapse FRET microscopy. Cell Death Discovery. 2018;**4**:45

[249] Majdi JA, Qian H, Li Y, Langsner RJ, Shea KI, Agrawal A, et al. The use of time-lapse optical coherence tomography to image the effects of microapplied toxins on the retina. Investigative Ophthalmology & Visual Science. 2014;**56**(1):587-597

[250] Pan YT, Wu ZL, Yuan ZJ, Wang ZG, Du CW. Subcellular imaging of epithelium with time-lapse optical coherence tomography. Journal of Biomedical Optics. 2007;**12**(5):050504

[251] Pan YT, Wu Q, Wang ZG, Brink PR, Du CW. High-resolution imaging characterization of bladder dynamic morphophysiology by time-lapse optical coherence tomography. Optics Letters. 2005;**30**(17):2263-2265

[252] Gabriel M, Balle D, Bigault S, Pornin C, Getin S, Perraut F, et al. Time-lapse contact microscopy of cell cultures based on non-coherent illumination. Scientific Reports. 2015;**5**: 14532

[253] Schiefer J, Kampe K, Dodt HU, Zieglgansberger W, Kreutzberg GW. Microglial motility in the rat facial nucleus following peripheral axotomy. Journal of Neurocytology. 1999;**28**(6):439-453

[254] Maurice DM, Zhao J, Nagasaki T. A novel microscope system for time-lapse observation of corneal cells in a living mouse. Experimental Eye Research. 2004;**78**(3):591-597

[255] Youssef S, Gude S, Radler JO. Automated tracking in live-cell time-lapse movies. Integrative Biology: Quantitative Biosciences From Nano to Macro. 2011;**3**(11):1095-1101

[256] Dewan MA, Ahmad MO, Swamy MN. Tracking biological cells in time-lapse microscopy: An adaptive technique combining motion and topological features. IEEE Transactions on Bio-Medical Engineering. 2011;**58**(6):1637-1647

[257] Kirby BB, Takada N, Latimer AJ, Shin J, Carney TJ, Kelsh RN, et al. In vivo time-lapse imaging shows dynamic oligodendrocyte progenitor behavior during zebrafish development. Nature Neuroscience. 2006;**9**(12):1506-1511

[258] Chen X, Zhou X, Wong ST. Automated segmentation, classification, and tracking of cancer cell nuclei in time-lapse microscopy. IEEE Transactions on Bio-Medical Engineering. 2006;**53**(4):762-766

[259] Meijering E, Dzyubachyk O, Smal I. Methods for cell and particle tracking. Methods in Enzymology. 2012;**504**:183-200

[260] Dzyubachyk O, van Cappellen WA, Essers J, Niessen WJ, Meijering E. Advanced level-set-based cell tracking in time-lapse fluorescence microscopy. IEEE Transactions on Medical Imaging. 2010;**29**(3):852-867

[261] Jaqaman K, Loerke D, Mettlen M, Kuwata H, Grinstein S, Schmid SL, et al. Robust single-particle tracking in live-cell time-lapse sequences. Nature Methods. 2008;**5**(8):695-702

[262] Huth J, Buchholz M, Kraus JM, Schmucker M, von Wichert G, Krndija D, et al. Significantly improved precision of cell migration analysis in time-lapse video microscopy through use of a fully automated tracking system. BMC Cell Biology. 2010;**11**:24

[263] Schoenauer Sebag A, Plancade S, Raulet-Tomkiewicz C, Barouki R, Vert JP, Walter T. A generic methodological framework for studying single cell motility in high-throughput time-lapse data. Bioinformatics. 2015;**31**(12):i320-i328

[264] Bray MA, Carpenter AE. CellProfiler tracer: Exploring and validating high-throughput, time-lapse microscopy image data. BMC Bioinformatics. 2015;**16**:368

[265] Brandes S, Mokhtari Z, Essig F, Hunniger K, Kurzai O, Figge MT. Automated segmentation and tracking of non-rigid objects in time-lapse microscopy videos of polymorphonuclear neutrophils. Medical Image Analysis. 2015;**20**(1):34-51

[266] Barbier de Reuille P, Routier-Kierzkowska AL, Kierzkowski D, Bassel GW, Schupbach T, Tauriello G, et al. MorphoGraphX: A platform for quantifying morphogenesis in 4D. eLife. 2015;**4**:05864

[267] Mankowski WC, Winter MR, Wait E, Lodder M, Schumacher T, Naik SH, et al. Segmentation of occluded hematopoietic stem cells from tracking. Conference proceedings: Annual International Conference of the IEEE Engineering in Medicine and Biology Society IEEE Engineering in Medicine and Biology Society Annual Conference. 2014;**2014**:5510-5513

[268] Tarnawski W, Kurtcuoglu V, Lorek P, Bodych M, Rotter J, Muszkieta M, et al. A robust algorithm for segmenting and tracking clustered cells in time-lapse fluorescent microscopy. IEEE Journal of Biomedical and Health Informatics. 2013;**17**(4):862-869

[269] Storr A, Venetis C, Cooke S, Kilani S, Ledger W. Time-lapse algorithms and morphological selection of day-5 embryos for transfer: A preclinical validation study. Fertility and Sterility. 2018;**109**(2):276-83.e3

[270] Liu Y, Feenan K, Chapple V, Matson P. Assessing efficacy of day 3 embryo time-lapse algorithms retrospectively: Impacts of dataset type and confounding factors. Human Fertility. 2018:1-9. https://doi.org/10.1080/14647273.2018.1425919

[271] Reignier A, Lammers J, Barriere P, Freour T. Can time-lapse parameters predict embryo ploidy? A systematic review. Reproductive Biomedicine Online. 2018;**36**(4):380-387

[272] Swain JE. Could time-lapse embryo imaging reduce the need for biopsy and PGS? Journal of Assisted Reproduction and Genetics. 2013;**30**(8):1081-1090

[273] Herrero J, Meseguer M. Selection of high potential embryos using time-lapse imaging: The era of morphokinetics. Fertility and Sterility. 2013;**99**(4):1030-1034

[274] Ergin EG, Caliskan E, Yalcinkaya E, Oztel Z, Cokelez K, Ozay A, et al. Frequency of embryo multinucleation detected by time-lapse system and its impact on pregnancy outcome. Fertility and Sterility. 2014;**102**(4):1029-33.e1

[275] Desai N, Goldberg JM, Austin C, Falcone T. Are cleavage anomalies, multinucleation, or specific cell cycle kinetics observed with time-lapse imaging predictive of embryo developmental capacity or ploidy? Fertility and Sterility. 2018;**109**(4):665-674

[276] Milewski R, Ajduk A. Time-lapse imaging of cleavage divisions in embryo quality assessment. Reproduction. 2017;**154**(2):R37-R53

[277] Chawla M, Fakih M, Shunnar A, Bayram A, Hellani A, Perumal V, et al. Morphokinetic analysis of cleavage stage embryos and its relationship to aneuploidy in a retrospective

time-lapse imaging study. Journal of Assisted Reproduction and Genetics. 2015;**32**(1): 69-75

[278] Campbell A, Fishel S, Laegdsmand M. Aneuploidy is a key causal factor of delays in blastulation: Author response to 'a cautionary note against aneuploidy risk assessment using time-lapse imaging'. Reproductive Biomedicine Online. 2014;**28**(3):279-283

[279] Daughtry BL, Chavez SL. Time-lapse imaging for the detection of chromosomal abnormalities in primate preimplantation embryos. Methods in Molecular Biology. 2018;**1769**:293-317

[280] Castello D, Motato Y, Basile N, Remohi J, Espejo-Catena M, Meseguer M. How much have we learned from time-lapse in clinical IVF? Molecular Human Reproduction. 2016;**22**(10):719-727

[281] Sterckx S, Cockbain J, Pennings G. Patenting time-lapse microscopy: The European story. Reproductive Biomedicine Online. 2014;**28**(2):146-150

[282] Sterckx S, Cockbain J, Pennings G. Patenting medical diagnosis methods in Europe: Stanford University and time-lapse microscopy. Reproductive Biomedicine Online. 2017;**34**(2):166-168

[283] Urbano LF, Masson P, VerMilyea M, Kam M. Automatic tracking and motility analysis of human sperm in time-lapse images. IEEE Transactions on Medical Imaging. 2017;**36**(3):792-801

[284] Wei F, Rong XX, Xie RY, Jia LT, Wang HY, Qin YJ, et al. Cytokine-induced killer cells efficiently kill stem-like cancer cells of nasopharyngeal carcinoma via the NKG2D-ligands recognition. Oncotarget. 2015;**6**(33):35023-35039

[285] Schiraldi C, Zappavigna S, D' Agostino A, Porto S, Gaito O, Lusa S, et al. Nanoparticles for the delivery of zoledronic acid to prostate cancer cells: A comparative analysis through time lapse video-microscopy technique. Cancer Biology & Therapy. 2014;**15**(11): 1524-1532

[286] Lin HD, Fong CY, Biswas A, Choolani M, Bongso A. Human Wharton's jelly stem cells, its conditioned medium and cell-free lysate inhibit the growth of human lymphoma cells. Stem Cell Reviews. 2014;**10**(4):573-586

[287] Bago JR, Pegna GJ, Okolie O, Mohiti-Asli M, Loboa EG, Hingtgen SD. Electrospun nanofibrous scaffolds increase the efficacy of stem cell-mediated therapy of surgically resected glioblastoma. Biomaterials. 2016;**90**:116-125

[288] Bago JR, Alfonso-Pecchio A, Okolie O, Dumitru R, Rinkenbaugh A, Baldwin AS, et al. Therapeutically engineered induced neural stem cells are tumour-homing and inhibit progression of glioblastoma. Nature Communications. 2016;**7**:10593

[289] Yasin R, Van Beers G, Riddle PN, Brown D, Widdowson G, Thompson EJ. An abnormality of cell behaviour in human dystrophic muscle cultures: A time-lapse study. Journal of Cell Science. 1979;**38**:201-210

[290] Weiger MC, Vedham V, Stuelten CH, Shou K, Herrera M, Sato M, et al. Real-time motion analysis reveals cell directionality as an indicator of breast cancer progression. PLoS One. 2013;**8**(3):e58859

[291] Tsujioka T, Matsuoka A, Tohyama Y, Tohyama K. Approach to new therapeutics: Investigation by the use of MDS-derived cell lines. Current Pharmaceutical Design. 2012;**18**(22):3204-3214

[292] Yano S, Tazawa H, Hashimoto Y, Shirakawa Y, Kuroda S, Nishizaki M, et al. A genetically engineered oncolytic adenovirus decoys and lethally traps quiescent cancer stem-like cells in S/G2/M phases. Clinical Cancer Research: An Official Journal of the American Association for Cancer Research. 2013;**19**(23):6495-6505

[293] Turani M, Banfalvi G, Peter A, Kukoricza K, Kiraly G, Talas L, et al. Antibiotics delay in vitro human stem cell regrowth. Toxicology In Vitro: An International Journal Published in Association with BIBRA. 2015;**29**(2):370-379

[294] Pulkkinen JO, Elomaa L, Joensuu H, Martikainen P, Servomaa K, Grenman R. Paclitaxel-induced apoptotic changes followed by time-lapse video microscopy in cell lines established from head and neck cancer. Journal of Cancer Research and Clinical Oncology. 1996;**122**(4):214-218

[295] Nakamura Y, Ishigaki Y. Immunostaining and time-lapse analysis of vinblastine-induced paracrystal formation in human A549 cells. Oncology Letters. 2014;**8**(6):2387-2392

[296] Furukawa T, Inoue H, Sugita K, Eguchi M, Sakakibara H, Sugiyama S, et al. Long-term cinemicrography of erythroblasts from a patient with congenital dyserythropoietic anemia type III: Direct observation of dysplastic erythroblast formation. Blood Cells. 1993;**19**(2):493-506. Discussion 7-8

[297] Xiang Z, Block M, Lofman C, Nilsson G. IgE-mediated mast cell degranulation and recovery monitored by time-lapse photography. The Journal of Allergy and Clinical Immunology. 2001;**108**(1):116-121

[298] Barretto RP, Ko TH, Jung JC, Wang TJ, Capps G, Waters AC, et al. Time-lapse imaging of disease progression in deep brain areas using fluorescence microendoscopy. Nature Medicine. 2011;**17**(2):223-228

[299] Dieterlen-Lievre F, Jaffredo T. Decoding the hemogenic endothelium in mammals. Cell Stem Cell. 2009;**4**(3):189-190

[300] Slocum HK, Parsons JC, Winslow EO, Broderick L, Minderman H, Toth K, et al. Time-lapse video reveals immediate heterogeneity and heritable damage among human ileocecal carcinoma HCT-8 cells treated with raltitrexed (ZD1694). Cytometry. 2000; **41**(4):252-260

6

Cell-Based Assays for Evaluation of Autophagy in Cancers

Siew-Wai Pang, Noel Jacques Awi,
Hooi-Yeen Yap and Sin-Yeang Teow

Abstract

Autophagy is a cellular mechanism that degrades damaged organelles and misfolded proteins to maintain cellular homeostasis. Autophagy in cancers is drawing increasing attentions due to its multifaceted roles in cancer development, progression, and treatment. There are several key autophagy effectors that are being extensively studied to understand the role of autophagy in cancer as well as their potential value as predictive and/or prognostic biomarkers and therapeutic target. These include ATG4A, ATG4B, Beclin-I, p62, LC3A, LC3B, LC3C, and LAMP. While having its own sophisticated pathway, autophagy has been reported to associate with multiple oncogenic pathways such as NF-kB, mTOR, and PI3K signaling. This chapter aims to provide a detailed protocol for researchers to investigate the role of autophagy using in vitro cell line as model. Here, we demonstrate several techniques including Western blot (WB), immunofluorescence (IF), and small-interfering RNA (siRNA) knockdown using colorectal cancer cell lines as samples. This chapter provides information to researchers especially those in their early- and mid-career to plan and design their experiments to study the autophagy events in their area of interests.

Keywords: autophagy, cancer, Western blot, immunofluorescence, cell

1. Introduction

1.1. Brief history of autophagy

The term autophagy, or sometimes known as autophagocytosis, comes from the Greek language for "self-eating." This term was coined by a Belgian biochemist, Christian de Duve in

1963. Prior to this term, however, autophagy was first observed, or at least hinted in as early as 1955 [1]. Kleinfeld and his colleagues found that the process involves three continuous stages of maturation and is seen as being used for the reutilization of cellular materials and organelle disposal. In Duve's definition, autophagy is a part of lysosomal function and glucagon being the main inducer of hepatic cell degradation. Together with his student, Russell Deter, they were the first to demonstrate that lysosomes play a central role for intracellular autophagy [2, 3]. Using Duve's work as reference, independent scientist groups discovered autophagy-related genes in yeast. In that period, Ohsumi and Michael Thumm studied on nonselective autophagy induced by starvation [4, 5]. At the same time, Klionsky discovered a form of selective autophagy called the cytoplasm-to-vacuole targeting (CVT) pathway [6, 7]. Not longer after, they discovered that their independent work actually revolves around the same pathway. Through collaborating, they published a paper titled "Cytoplasm-to-vacuole targeting and autophagy employ the same machinery to deliver proteins to the yeast vacuole" [8]. In 2003, a unified nomenclature was advocated by scientists in the field to use ATG for autophagy-related genes [9]. More than a decade later, in 2016, Ohsumi was finally awarded a Nobel Prize in Physiology or Medicine for his contribution toward the field of autophagy. While it is undeniable that Ohsumi deserves the Nobel Prize, some individuals have pointed out that the prize should have been more inclusive of other researchers who made Ohsumi's work possible [10]. In the second millennium, there was an accelerated growth of research in autophagy thanks to the work and contribution by these scientists on ATG genes. With the fundamentals on autophagy set strong, scientists began to study its association with human health and diseases. The first breakthrough discovery associating autophagy with cancer

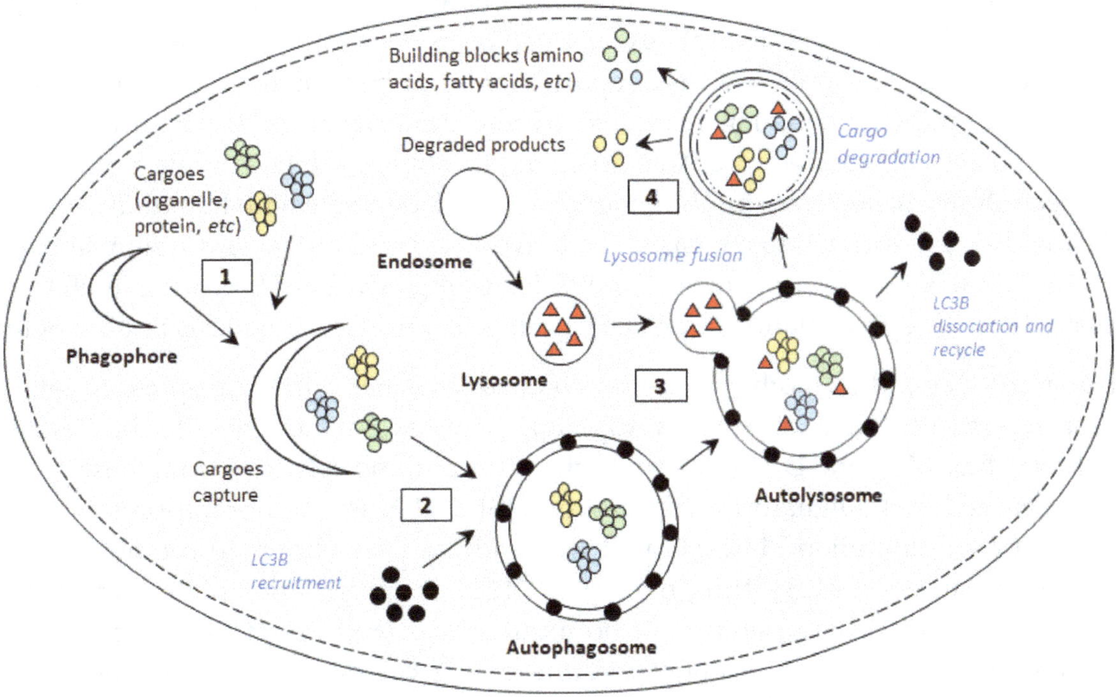

Figure 1. Cellular autophagy processes. (1) Phagophore forms and elongates to package the cargoes comprising damaged organelles and proteins. (2) LC3B proteins are then recruited and together with phagophore, it forms autophagosome. (3) LC3B proteins are then dissociated, and the cytoplasmic cargo is sequestered to fuse with endosome-derived lysosome to form autolysosome. (4) Hydrolytic enzymes then degrade the cargoes and release metabolic products and building blocks such as amino acids and fatty acids for nutrient recycling.

was landmarked by Beth Levine's group in 1999 [11]. To date, the link between cancer and autophagy remains to be the top focus of autophagy researchers.

1.2. Autophagy pathway

Autophagy is an evolutionary conserved mechanism involved in maintaining homeostasis and metabolism at the cellular level by degrading proteins with long half-life and clearing cytoplasmic organelles through lysosomes. Lipids, nucleotides, and glycogen are also subjected to lysosomal degradation via autophagy. Like other pathways involved in homeostasis, the cellular pathways are highly regulated. There are evidences that the recycling of proteins and other macromolecules may contribute to protective roles in normal development, senescence, cell death, and as a defense mechanism against intracellular infections. **Figure 1** shows the detailed processes involved in the autophagy pathway. The dysregulation of autophagy has been shown to cause cancer, inflammatory, metabolic, and neurodegenerative diseases [12, 13].

2. Role of autophagy in cancers

2.1. Autophagy as tumor suppressor

The suppression of autophagy has also been linked to an increase in oxidative stress, genome instability, and activation of the DNA damage response. The increase in oxidative stress leads to a cascade of reaction which may promote tumor growth [14]. A lack of autophagy in hepatocytes can also cause cell death and inflammation, which are known to progress to liver cancer [14]. Deficiency in autophagy has also been shown to promote the accumulation of p62, an autophagy substrate used as a reporter for autophagy activity [15, 16]. The aberrant accumulation of p62 is linked to increase toxicity and tumorigenesis [15]. In other conditions, the expression of p62 increased oxidative stress and tumor growth [17], while a suppression of p62 has been shown to suppress tumorigenesis in modified mouse with hereditary lung cancer [18]. Besides being associated to increased oxidative stress, p62 also acts as a signaling adaptor for the regulation of many oncogenic pathways, including NRF2, mTOR, and NF-κB [19]. That being said, how the dysregulation of p62 contributes toward tumorigenesis is still not known.

Other than BECN1 and p62, other well-known oncogenes and tumor suppressor genes have also been reported to interrupt upon autophagic pathways, which include the PTEN/PI3K/Akt pathway, Ras, Myc, and DAPk. The autophagy-impinging activities from these genes may have been caused by malignant transformation [20]. PTEN is a tumor suppressor gene that inhibits the pro-proliferation PI3K/Akt pathway and has been shown to promote autophagy in HT-29 colon cancer cells [21]. Thus, the loss of PTEN or upregulation of PI3K can contribute to malignancy through the inhibition of both autophagy and apoptosis. Myc on the other hand has been shown to promote apoptosis, autophagic cell death, and even oncogenesis. In rat fibroblasts, the overexpression of Myc improved autophagic activities [22]. It is interesting to note that the autophagic-inducing domain of the Myc gene is different to that of apoptosis and oncogenesis. The oncogenic Ras protein, including KRAS and NRAS, has been implicated in the promotion of p53-independent non-apoptotic autophagic cell death, which cannot be blocked by Bcl-2 overexpression. This is exemplified in a study where non-apoptotic neuroblastoma degeneration undergoing autophagic cell death showed high expressions of RAS [23]. Similarly, this phenomenon was also observed in neuroblastoma and HT-29 colon cancer cell lines in vitro [20]. DAPk was initially identified as a cell death-promoting protein before its role in mediating autophagy and tumor suppression was characterized [24]. Like other cases of tumor suppressors, DAPk expression was found to be reduced in several cancer cell lines including but not limited to bladder, breast, renal, lung, ovarian, cervix, colon, head and neck, prostate, and brain cancers. The reintroduction of DAPk into highly metastatic cell lines ameliorated metastasis and tumorigenesis and improved cell death. These data suggest that DAPk has antimetastasis and tumor suppressor properties [20].

2.2. Autophagy as tumor promoter

Contrary to what was discussed above, cancer cells also rely on autophagy for survival and in many cases can be more dependent than normal cells. This is possibly due to the heavy reliance of tumors on nutrient supplies for the maintenance of rapid cell proliferation. In tumors inflicted with hypoxia, autophagy is essential for tumor survival [25]. In RAS-transformed cancer cells, autophagy was shown to be upregulated, and this promotes their survival, growth, invasion, and metastasis. The upregulation of autophagy in RAS-driven cancers has also been reported to ameliorate mitochondrial metabolic defects and the resulting sensitivity to stress. The genetic mechanism on how autophagy dependency arises is still at its infant stage [15].

In a study specific on the role of autophagy in GEM models for RAS-driven non-small-cell lung cancer (NSCLC), the essential autophagy gene ATG7 was deleted together with mutant KRAS in tumor cells. They observed that the absence of ATG7 causes an aberrant accumulation of defective mitochondria, and this leads to the accelerated activation of p53, cell growth arrest, and cell death. ATG7 loss also results in reduced tumor burden by facilitating the conversion of adenomas and carcinomas to a rare form of benign neoplasms called oncocytomas, characterized by the accumulation of nonfunctional mitochondria [26]. While it may seem beneficial to reduce the expression of ATG7 in cancer cells, there is no extension to the life-span of the mice as they die of pneumonia, possibly triggered by inflammation caused by autophagy defects [15]. The deletion of an alternative autophagy gene ATG5 in the same setting also resulted in a similar reduction of tumorigenesis, suggesting that tumorigenesis is mediated by autophagy itself and not just by ATG7 alone. Similarly, the activation of autophagy in GEM models of RAS-driven pancreatic cancer also suppresses p53 activation [27]. In a separate study, it was found that the allelic loss of BECN1 promotes the activation of tumor suppressor p53, and this in turns reduced tumorigenesis [28]. However, large-scale genomic analysis to date has not been successful in identifying recurrent mutations in essential autophagy genes such as BECN1 [29].

2.3. Potential targets and biomarkers in cancers

Oftentimes, the pathways of important cellular processes such as apoptosis and inflammation that can be linked to diseases are studied extensively in hopes of searching for biomarkers or potential drug targets. Autophagy is no exception, as its role in cancer has been demonstrated in the previously mentioned studies; several groups are investigating their predictive and prognostic values. As of 2017, most of the characterized autophagy-related protein markers are prognostic in nature but are not approved for use in clinical settings. To date,

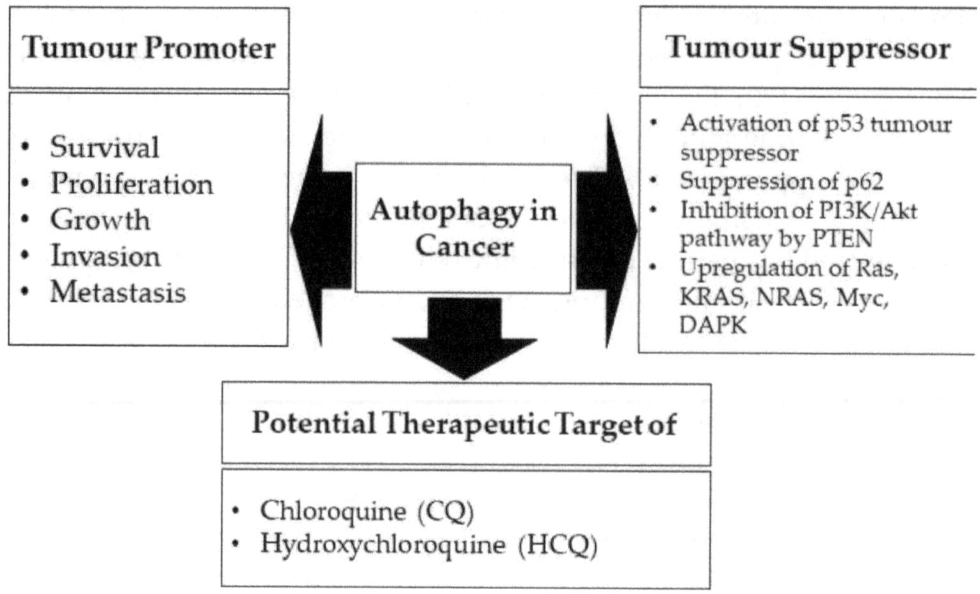

Figure 2. The role of autophagy in cancers and its therapeutic value.

drug development wecules targeting autophagy proteins. An example of lysosomal inhibitors will be the chloroquine (CQ) and hydroxychloroquine (HCQ), both of which are the main characters in clinical trials. The derivative of chloroquine Lys05 is currently going through optimization for clinical trials [30]. LC3B and p62 are common autophagy markers generally used in some assays to assess autophagy turnover, and these two can be useful to monitor the effectiveness of autophagy treatment. Potential targets for autophagy proteins to be discussed in the next section include LC3A, LC3B, p62, ULK, ATG4, ATG7, and Beclin-1. **Figure 2** summarizes the tumor promoting and suppressing roles of autophagy in cancers as well as their therapeutic potentials.

3. Autophagy markers

3.1. LC3B and LC3A

LC3B is an extensively studied autophagy-related protein and is the shortened name for microtubule-associated protein 1 light chain 3B. Together with p62, LC3B has been used as an autophagy marker in several laboratory assays to measure autophagy flux. While in general a high expression of LC3B in cancer patients represents a bad overall prognosis, there are exceptions. This is especially apparent in breast cancer, where the genetic disposition of the cancer subtype will determine the prognostic values of LC3B expression. In the context of colorectal cancer, poor prognosis with high LC3B expression was observed only in KRAS-mutated samples but showed no changes in KRAS-WT specimens [31]. Interestingly, a high expression of LC3B was found to be in favor of the patient's prognosis in NSCLC [31]. With these different findings, LC3B cannot be used as a general prognostic biomarker across all cancer types or even within cancer subgroups. It will be important to study and identify its prognostic value with attention to different cancer types and subtypes.

LC3A on the other hand is another autophagy marker that is being studied, albeit not as extensive as LC3B. IHC staining of LC3A can be defined into three distinct staining patterns: juxtanuclear staining, staining of "stone-like" structures (SLS), and diffuse cytoplasmic staining [32]. All three patterns translate to different prognostic values. For example, a high staining pattern of juxtanuclear LC3A in colorectal [33] and breast cancers [32] correlates with good prognosis; however, the opposite is true if it was SLS. Accumulation of SLS is also linked to poor prognosis in other cancer types including gastric [34], NSCLC [35], hepatocellular carcinoma [36], and clear cell ovarian carcinoma [30, 37].

3.2. p62/SQSTM1

p62/SQTM1 was initially characterized as the mediator of NFκB signaling and is now known as the signaling hub for a variety of cellular events including oxidative stress response and amino acid sensing [37]. As discussed previously, p62 is a signaling adaptor which links autophagic substrates to autophagic machineries. During autophagy, the degradation of p62 serves as a marker for functional autophagic activities. Thus, in a Western blot, the accumulation of p62 protein can be indicative of autophagy inhibition. While the principle of p62 assays seems straightforward, caution has to be taken during the interpretation by taking into account the regulation of p62 concentration at the transcriptional and post-translational stages. Taking autophagy out of the context, compilations of reports and studies demonstrate the link between high cytoplasmic but low nuclear expression levels with high tumor aggression and poor prognosis. The cancer types studied include endometrial cancer, oral squamous cell carcinoma, NSCLC, and epithelial ovarian cancer. Specifically, in breast cancer, high cytoplasmic expression levels of p62 have been shown to correlate with cancer grade, metastatic status, and reduction in five-year survival. Additionally, it was also significantly associated to high levels of epidermal growth factor receptor (EGFR) and human epidermal growth factor receptor-2/3/4 (HER2/3/4) expression. Poor prognosis due to high levels of cytoplasmic p62 expression also holds true in triple-negative breast cancer [38]. In general, the studies available to date indicate high cytoplasmic p62 expression as a marker for poor prognosis. More studies will have to be done on nuclear-localized p62 across cancer types to understand its prognostic values.

3.3. Beclin-1/VPS34

The Beclin-1/VPS34 complex is also one of the most popularly studied autophagy-related genes as it is the central coordinator of autophagy. Normally, Beclin-1 inhibits the anti-apoptotic Bcl-2 proteins; thus, the reduction of Beclin-1 expression can lead to enhanced anti-apoptotic pathways, consequently translating to poor prognosis. Overexpression on the other hand induces tumor hypoxia and acidity, which is likely to promote autophagy. Such events may increase aggressive tumor behavior [39]. Beclin-1 is a favorable prognostic marker in several cancers including lymphoma, lung, breast, and gastric cancer. In the context of colorectal cancer, the prognostic values can be split between poor and favorable. Poor prognosis in colorectal cancer patients was demonstrated in a study by Han et al. [40], and this finding was supported by additional studies by other researchers. The contradiction between colorectal cancer and other cancer types was addressed by Han et al. when he found that high expression level was associated to poor survival in patients undergoing chemotherapy, whereas patients without chemotherapy showed higher overall survival. Since most chemotherapeutic drugs actively induce autophagy [41], it is possible that the increased autophagic rate caused by high expression levels promoted resistance to chemotherapy. As with p62, expression levels of Beclin-1 in different histologic and genetic subtypes of breast cancer must be examined. To date, Beclin-1 expression was shown to be significantly correlated to estrogen receptor (ER) negativity. Expression levels vary between subtypes, with maximum expression found in triple-negative breast cancer [42].

3.4. ULK-1/ULK-2

ULK-1 and ULK-2 are serine/threonine kinases that are the most upstream components of the autophagy pathway and as such became attractive drug targets. While ULK-1/ULK-2 small molecule inhibitors are being developed, the data on the prognostic value of ULK-1/ULK-2 in different cancers is contradictory and limited. In breast cancer [40] and gastric cancer [43], high expression of ULK-1 appears to associate to good patient prognosis. This contradicts the findings in colorectal cancer [44]. A separate study in colorectal cancer did not demonstrate any correlation to survivability, even after taking into account the status of KRAS. However, they found that high expression levels were linked to lymph node metastasis. High levels of ULK-1 expression were also translated to adverse prognosis in esophageal squamous cell carcinoma [45] and nasopharyngeal carcinoma [46]. As for ULK-2, high expression levels were reported in prostate cancer tissue compared to adjacent healthy prostate tissue, but no correlation to prognosis was possible from this study [47]. With these contradictions in mind, additional studies with bigger patient cohorts will be required to further understand the prognostic values of ULK-1 and ULK-2 in cancer.

3.5. ATG4B

ATG4B is a cysteine protease that plays a central role in autophagy. While it is one of the autophagy proteins that is being targeted as a potential therapeutic protein, not much is known about its prognostic value. Some studies showed elevated expression levels in chronic myeloid leukemia, colorectal, and lung cancer; however, there is no mention of its prognostic value in cancers, including members of the ATG4 family [30].

3.6. GABARAP

GABARAP is a part of the ATG8 protein homologs that consist of the subfamily LC3s and GABARAPs. While LC3 has been widely studied as the marker for autophagy, little is known about the GABARAP. GABARAP subfamily consists of GABARAP, GABARAPL1, and GABARAPL2 (GATE16) [48]. While LC3B is responsible in the elongation of autophagosome, Joachim et al. suggests that GABARAP lipidation onto the phagosome keeps the ULK-1 in an activated state until the dissociation of ULK-1 complex and the closing of the phagosome [49]. GABARAPL1 has been implicated in the formation of tumor in mice exposed to genotoxic DMBA [50]. Knockout GABARAPL1 mice showed a reduced expression of TGF-β1, which acts as a suppressor of T helper 1 cells (T_H1). The reduced suppression on T_H1 results in increased expression of cytokines IL-2 and IFN-γ, which induces an immune response during the exposure of DMBA on the knockout mice. Salah et al. also suggest that the cytokines promote the expression of Xaf1, which inhibits Xiap, a negative regulator of apoptosis, thus promoting cell death and preventing formation of cancerous cells. On the other hand, Berthier et al. found that GABARAPL1 inhibits the growth of breast cancer cell line, MCF-7, and the expression of *GAPARAPL1* gene is downregulated in breast cancer tissues [51].

4. In vitro studies of cellular autophagy

4.1. 2D and 3D establishment using CRC cell lines

CRC cell lines such as HT29, SW48, HCT116, and SW480 are adherent cell lines. By default, they will grow in normal tissue-cultured flat-bottom plates or flask in a monolayer, which is a 2D structure. However, monolayer morphology is not the natural appearance of all the established cell lines. Therefore, scientists have developed 3D models to bridge the difference between in vitro and in vivo experiments. For 3D model establishment, the cells were seeded into a round-bottom ultra-low-attachment 96-well plate and allowed to settle down and grow. The cells were observed every day until clear spheroids can be seen. It is important to capture images every day until clear spheroids can be seen. Some cell lines such as SW480 might not form clear rounded spheroids. Once the spheroids are formed, it is important to not disrupt spheroid formation when handling it in situations such as changing media. This is because the cells were seeded in ultra-low-attachment plate, which means they no longer attach to the bottom of the well. Careless handling of the spheroids might cause disruption to it or worse, the loss of spheroids. Although the process of 3D formation is straight forward, not all cell lines are able to form spheroids. Such example is the SW48, which when seeded into the well, remained in multiple clusters instead of forming spheroids as shown in **Figure 3**. Through the formation of 3D models, scientists can better study the gene expressions and cell behaviors [52], as well as carry out drug response experiments on a model that better mimics the natural

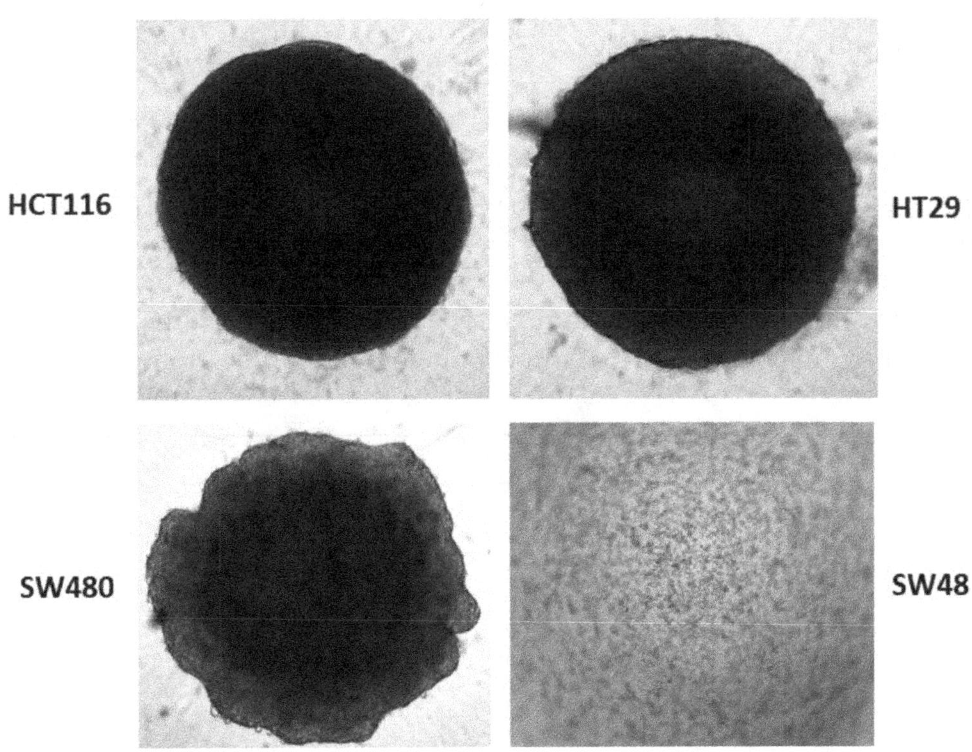

Figure 3. Spheroid formation of colorectal cancer (CRC) cell lines. 1 × 10^5 cells were seeded into round-bottom ultra-low-attachment 96-well plate and incubated at 37°C at 5% CO_2 with 95% humidity. The cells were observed every day until spheroids were formed. The images above were taken at 72 hours after seeding. HCT116 and HT29 both formed a round spheroid in the well, while SW480 formed an irregular spheroid. No spheroid was formed 72 hours after seeding.

tumor morphology in the human body [53]. 3D models can also be cocultured with other cells to emulate the tumor microenvironment and investigate cell-cell interactions.

4.2. Effect of starvation on autophagy

Starvation is one of the well-known methods that can positively induce autophagy for autophagy-related studies. Common methods of starvation include removal of amino acids as well as serum from the media. The effect of autophagy can be detected as early as 1 hour after removal of the amino acids from the media [54]. This effect can be studied using either Western blot or immunofluorescence (i.e. p62 accumulation).

4.3. Autophagy inhibitors and treatments

Autophagy-related activities can be inhibited by treating cell lines with inhibitors. In the market, there are multiple types of inhibitors that are manufactured to target specific pathways that are related to autophagy in cells. As shown in **Table 1** below, these inhibitors with known mechanisms can be used in vitro.

Cytotoxicity experiments should be carried out by researchers when using such inhibitors as they can reduce the viability of cells tested on. This can be carried out through serial-diluted inhibitors using MTT or any luminescence-based viability assays. The effect of inhibition can then be determined using the safe range of concentration obtained through the cytotoxicity assay followed by Western blot or immunofluorescence.

4.4. siRNA knockdown for autophagy

siRNAs have been developed by companies that specifically targets different genes that produce important proteins that play roles in the activation of autophagy. Some examples of currently available siRNAs are those that target Atg3, Atg5, Atg7, Atg10, Atg12, Atg13, Atg14, and Atg101. When researchers first receive the siRNA, the user should run an optimization experiment to determine a few characteristics of the siRNA. The toxicity of such siRNA should first be tested, and the IC50 should be determined to better understand what concentration range might work the best and have the least impact on cell viability. Cytotoxicity assays such as

Name	Mechanism	References
LY294002	PI3K inhibitor	[55]
3-methyladenine	PI3K inhibitor	[56]
Wortmannin	PI3K inhibitor	[57]
SBI-0206965	ULK-1 inhibitor	[58]
Spautin-1	USP10 and USP13 inhibitors	[59]
SAR405	Vsp18 and Vsp34 inhibitor	[60]
NSC195058	ATG4 inhibitor	[61]

Table 1. Autophagy inhibitors that can be used in vitro.

MTT or any real-time luminescence-based assays can be carried out for this purpose. After that, the user should run the silencing experiments using serial-diluted siRNA followed by Western blot to determine what is the minimum effective siRNA concentration for the best gene-silencing results. These steps should be repeated for every cell line that is going to be tested using the siRNA because there will be some differences between cell lines.

4.5. Evaluation of autophagy by immunofluorescence

Immunofluorescence is a common method of immunostaining. This technique uses the specificity of antibodies to their antigen, a specific biomolecule target within or around a cell for the visualization of the distribution of the target molecules in the cells. Through this technique, researchers can visualize the location of the desired targets as well as qualitatively analyze protein concentration in the cells. The following protocols are carried out in tissue-culture flat-bottom 96-well plate.

4.5.1. Materials

1. LC3B (D11) XP® Rabbit mAb—Cell Signaling Technology, #3868
2. Anti-rabbit IgG (H + L), $F(ab')_2$ Fragment (Alexa Fluor® 488 conjugated)—Cell Signaling Technology, #4412
3. DAPI—Cell Signaling Technology, #4083
4. Phosphate buffered salts (PBS) tablets—Takara, #T900
5. Tissue culture-treated 96-well flat-bottom plate—TPP, #92096
6. Albumin, bovine serum, fraction V, low heavy metals—Merck, #12659-100GM
7. Methanol, methyl alcohol, Grade AR—Riendemann Chmidt, #M2097-1-2500
8. Triton X-100 (for molecular biology)—Sigma, #T8787

4.5.2. Methods

1. Cells were seeded into 96-well plates and grown to a confluency of ~50%.
2. Treatments such as starvation and autophagy inhibition were carried out in the well prior to fixation.
3. The spent media were removed and the cells were washed with 1X PBS once, before 200 uL ice-cold 100% methanol was added into each well, followed by incubation in –20°C for 15 minutes.
4. The methanol was removed from each well, and the wells were washed for three times with 200 uL 1X PBS incubated for 5 minutes between each wash on the bench.
5. 5% bovine serum albumin (BSA) dissolved in 1X PBS with 0.3% Triton X-100 was prepared, and 50 uL is added into each well for blocking. The cells were incubated for 1 hour at room temperature.
6. Antibody diluent buffer was prepared by dissolving 1% BSA in 1X PBS with 0.3% Triton X-100 while the cells are blocked.

7. Primary antibody was diluted in antibody diluent buffer at the recommended concentration by the manufacturer. LC3B was diluted 1:200 for this experiment.

8. After blocking, the blocking buffer was removed, and 50 uL of the diluted antibody was added into each well. The plate was incubated at 4°C overnight.

9. The primary antibody was removed the next day, and the wells were washed as mentioned in Step 4.

10. Fluorophore-conjugate secondary antibody targeting the primary antibody is diluted in antibody diluent buffer (Step 6) at the recommended concentration by the manufacturer. For this experiment, Alexa Flour 488-conjugated anti-Rabbit IgG antibody was diluted 1:1000. 50 uL of diluted secondary antibody was added into each well and incubated at room temperature for 2 hours in the dark.

11. The Secondary Antibody was removed and the wash step in Step 4 was repeated 5 times.

12. DAPI was diluted to a concentration of 0.5 ug/mL in 1X PBS, and 50 uL was added into each well in the dark. The wells were incubated for 5 minutes at RT in the dark.

13. The wells were washed 1 time according to Step 4 and viewed using a fluorescence microscope in the dark.

Cells were treated with LY294002, which is a cell-permeable inhibitor for phosphoinositide 3-kinase (PI3K) that acts on the enzyme ATP binding site (**Table 1**). As PI3K is strictly required for autophagy, PI3K inhibition sequentially leads to autophagy inhibition. As shown in **Figure 4**, both CCD112 and HT29 showed an increase in LC3B signal after starvation and LY294002 treatment. The signal is correlated to the increase in autophagosome formation after the treatment. However, such result was not obvious for HCT116, which has a high autophagosome formation in the untreated cell lines. This may be a result of continuous activation of autophagy pathway due to KRAS mutation.

4.6. Evaluation of autophagy by Western blot analysis

One of the most widely used method for the examination of autophagy activity is by elucidating the protein expression of the autophagy markers through immunoblotting. The fluctuation in the expression can help in showing the effect of different interventions such as gene silencing or inhibition on autophagy activity. The following protocol for determination of autophagy marker expressions is adapted from the general protocol for Western blotting (Bio-Rad).

4.6.1. Materials

1. Cell lines (ATCC):
   a. HT-29 (ATCC HTB-38)
   b. HCT 116 (ATCC CCL-247)
   c. CCD-112CoN (ATCC CRL-1541)
2. RIPA lysis buffer, 10X—Cell Signaling Technology, #9806S

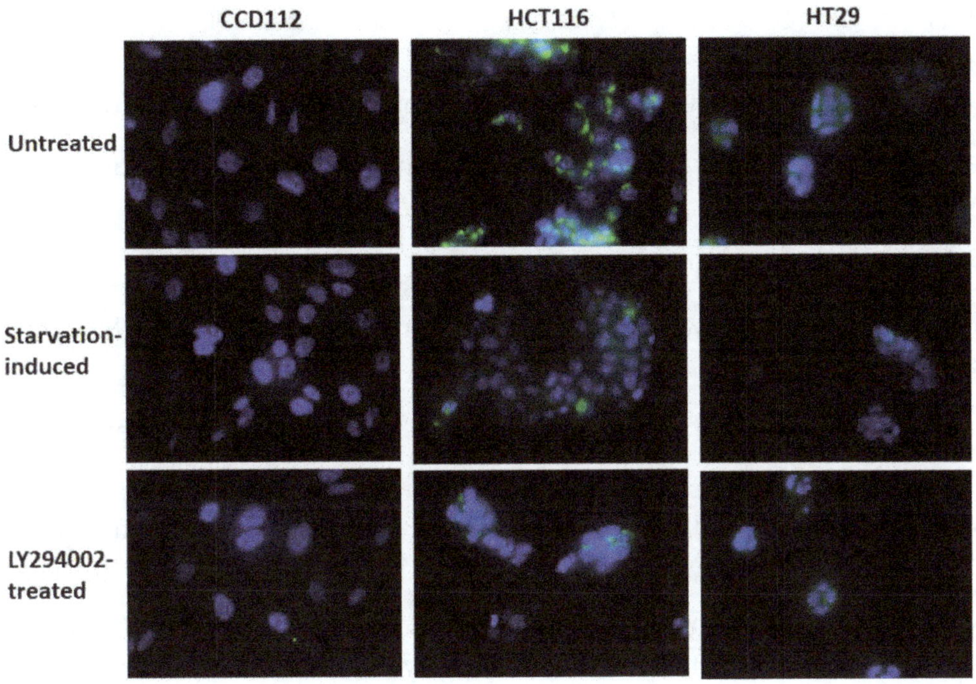

Figure 4. Immunofluorescence analysis of autophagy marker LC3B expression in colon cells after treatment. Immunofluorescence analysis of starvation-induced and PI3K-inhibited colorectal cancer cell lines: HT29, HCT116, and CCD112. Cells were seeded at 70,000 per well and incubated for 24 hours prior to treatment. Each cell line was subjected to two different treatments: starvation in serum-free media and inhibition of PI3K in media containing 50 µM of LY294002. Cells were incubated for 24 hours prior to fixation and staining. The DAPI-stained nuclei is in blue, while the LC3B expression is in green (Alexa Fluor 488 conjugated antibody).

3. Protease inhibitor cocktail, 100X—Cell Signaling Technology, #5871
4. Pierce BCA protein assay kit—ThermoFisher Scientific, #23227
5. 2-mercaptoethanol—Merck, 60–24-2
6. Tris/Glycine/SDS running buffer, 10X—Bio-Rad, #1610772
7. 4–20% Mini-PROTEAN® TGX™ Precast Protein Gels, 15-well—Bio-Rad #4561096
8. FlashBlot transfer buffer—Advansta, 150,421–95
9. Immun-Blot PVDF membrane—Bio-Rad, #1620177
10. Blotting Grade Blocker—Bio-Rad, #170–6404
11. Primary antibody (Cell Signaling Technology)
    a. β-Actin—#3700
    b. LC3B—#4599
12. Secondary antibody (Cell Signaling Technology)
    a. Anti-mouse IgG, HRP-linked antibody—#7076S
    b. Anti-rabbit IgG, HRP-linked antibody—#7074S
13. WesternBright Sirius HRP substrate—Advansta, 170,501–39

4.6.2. Methods

1. In a 24-well plate, the cells were seeded at 50% confluency and left to incubate overnight.
2. The media were then changed accordingly, with normal media, serum-free media, and media with the inhibitor.
3. After 24-hour incubation, the cells were collected and rinsed with PBS.
4. The cells were then lysed with lysis cocktail consisting of 1X RIPA lysis buffer and 1X protease inhibitor cocktail.
5. The lysates were vortexed briefly every 5 minutes for 30 minutes and kept in ice in between mixing.
6. The lysates were then centrifuged at 16000×g for 15 minutes at 4°C, and the supernatants were kept.
7. The lysates were then quantified by using BCA protein assay and immediately diluted with deionized water to obtain 13 μL of 20 μg lysate and then added to 5 μL of 4X SDS loading buffer and 2 μL of 2-mercaptoethanol.
8. The samples were boiled at 90°C for 10 minutes, and 10 μL of the samples were loaded into the well of a gradient gel.
9. The lysates were separated by SDS electrophoresis by using Tris/Glycine/SDS running buffer at 150 V for approximately 45 minutes and transferred to a PVDF membrane by using FlashBlot transfer buffer at 55 V for 1 hour.
10. The membrane was then blocked by using 5% Blotting Grade Blocker in TBST for 1 hour to prevent non-specific binding of antibody.
11. The membrane was then incubated with primary antibody diluted with the Blocker in TBST at 1:2000 dilution.
12. The incubation was done overnight at 4°C or 1 hour at room temperature.
13. Any unbound antibodies were then removed by washing with TBST for 10 minutes once and 5 minutes for another 2 times.
14. Secondary antibodies conjugated with horseradish peroxidase (HRP) were then prepared by diluting the antibody in the Blocker in TBST at 1:3000 dilution.
15. The membrane was then incubated in the secondary antibody:
    a. Anti-rabbit secondary antibody: LC3B
    b. Anti-mouse secondary antibody: β-Actin
16. The membrane was incubated for 1 hour at room temperature, and unbound antibodies are washed away with TBST once for 10 minutes and for 5 minutes for another 4 times.
17. The WesternBright Sirius HRP substrate was then dropped on the membrane, and the proteins were viewed by using ImageQuant LAS 500.
18. The blot images were then analyzed using ImageJ software for densitometry analysis.

As shown in **Figure 5**, the level of the LC3B-II proteins is generally lower than LC3B-I in cells grown at normal condition. However, following the inhibition with LY294002, the conversion of the LC3B-I to LC3B-II increases dramatically especially for the cancer cell lines. This shows an increase in the formation of the autophagophore in response to the inhibition. This is similar to the findings of Luo et al. where the inactivation of the PI3K/Akt pathway results in an increase in expression of LC3B-II proteins [62]. Meanwhile, following starvation, the cell lines also showed a marked increase in conversion of LC3B-I to LC3B-II indicating increase in autophagy activity. The increase in formation of autophagosome in cells undergoing starvation and inhibition by LY294002 is due to the effect of the PI3K pathway on the mammalian target of rapamycin complex 1 (mTORC1). In nutrient-deprived cells, IκB kinase (IKK) expression has been shown to be upregulated, while p85 regulatory subunit of PI3K has been shown to be a substrate of IKK. During starvation, the increase in IKK expression leads to the increase in phosphorylation of the p85 subunit of PI3K leading to inactivation of PI3K pathway [63]. The inactivation of the PI3K pathway inactivates mTORC1, which has an inhibitory effect on ULK1–Atg13–FIP200 complex, an autophagy initiation complex [64, 65]. So the inactivation of the PI3K pathway in both starvation and LY294002 treatment leads to the activation of autophagy by ULK-1 complex. However, the monitoring of only one protein marker is not enough to conclusively indicate the effect of the treatment.

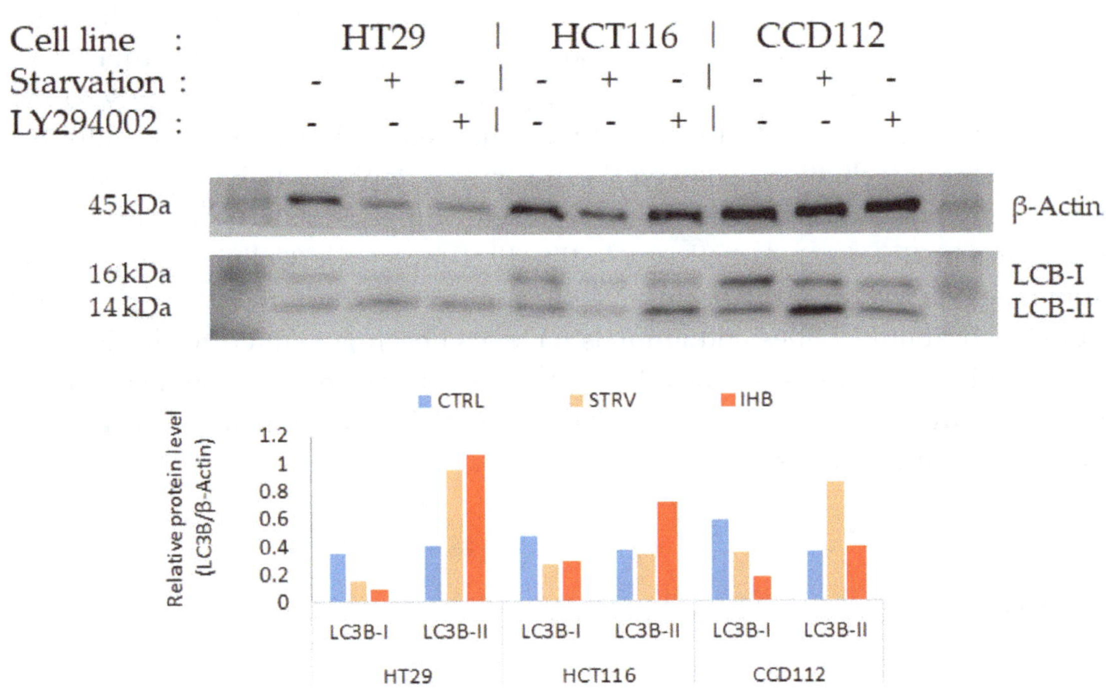

Figure 5. Western blot analysis of LCB-I and LCB-II. Top panel: Western blot analysis of starvation-induced and PI3K-inhibited colorectal cancer cell lines: HT29, HCT116, and CCD112. Cells were seeded at 70,000 per well and incubated for 24 hours prior to treatment. Each cell line was subjected to two different treatments: Starvation in serum-free media and inhibition of PI3K in media containing 50 μM of LY294002. Cells were incubated for 24 hours prior to harvesting. Cell lysates were run on 4–20% gradient gels under reducing conditions, and proteins were immunodetected on a PVDF membrane with rabbit anti-LC3B Mab (#3868) and mouse anti-β-actin Mab (#3700) from cell signaling technology. Both antibodies were diluted to 1:2000 with 5% milk in TBST. The bands were subsequently visualized with HRP-labeled anti-rabbit IgG antibodies (#7074) for LC3B and anti-mouse IgG antibodies (#7076) for β-actin (cell signaling technology) diluted at 1:3000 with 5% milk in TBST. Bottom panel: Densitometry analysis of protein bands. The analysis was done by using image J, and the relative protein levels were calculated by dividing absolute protein level of LC3B with β-actin.

5. Notes and limitations

The enclosed protocols in this chapter focus on evaluating autophagy using in vitro cancer cell lines but not limited to them. It should be noted that this fundamental cellular mechanism can be detected and studied in other cell types such as leucocytes, fibroblasts, stem cells, and so on. Autophagy is also extensively studied in fixed and live tissues (which are not discussed here) with regard to cancers and other diseases. We have only included Western blot and immunofluorescence protocols because of their simplicity and cost-effectiveness. Due to availability, colorectal cancer cell lines were used as study models for autophagy in this chapter. It must be noted that the expression pattern of studied proteins may vary among cell lines and across different cell types. Hence, the enclosed data should only be used as a reference. Researchers are advised to perform their own optimization experiments and baseline studies based on the given protocols. There are numerous varying parameters that may contribute to varying outcomes including brand and manufacturer of reagents and consumables, ambient conditions, personnel, instrumentation, and so on. Here, we have only targeted one of the autophagy effectors, LC3B for demonstration. It should be noted that there is a list of autophagy-related proteins/mRNA/DNA (described in Section 3) that can be studied according to the researchers' target of interest with respect to the nature of the research project. In addition, a plethora of autophagy-associated inducer or inhibitor (described in Section 4) can be chosen to study a specific protein/mRNA or pathway in autophagy. Last but not least, to further understand how autophagy functions and its association with a disease or disorder, it is always more favorable to study two or more autophagy-related targets concurrently to maximize the gained output and cost-effectiveness. Our enhanced understanding on autophagy and the development of technology allowed the study of autophagy to be made easier through panel assays such as Autophagy Regulators Panel (Millipore), CYTO-ID Autophagy detection kit (Enzo Life Sciences), Autophagy Antibody Sampler Kit (Cell Signaling Technology), and Autophagy Detection Kit (Abcam). Newly engineered study models such as ATG, p62, and ULK-1 knockout cell lines and animals have also been generated for in-depth study of autophagy pathway. In addition, the advancement in bioinformatics also helps in data organization and analysis as well as deciphering the potential interaction of autophagy with other unexplored cellular pathways.

6. Conclusions

The phenomenon of autophagy has been hinted since decades ago and has been a hot area of research ever since the 1990s. Extensive studies have been done on the characterization, mechanism, function, and its association with a multitude of diseases including cancer. While a simple search of autophagy in Google may yield an insurmountable amount of information, its role, mechanism, and function in relation to cancer are still not fully understood and present multiple contradictions between and within different cancer types. The simplified literature review and the protocols enclosed in this book chapter will hopefully help researchers in further understanding the roles and mechanisms of autophagy in different cancer cell types, identifying new therapeutic targets and predictive/prognostic biomarkers, and developing diagnostic assays and therapeutic drugs.

Acknowledgements

We thank Sunway Internal Research Grant INT-2018-SHMS-SIHD-01 from Sunway University and National Cancer Council Malaysia (MAKNA) Cancer Research Award (CRA) 2016 (EXT-SIDS-SIHD-MAKNA-2017-01) for partly supporting this work. This work is also partly supported by Sunway Medical Centre Research Funds (SRC/002/2017/FR and SRC/003/2017/FR). Siew-Wai Pang, Noel Jacques Awi, and Hooi-Yeen Yap are recipients of Sunway University Research Scholarships.

Conflict of interest

The authors declare no conflict of interests for this article.

List of Abbreviations

ATG	autophagy-related
ATP	adenosine triphosphate
CQ	chloroquine
EGFR	epidermal growth factor
ER	estrogen receptor
HCQ	hydroxychloroquine
HER	human epidermal growth factor receptor
HRP	horseradish peroxidase
IF	immunofluorescence
IHC	immunohistochemistry
IKK	IκB kinase
mTORC1	mammalian target of rapamycin complex 1
NF-κβ	nuclear factor kappa beta
NSCLC	non-small-cell lung cancer

PI3K	phosphoinositide 3-kinase
SLS	stone-like structure
WB	Western blot
WT	wild type

Author details

Siew-Wai Pang, Noel Jacques Awi, Hooi-Yeen Yap and Sin-Yeang Teow*

*Address all correspondence to: ronaldsyeang@gmail.com

Department of Medical Sciences, School of Healthcare and Medical Sciences, Sunway University, Jalan Universiti, Subang Jaya, Selangor Darul Ehsan, Malaysia

References

[1] Hruban Z, Spargo B, Swift H, Wissler RW, Kleinfeld RG. Focal cytoplasmic degradation. The American Journal of Pathology. 1963;**42**(6):657-683

[2] Deter RL, De Duve C. Influence of glucagon, an inducer of cellular autophagy, on some physical properties of rat liver lysosomes. The Journal of Cell Biology. 1967;**33**(2):437-449

[3] Deter RL, Baudhuin P, De Duve C. Participation of lysosomes in cellular autophagy induced in rat liver by glucagon. Journal of Cell Biology. 1967;**35**(2):C11-C16

[4] Takeshige K, Baba M, Tsuboi S, Noda T, Ohsumi Y. Autophagy in yeast demonstrated with proteinase-deficient mutants and conditions for its induction. The Journal of Cell Biology. 1992;**119**(2):301-312

[5] Thumm M, Egner R, Koch B, Schlumpberger M, Straub M, Veenhuis M, et al. Isolation of autophagocytosis mutants of *Saccharomyces cerevisiae*. FEBS Letters. 1994;**349**(2):275-280

[6] Klionsky DJ, Cueva R, Yaver DS. Aminopeptidase I of *Saccharomyces cerevisiae* is localized to the vacuole independent of the secretory pathway. The Journal of Cell Biology. 1992;**119**(2):287-300

[7] Harding TM, Morano KA, Scott SV, Klionsky DJ. Isolation and characterization of yeast mutants in the cytoplasm to vacuole protein targeting pathway. The Journal of Cell Biology. 1995;**131**(3):591-602

[8] Harding TM, Hefner-Gravink A, Thumm M, Klionsky DJ. Genetic and phenotypic overlap between autophagy and the cytoplasm to vacuole protein targeting pathway. The Journal of Biological Chemistry. 1996;**271**(30):17621-17624

[9] Klionsky DJ, Cregg JM, Dunn WA, Emr SD, Sakai Y, Sandoval IV, et al. A unified nomenclature for yeast autophagy-related genes. Developmental Cell. 2003;**5**(4):539-545

[10] Van Noorden R, Ledford H. Medicine nobel for research on how cells 'eat themselves'. Nature. 2016;**538**(7623):18-19

[11] Liang XH, Jackson S, Seaman M, Brown K, Kempkes B, Hibshoosh H, et al. Induction of autophagy and inhibition of tumorigenesis by Beclin 1. Nature. 1999;**402**(6762):672-676

[12] Mizushima N, Komatsu M. Autophagy: Renovation of cells and tissues. Cell. 2011;**147**(4): 728-741

[13] Strohecker AM, Guo JY, Karsli-Uzunbas G, Price SM, Chen GJ, Mathew R, et al. Autophagy sustains mitochondrial glutamine metabolism and growth of BrafV600E-driven lung tumors. Cancer Discovery. 2013;**3**(11):1272-1285

[14] Sun B, Karin M. Inflammation and liver tumorigenesis. Frontiers of Medicine. 2013;**7**(2):242-254

[15] White E. The role for autophagy in cancer. The Journal of Clinical Investigation. 2015; **125**(1):42-46

[16] Liu WJ, Ye L, Huang WF, Guo LJ, Xu ZG, Wu HL, et al. P62 links the autophagy pathway and the ubiqutin-proteasome system upon ubiquitinated protein degradation. Cellular & Molecular Biology Letters. 2016;**21**:29

[17] Mathew R, Karp CM, Beaudoin B, Vuong N, Chen G, Chen HY, et al. Autophagy suppresses tumorigenesis through elimination of p62. Cell. 2009;**137**(6):1062-1075

[18] Duran A, Linares JF, Galvez AS, Wikenheiser K, Flores JM, Diaz-Meco MT, et al. The signaling adaptor p62 is an important NF-κB mediator in tumorigenesis. Cancer Cell. 2008;**13**(4):343-354

[19] Moscat J, Diaz-Meco MT. P62: A versatile multitasker takes on cancer. Trends in Biochemical Sciences. 2012;**37**(6):230-236

[20] Gozuacik D, Kimchi A. Autophagy as a cell death and tumor suppressor mechanism. Oncogene. 2004;**23**(16):2891-2906

[21] Arico S, Petiot A, Bauvy C, Dubbelhuis PF, Meijer AJ, Codogno P, et al. The tumor suppressor PTEN positively regulates macroautophagy by inhibiting the phosphatidylinositol 3-kinase/protein kinase b pathway. The Journal of Biological Chemistry. 2001;**276**(38):35243-35246

[22] Tsuneoka M, Umata T, Kimura H, Koda Y, Nakajima M, Kosai K, et al. C-myc induces autophagy in rat 3Y1 fibroblast cells. Cell Structure and Function. 2003;**28**(3):195-204

[23] Kitanaka C, Kato K, Ijiri R, Sakurada K, Tomiyama A, Noguchi K, et al. Increased Ras expression and caspase-independent neuroblastoma cell death: Possible mechanism of spontaneous neuroblastoma regression. Journal of the National Cancer Institute. 2002;**94**(5):358-368

[24] Singh P, Ravanan P, Talwar P. Death associated protein kinase 1 (DAPK1): A regulator of apoptosis and autophagy. Frontiers in Molecular Neuroscience. 2016;**9**:46

[25] Degenhardt K, Mathew R, Beaudoin B, Bray K, Anderson D, Chen G, et al. Autophagy promotes tumor cell survival and restricts necrosis, inflammation, and tumorigenesis. Cancer Cell. 2006;**10**(1):51-64

[26] Gasparre G, Romeo G, Rugolo M, Porcelli AM. Learning from oncocytic tumors: Why choose inefficient mitochondria? Biochimica et Biophysica Acta, Bioenergetics. 2011; **1807**(6):633-642

[27] Rosenfeldt MT, O'Prey J, Morton JP, Nixon C, Mackay G, Mrowinska A, et al. P53 status determines the role of autophagy in pancreatic tumour development. Nature. 2013;**504**:296-300

[28] Huo Y, Cai H, Teplova I, Bowman-Colin C, Chen G, Price S, et al. Autophagy opposes p53-mediated tumor barrier to facilitate tumorigenesis in a model of PALB2 -associated hereditary breast cancer. Cancer Discovery. 2013;**3**(8):894-907

[29] Lawrence MS, Stojanov P, Mermel CH, Robinson JT, Garraway LA, Golub TR, et al. Discovery and saturation analysis of cancer genes across 21 tumour types. Nature. 2014; **505**(7484):495-501

[30] Bortnik S, Gorski SM. Clinical applications of autophagy proteins in cancer: From potential targets to biomarkers. International Journal of Molecular Sciences. 2017;**18**(7):1496

[31] McAfee Q, Zhang Z, Samanta A, Levi SM, Ma XH, Piao S, et al. Autophagy inhibitor Lys05 has single-agent antitumor activity and reproduces the phenotype of a genetic autophagy deficiency. Proceedings of the National Academy of Sciences USA. 2012;**109**(21):8253-8258

[32] Sivridis E, Koukourakis MI, Zois CE, Ledaki I, Ferguson DJP, Harris AL, et al. LC3A-positive light microscopy detected patterns of autophagy and prognosis in operable breast carcinomas. The American Journal of Pathology. 2010;**176**(5):2477-2489

[33] Giatromanolaki A, Koukourakis MI, Harris AL, Polychronidis A, Gatter KC, Sivridis E. Prognostic relevance of light chain 3 (LC3A) autophagy patterns in colorectal adenocarcinomas. Journal of Clinical Pathology. 2010;**63**(10):867-872

[34] Liao W, Sun L, Wang C, Huang H, Liu J, Liao W, et al. LC3A-positive 'stone-like' structures predict an adverse prognosis of gastric cancer. The Anatomical Record. 2014;**297**(4):653-662

[35] Karpathiou G, Sivridis E, Koukourakis MI, Mikroulis D, Bouros D, Froudarakis ME, et al. Light-chain 3A autophagic activity and prognostic significance in non-small cell lung carcinomas. Chest. 2011;**140**(1):127-134

[36] Lee YJ, Ha YJ, Kang YN, Kang KJ, Hwang JS, Chung WJ, et al. The autophagy-related marker LC3 can predict prognosis in human hepatocellular carcinoma. PLoS One. 2013;**8**:e81540

[37] Spowart JE, Townsend KN, Huwait H, Eshragh S, West NR, Ries JN, et al. The autophagy protein LC3A correlates with hypoxia and is a prognostic marker of patient survival in clear cell ovarian cancer. The Journal of Pathology. 2012;**228**(4):437-447

[38] Rolland P, Madjd Z, Durrant L, Ellis IO, Layfield R, Spendlove I. The ubiquitin-binding protein p62 is expressed in breast cancers showing features of aggressive disease. Endocrine-Related Cancer. 2007;**14**:73-80

[39] Koukourakis MI, Giatromanolaki A, Sivridis E, Pitiakoudis M, Gatter KC, Harris AL. Beclin 1 over- and underexpression in colorectal cancer: Distinct patterns relate to prognosis and tumour hypoxia. British Journal of Cancer. 2010;**103**(8):1209-1214

[40] Han Y, Xue X, Shen H, Guo X, Wang X, Guo X, et al. Prognostic significance of Beclin-1 expression in colorectal cancer: A meta-analysis. Asian Pacific Journal of Cancer Prevention. 2014;**15**(11):4583-4587

[41] Thorburn A, Thamm DH, Gustafson DL. Autophagy and cancer therapy. Molecular Pharmacology. 2014;**85**(6):830-838

[42] Cha YJ, Kim YH, Cho NH, Koo JS. Expression of autophagy related proteins in invasive lobular carcinoma: Comparison to invasive ductal carcinoma. International Journal of Clinical and Experimental Pathology. 2014;**7**(6):3389-3398

[43] Cao QH, Liu F, Yang ZL, Fu XH, Yang ZH, Liu Q, et al. Prognostic value of autophagy related proteins ULK1, Beclin 1, ATG3, ATG5, ATG7, ATG9, ATG10, ATG12, LC3B and p62/SQSTM1 in gastric cancer. American Journal of Translational Research. 2016;**8**(9): 3831-3847

[44] Zou Y, Chen Z, He X, He X, Wu X, Chen Y, et al. High expression levels of unc-51-like kinase 1 as a predictor of poor prognosis in colorectal cancer. Oncology Letters. 2015;**10**(3):1583-1588

[45] Jiang L, Duan BS, Huang JX, Jiao X, Zhu XW, Sheng HH, et al. Association of the expression of unc-51-like kinase 1 with lymph node metastasis and survival in patients with esophageal squamous cell carcinoma. International Journal of Clinical and Experimental Medicineis. 2014;**7**(5):1349-1354

[46] Yun M, Bai HY, Zhang JX, Rong J, Weng HW, Zheng ZS, et al. ULK1: A promising biomarker in predicting poor prognosis and therapeutic response in human nasopharyngeal carcinoma. PLoS One. 2015;**10**:e0117375

[47] John Clotaire DZ, Zhang B, Wei N, Gao R, Zhao F, Wang Y, et al. MiR-26b inhibits autophagy by targeting ULK2 in prostate cancer cells. Biochemical and Biophysical Research Communications. 2016;**472**(1):194-200

[48] Lee YK, Lee JA. Role of the mammalian ATG8/LC3 family in autophagy: Differential and compensatory roles in the spatiotemporal regulation of autophagy. BMB Reports. 2016;**49**(8):424-430

[49] Joachim J, Jefferies HBJ, Razi M, Frith D, Snijders AP, Chakravarty P, et al. Activation of ULK kinase and autophagy by GABARAP trafficking from the centrosome is regulated by WAC and GM130. Molecular Cell. 2015;**60**(6):899-913

[50] Salah FS, Ebbinghaus M, Muley VY, Zhou Z, Al-Saadi KRD, Pacyna-Gengelbach M, et al. Tumor suppression in mice lacking GABARAP, an Atg8/LC3 family member implicated in autophagy, is associated with alterations in cytokine secretion and cell death. Cell Death & Disease. 2016;**7**:e2205

[51] Berthier A, Seguin S, Sasco AJ, Bobin JY, De Laroche G, Datchary J, et al. High expression of gabarapl1 is associated with a better outcome for patients with lymph node-positive breast cancer. British Journal of Cancer. 2010;**102**(6):1024-1031

[52] Lin RZ, Chang HY. Recent advances in three-dimensional multicellular spheroid culture for biomedical research. Biotechnology Journal. 2008;**3**(9-10):1172-1184

[53] Perche F, Torchilin VP. Cancer cell spheroids as a model to evaluate chemotherapy protocols. Cancer Biology & Therapy. 2012;**13**(12):1205-1213

[54] Mizushima N, Yoshimori T, Levine B. Methods in mammalian autophagy research. Cell. 2010;**140**(3):313-326

[55] Avni D, Glucksam Y, Zor T. The phosphatidylinositol 3-kinase (PI3K) inhibitor LY294002 modulates cytokine expression in macrophages via p50 nuclear factor kappa B inhibition, in a PI3K-independent mechanism. Biochemical Pharmacology. 2012;**83**(1):106-114

[56] Wu YT, Tan HL, Shui G, Bauvy C, Huang Q, Wenk MR, et al. Dual role of 3-methyladenine in modulation of autophagy via different temporal patterns of inhibition on class I and III phosphoinositide 3-kinase. The Journal of Biological Chemistry. 2010;**285**(14): 10850-10861

[57] Hansen SH, Olsson A, Casanova JE. Wortmannin, an inhibitor of phosphoinositide 3-kinase, inhibits transcytosis in polarized epithelial cells. The Journal of Biological Chemistry. 1995;**270**(47):28425-28432

[58] Egan DF, Chun MGH, Vamos M, Zou H, Rong J, Miller CJ, et al. Small molecule inhibition of the autophagy kinase ULK1 and identification of ULK1 substrates. Molecular Cell. 2015;**59**(2):285-297

[59] Liu J, Xia H, Kim M, Xu L, Li Y, Zhang L, et al. Beclin 1 controls the levels of p53 by regulating the deubiquitination activity of USP10 and USP13. Cell. 2011;**147**(1):223-234

[60] Ronan B, Flamand O, Vescovi L, Dureuil C, Durand L, Fassy F, et al. A highly potent and selective Vps34 inhibitor alters vesicle trafficking and autophagy. Nature Chemical Biology. 2014;**10**(12):1013-1019

[61] Akin D, Wang SK, Habibzadegah-Tari P, Law B, Ostrov D, Li M, et al. A novel ATG4B antagonist inhibits autophagy and has a negative impact on osteosarcoma tumors. Autophagy. 2014;**10**(11):2021-2035

[62] Luo M, Liu Q, He M, Yu Z, Pi R, Li M, et al. Gartanin induces cell cycle arrest and autophagy and suppresses migration involving PI3K/Akt/mTOR and MAPK signalling pathway in human glioma cells. Journal of Cellular and Molecular Medicine. 2017;**21**(1):46-57

[63] Comb W, Hutti J, Cogswell P, Cantley L, Baldwin A. p85α SH2 domain phosphorylation by IKK promotes feedback inhibition of PI3K and Akt in response to cellular starvation. Molecular Cell. 2012;**45**(6):719-730

[64] Dibble C, Cantley L. Regulation of mTORC1 by PI3K signaling. Trends in Cell Biology. 2015;**25**(9):545-555

[65] Hosokawa N, Hara T, Kaizuka T, Kish C, Takamura A, Miura Y, et al. Nutrient-dependent mTORC1 association with the ULK1–Atg13–FIP200 complex required for autophagy. Molecular Biology of the Cell. 2009;**20**(7):1981-1991

7

Air Pouch Model: An Alternative Method for Cancer Drug Discovery

Moisés Armides Franco-Molina,
Silvia Elena Santana-Krímskaya and
Cristina Rodríguez-Padilla

Abstract

The tumor microenvironment (TME) is composed of cancer, immune, and stromal cells that interact through cell-to-cell contact and a diverse milieu of cytokines, chemokines, growth factors, and proteases. Several reports have linked the presence of specific cell subtypes with tumor stages, prognosis, and patient survival. Understanding cancer cell behavior and their response to treatment within the tumor microenvironment is essential to prevent establishment, growth, and progression of tumors. Many synthetic and biological agents have been tested using cell-based assays, which do not provide reliable predictive capacity for drug candidate performance in vivo. In this chapter, we discuss about the benefits of an air pouch tumor model, in which tumor cells are inoculated inside an air pouch created on the back of the mouse. The air pouch cancer model serves as a confined/localized tumor microenvironment, where direct contact of drug candidates and tumor cells is guaranteed in a tumor microenvironment context. Therefore, the efficacy of the therapeutic agent can be accurately assessed in vivo.

Keywords: tumor, air pouch, mice, in vitro, in vivo, antitumor

1. Introduction

Cancer research requires a significant amount of in vitro and in vivo preclinical studies. In vitro cancer cell line cultures are routinely used as the first step for evaluating potential efficacy for cancer drugs and therefore determine the "stop/go" decision for drug development, followed by animal and finally human trials [1]. However, only about 5% of anticancer drugs finally get approved for clinical use due to lack of clinical efficacy or intolerable toxicity [1, 2].

In vitro approach contribution for biotechnological development is undeniable; however, cell lines are maintained in nonphysiological conditions that do not resemble body temperature, electrolyte concentration, extracellular matrix contact, cell density, and heterogeneity. Culture conditions also implicate sudden changes such as media exchange and nutrient depletion. Furthermore, rapid cell growth is desirable and induced, driving cell subpopulations not to differentiate. All of these factors alter cell signaling and favor specific subpopulations of cells that adapt to these artificial conditions and loose some of their original characteristics [3].

To optimize cancer drug screening, it is necessary to include the biological and genetic components that influence cancer treatment outcome. Therefore, to study the cancer cell in vivo, it is essential to understand how to prevent the establishment, growth, progression, and metastasis of cancers and how to modulate the tumor microenvironment (TME) for therapeutic gain [4].

Mouse models have been used to assess toxicology and efficacy of anticancer drugs, as well as for the study of tumor induction (establishment), progression, and metastasis; these traditional mouse models have been successfully used to identify cytotoxic drugs that are still the main anticancer treatment in therapy today [4, 5].

In this chapter, we discuss about the benefits of a tumor model in which tumors can grow inside an air pouch created in the dorsal part of the mouse (**Figure 1**). The air pouch cancer model serves as the local microenvironment, which can be modulated to study establishment, progression, and metastasis as well as the efficacy of therapeutic agents.

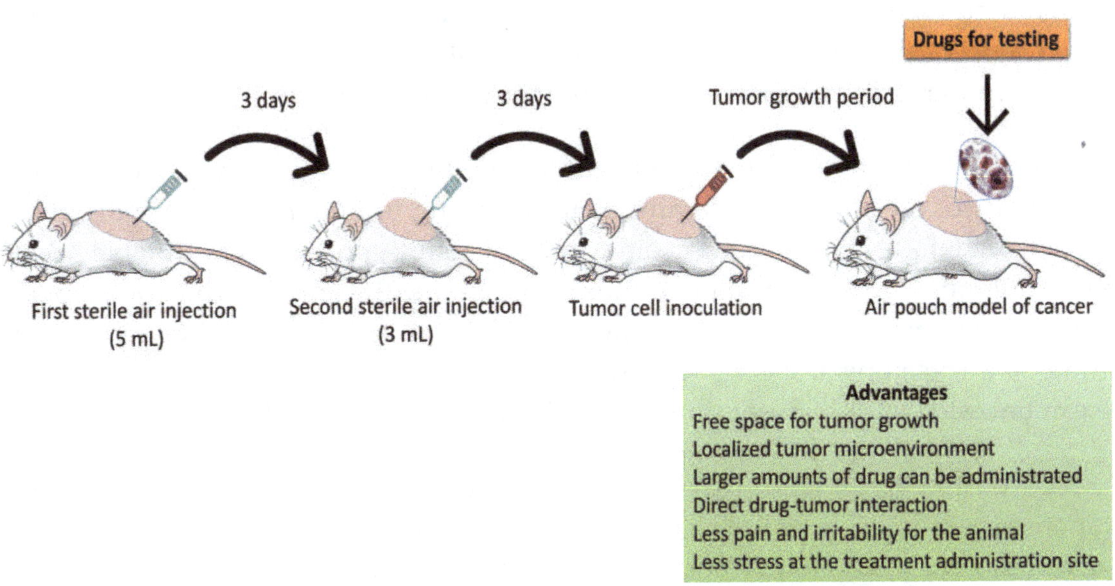

Figure 1. Mouse air pouch model for the study of cancer scheme. The general methodology to induce the air pouch model for the study of cancer and its advantages is depicted. Briefly, 5 milliliters of sterile air is injected into the shaved back of a mouse; after 3 days the pouch is reinflated with 5 milliliters of sterile air. After 3 days tumor cells are inoculated and allowed to grow prior to drug administration.

2. Model review

1953: The air pouch model was described by Selye for the study of the mechanism of action of hydrocortisone. Selye termed his model the granuloma pouch model and described it as a "procedure designed to permit the objective, quantitative analysis of factors regulating inflammation and wound-healing" [6].

1954: Kleinfeld and Habif evaluated the anti-inflammatory effect of trypsin and chymotrypsin inside a granuloma pouch. According to the authors, this method provides a standardized and reproducible inflammatory lesion. They evaluated the volume of the exudate and the weight of the pouches. The authors concluded that parenteral administration of trypsin and chymotrypsin does not affect the formation of granulation tissue in a chronic sterile inflammation environment [7].

1956: Hewitt observed that the number of viable cancer cells needed to induce tumors in half of an experimental group of adult mice was 1641 and in newborn mice was 10. He hypothesized that tumorigenesis could be affected by the dispersion of cancer cells and by the vascularity of the injection site. To test this hypothesis, he injected tumor cells into subcutaneous tissue and into an air pouch, and to evaluate the vascularity effect, he induced hyperemia in air pouches with a formic acid solution. Hewitt concluded that neither cancer cell dispersion nor vascularity affected the capacity of cancer cells to induce a tumor. He also suggested the use of the air pouch technique for further tumor transplantation studies [8].

1957: Selye wrote another article this time addressing the question: is inflammation good or bad for cancer development? He induced two air pouches on the back of the same animal, administered croton oil into one of the pouches to establish an inflammatory environment, inoculated cancer cells into both pouches, and evaluated which tumor developed faster. The results showed that inflammation can promote tumor development [9].

1957: A scientific report stated that the air pouch model presented several advantages over other methods for the study of inflammation. According to Robert and Nezamis, the granuloma pouch technique is simple and yields uniform results, and it is quantifiable because the degree of inflammation is reported in grams of the tissue, thickness of the pouch membrane, or volume of exudate. There is no systemic response, and furthermore they recommend this technique as a screening test for unknown compounds because minimal amounts of substance can be tested with high sensitivity in a short period of time [10].

1957: Based on the observation that two concomitant tumors in the same animal inhibited or promoted their development, Hans Richer inoculated nonviable sarcoma cells and viable Walker tumor cells into a mouse air pouch, to determine if the growth inhibition between tumors was due to nutrient competition. The air pouch model was used in this study because, according to the author, it is an ideal method to test a localized effect. He concluded that nonviable sarcoma cells delayed tumor growth when injected previously or concomitantly with the viable Walker tumor [11].

1958: It was known that cortisol retarded the growth of Walker tumor in rats; to determine if the effect was due to a local or systemic action of the hormone, Selye used his air pouch technique. According to him, the "granuloma pouch technique forces the malignant tissue to grow in the form of a thin lining on the wall of a regularly shaped ellipsoid cavity to which accurately measured amounts of hormones can be applied." Selye concluded that tumor growth inhibition was considerably more pronounced when cortisol was administered inside the air pouch containing the tumor as compared to administration subcutaneously in another location. In the same article, the author suggests that the hormones probably do not act directly upon cancer cells, but rather modify the stroma and the vascularization [12].

1958: Another adaptation of the air pouch model was used by Ship et al. to determine whether administration of local cytotoxic agents could prevent tumor recurrence after surgical interventions. For a better recreation of a surgical wound environment, Ship removed the air out of the pouches after tumor inoculation. He termed his modified model "the air bubble technique" and demonstrated the effectiveness of formaldehyde against tumor cells implanted in vivo [13].

1958: The granuloma pouch was used to study the role of histamine as a mediator of acute inflammation. The air pouch model was employed because, according to previous reports, croton oil used to form the granuloma pouch disrupted mast cells, therefore releasing histamine. The histological changes associated with inflammation were not observed when histamine was depleted. The author concludes that histamine is closely related to early inflammatory reactions [14].

1961: Thoracic duct lymph cells from donor rats were injected into an air pouch of a receptor rat, for cytological and histological studies of the lymphocyte. The inoculation of thoracic lymph duct cells into the air pouch also allows simultaneous injection of antigen, fatty substances, and other materials to stimulate development of plasma cells and macrophages, among others, and either donor or pouch animal can be pretreated to alter the lymph cells or their tissue environment [15].

1967: Chang, Gibley, and Ichinoe used the air pouch to study the histological features of early passages of a chemically induced hepatocarcinoma. The authors stated that the air pouch method provides easy access to the tumor for faster processing of the tissue for histochemical, ultrastructural, or biochemical studies [16].

1967: The effect of the hormones progestogen, estrogen, and chorionic gonadotropin on the inflammatory response was studied. The granuloma pouch technique (air pouch with croton oil injection) in rats was used as the inflammation model. The thickness of the granuloma pouch was the parameter used to measure inflammation. It was concluded in this study that progestogen, estrogen, and chorionic gonadotropin act as antiphlogistic substances on acute inflammation [17].

1968: Orchidectomized adult male rats received a subcutaneous transplant of ovarian grafts. The experiment determined that daily injections of progesterone (0.1 milligrams) and continuous mild stress (air pouch with croton oil) or higher doses of progesterone (0.25 milligrams) induce luteinization [18].

1968: Another adaptation of the air pouch model was employed by Toto et al. to determine if artificially produced cell culture media could induce cell proliferation in vivo. This group administered tritiated thymidine into the mouse to evaluate the proliferation rate. Their results suggest that T199 tissue culture medium supports and promotes cell growth in vivo [19].

1969: The air pouch model was used to examine the role of polymorphonuclear leukocytes in tissue repair. Briefly, an air pouch was formed in the back of mice, and turpentine was injected to cause acute inflammation. Polymorphonuclear leukocytes were extracted from this air pouch, labeled with tritiated thymidine, and inoculated into the pouch of a recipient mouse. Leukocyte cytolysis was detected and the tritiated thymidine was incorporated by DNA synthetizing fibroblasts. The authors concluded that the neutrophil degradation products are used by fibroblast to proliferate and repair damaged tissue [20].

1969: Willis used the air pouch model to obtain large volumes of exudate to extract and quantify prostaglandins for further assays. He suggested that pharmacological activity of inflammatory exudates was attributed to E-type prostaglandins [21].

1970: Anti-inflammatory activity of topically applied hydrocortisone acetate, methylprednisolone acetate, betamethasone 17-valerate, triamcinolone 16,17-acetonide, betamethasone, and fluocinolone acetonide was tested in an air pouch model. It was concluded that the air pouch model served as an application area and was used for the evaluation of commercially available corticosteroids [22].

1970: An ascitic variant of an induced hepatoma cell line was established. The original tumor was inoculated inside an air pouch. Tumor cells grew attached to the pouch walls, and in addition some cells accumulated in the fluid inside the pouch. The fluid was extracted and injected intraperitoneally in female mice and maintained for more than six passes. This method produces high amount of tumor cells free of non-tumor cells and connective tissue [23].

1973: Sensitivity of the capillary endothelium to radiation was tested. Briefly, an air pouch was induced in rats, then the area was depleted of blood vessels by freezing, and vascular proliferation was later induced employing uric acid and lithium lactate. The area becomes revascularized within 12 days. It was observed that in irradiated preparations, revascularization decreased in a dose-dependent manner [24].

1974: On the 30th symposium of the Society for Developmental Biology, the air pouch model was listed as one of the bioassays to evaluate the tumor angiogenesis factor [25].

1974: The air pouch model was used to study the binding of carcinogen 7,12-dimethylbenz[a]anthracene (DMBA) to the DNA of the mouse epidermis. The results showed that it is possible to study molecular events, such as the binding of DMBA to replicative or non-replicative DNA in vivo. It was concluded that DMBA binds preferentially to non-replicating DNA [26].

1974: An air pouch in the rat was the technique used to demonstrate that tumors can induce endothelial mitosis from 3 to 5 mm distance [27].

1974: Bladder fragments and air were subcutaneously injected in the backs of Wistar rats; within a month a cyst completely lined by urothelial cells was formed and persisted for at least 6 months. This heterotypic bladder in the rat was described by the authors as the first step to study the implantation, progression, and spread of bladder cancer [28].

1974: The healing of tissues damaged by X-ray radiation was studied using the air pouch model. The measurement of radiation damaged was based on the DNA content of the granuloma tissue and the volume of blood produced by angiogenesis and by microcolony count. It was concluded that a single high dose of local radiation (2000–4000 rads) destroys the inner layer of rapidly growing cells, while outer layers of differentiated cells show less damage [29].

1976: Differences between the early and late stages of inflammation were studied in a carrageenan inflammation air pouch model. The results showed that vascular permeability is increased by histamine injection on the early stage, while prostaglandins increase vascular permeability on the chronic stage [30].

1976: Hypersensitivity to tobacco was studied in a mouse model. To determine the involvement of mast cell response, an air pouch was induced on the back of mice, tobacco extract was administered into the pouch, and the tissue was examined microscopically to determine the presence of degranulated and non-degranulated mast cells. It was concluded that mice serve as a good model to study tobacco hypersensitivity and that mast cell is sensitized by tobacco-specific IgG and IgE [31].

1977: The air pouch model with carrageenan was used to evaluate the effect of flurbiprofen, a nonsteroidal anti-inflammatory drug, on prostaglandin production and leucocyte migration. It was concluded that optimal doses of flurbiprofen may inhibit prostaglandin synthesis and polymorphonuclear leucocyte migration into the site of inflammation [32].

1979: It was observed that oxygenation of tumor tissues occurred after X-ray radiation in some tumors; this phenomenon was termed reoxygenation, and it was believed to affect cancer cell survival. Reinhold, Blachiewicz, and Berg-Blok modified the air pouch model to obtain a "sandwich chamber" on the back of rats, to evaluate the oxygenation effect on tumors with or without the previous radiation. They concluded that oxygen levels in the tumor microenvironment of rhabdomyosarcoma BA 1112 can temporarily increase after 20 Gy radiation [33].

1980: N-Methyl-N′-nitro-N-nitrosoguanidine (MNNG) and benzo[a]pyrene (BP) were administered into air pouches of adult male rats. Fort-eight hours later, the granulation tissue was excised, and the density of mutated cells resistant to 6-thioguanine was determined in vitro. There was a dose-dependent increase in the mutation rate for both compounds [34].

1980: L. van den Boocaard studied the antitumor effect of heat and irradiation combined. He induced an air pouch on the back of rats and inoculated tumor cells inside the pouch. Then, the dorsal skin of the rat was immersed in a water bath at 42°C, and the air pouch prevented the excessive heating of the body, while tumor was about the same as the water bath. Tumor heating was performed for 2 hours 1 day before irradiation. L. van den Boocaard concluded that the combined therapy induced tumor regression and a longer tumor-free interval [35].

1982: An air pouch was formed on the back of rats to induce rapidly proliferating granuloma tissue. Two days later the test compounds, 5-bromo-2′-deoxyuridine (BrdU), mitomycin C (MMC), colchicine, and cyclophosphamide (CP), were administered inside the pouch;

24 hours later, the tissue was excised, and sister chromatid exchange frequencies were determined. It was concluded that mitomycin C and cyclophosphamide induce sister chromatid exchange [36].

1983: The carcinogenic potential of procarbazine was evaluated in an air pouch model in rats. It was concluded that high doses of procarbazine (300 mg/Kg) injected into the subcutaneous tissue do not induce tumors, unless an inflammatory environment had been previously established with a croton oil injection [37].

1984: An air pouch model was induced beneath the mammary fat pad of Wistar rats and injected with 20 milligrams of DMBA (7,12-dimethylbenz[a]anthracene). This technique allowed the induction of localized, transplantable, and estrogen-dependent adenocarcinomas in 67–80 days as compared to 150–180 days required by other techniques. The authors suggested that this model might be useful to study the biochemical mechanisms involved in estrogen-dependent mammary cancers [38].

1986: Sedgwick and Lees compared three different acute inflammation models: a 6-day air pouch, polyester sponge, and pleurisy. Overall, they concluded that the air pouch model was the most sensitive to assess the effect of steroids [39].

1987: Pretreatment of mice with methotrexate inhibits neutrophil chemotaxis induced by B4 and the complement protein C5a administration inside an air pouch. These results indicated that the anti-inflammatory activity of methotrexate observed in clinical trials with arthritis patients may be due to neutrophil chemotaxis inhibition [40].

1987: The air pouch model was used to study the effect of inflammation over pH. The author concluded that inflammation slightly decreases pH (about 0.5 units) and suggests that the pH variation could affect the effect of anesthetics [41].

1987: The effect of a fish oil-supplemented diet over inflammation and immune response was studied using a rat air pouch model. The rats were given 500 mg/kg/day of eicosapentaenoic acid and 333 mg/kg/day of docosahexaenoic acid, and control groups received water, oleic acid, or safflower oil for a period of 50 days. After this period, acute and chronic inflammation responses were induced and evaluated in an air pouch model with bovine serum albumin antigen. The results showed that the fish oil-supplemented diet decreases prostaglandin E2 and leukotriene B4 and increases leukocyte infiltration during the chronic inflammation state. No difference was observed on the volume of inflammatory exudate, the amount of protein in the exudate, and connective tissue proliferation. The authors concluded that fish oil-derived fatty acids can probably modulate the chronic inflammation phase and cellular immune response, by inhibition of leukotriene B4 and prostaglandin E2 synthesis [42].

1987: The tumor targeting the potential of liposomes encapsulating the radioisotope Ga-67 and an antibody against DLAA (Dalton's lymphoma-associated antigen) was evaluated in an air pouch model. The results showed that anti-DLAA improves Ga-67 uptake by tumor cells and that negative and neutral charge liposomes increase the accumulation of GA-67 to tumor tissue as compared to positively charged liposomes. Liposomes with Ga-67 but without anti-DLAA had minimal accumulation in tumor tissue and maximal accumulation in the liver and spleen [43].

1988: Bottomley, Griffiths, Rising, and Steward implanted a rat cartilage wrapped in cotton inside an air pouch model in mice. The authors observed loss of peptidoglycan and collagen, and formation of a granuloma, and tested different types of anti-inflammatory and antirheumatic drugs. They concluded that the air pouch model is better than other animal models to predict therapeutic efficacy in men [44].

1989: The distribution of latex microspheres into inflamed tissues and inflammatory exudate was investigated. Briefly, an air pouch model with carrageenan was induced in rats, and microspheres were administrated either orally or intravenously. The authors concluded that microspheres could be used as a vehicle for inflammation-targeted treatment [45].

1996: Appleton et al. evaluated the role of vascular endothelial growth factor in inflammation-mediated angiogenesis with the air pouch model. They concluded that the vascular endothelial growth factor may be an important regulator of angiogenesis during the inflammation process [46].

1999: The air pouch model was used to evaluate the chemopreventive activity of celecoxib and indomethacin. Briefly, hairless mice were fed with celecoxib or indomethacin, and air pouches were exposed to ultraviolet light. Celecoxib and indomethacin prevented tumor formation in 89% and 78% of the cases, respectively [47].

2000: The activation of NF-κB in the carrageenan air pouch model was studied and also the effect of dexamethasone over NF-κB activation. Results showed that NF-κB activation starts on the first day of inflammation and increases as inflammation progresses and decreases with dexamethasone treatment [48].

2000: Hooper et al. evaluated leukocyte populations, oxygen reactive species release, and phagocytosis, in response to the biomaterials poly(tetrafluoroethylene) (ePTFE), silicone, low-density polyethylene (LDPE), poly(L-lactic acid) (PLLA), poly(desaminotyrosyl-tyrosine ethyl carbonate) [poly(DTE carbonate)], and poly(desaminotyrosyl-tyrosine benzyl carbonate) [poly(DTBzl carbonate)]. Poly(DTE carbonate), ePTFE, LDPE, or poly[DTBzl carbonate increased the levels of superoxide anions inside the pouches. The authors stated that the air pouch method is a highly sensitive method to test the response of inflammatory cells to biomaterials [49].

2001: Retroviral vectors expressing a human interleukin-1 receptor antagonist and a soluble tumor necrosis factor receptor, where injected inside an air pouch model of inflammation. The pouch tissues and exudates were collected after 48 or 72 hours. Gene transfection was corroborated by PCR analysis. And, decrease of inflammation was observed in transfected mice by histological analysis. The authors concluded that the air pouch model is useful to evaluate gene therapy [50].

2004: An air pouch model was used to determine that acute neutrophilic inflammation in response to urate crystals is dependent of chemokines that bind the mIL-8RH receptor, the mice homolog of CXCR-2 chemokine 8 receptor [51].

2007: A study evaluated the antitumor effect of liposome-encapsulated anatase particles of titanium dioxide (LT). The air pouch was inoculated with NBT-II bladder cancer cells, to simulate bladder cancer, and treated with LT injections followed by UVA radiation. Tumor growth inhibition and increased survival were observed particularly in the LT + UVA radiation group. The results suggest that LT is probably more effective than not encapsulated anatase particles of titanium dioxide for the treatment of bladder cancer [52].

2009: An air pouch model was induced to determine if interleukin-17 could initiate an inflammatory response. It was concluded that interleukin-17 cannot start an inflammatory response, but it is able to increase inflammation in its early stages [53].

2009: Kourelis et al. used the air pouch model for the evaluation of early immune response. This group determined the immunoregulatory properties of *Lactobacillus paracasei* subsp. *paracasei* B112, DC205, DC215, and DC412 strains [54].

2013: The air pouch model has also been validated for nanoparticle evaluation by Vandooren et al. This group evaluated biocompatibility, toxicity, and inflammatory and adaptive immunological response to nanoparticles designed for nanomedicine. They reported that this technique yields reproducible data, with three mice per test [55].

2014: Eteraf-Oskouei et al. evaluated the antiangiogenic effect of honey in vivo. This group induced an inflammation air pouch model in Wistar rats, by injecting 1 milliliter of carrageenan (1%) on day 6. Honey was injected into the pouch at the same time as carrageenan and then for 2 consecutive days. The evaluation parameters included hemoglobin concentration, vascular endothelial growth factors and prostaglandin E2 levels, and granulomatous tissue weight. The results showed a decrease of angiogenesis, and it was concluded that honey might be useful in the treatment of granulomatous inflammatory conditions [56].

2015: Another report published by Eteraf-Oskouei et al. evaluated the antiangiogenic effect of *Ficus carica* leaf extract. The volume of exudates, the cell number, and TNFα, PGE2, and VEGF levels contained in carrageenan air pouches were measured, and for angiogenesis determination, they measured hemoglobin quantity. The extract significantly decreased the volume of exudate and leukocyte accumulation and levels of TNFα, PGE2, and VEGF; it also inhibited angiogenesis [57].

2015: A study evaluated the role of interleukin-17 in invasive breast cancer tumor pathogenesis. Air pouches were created on the back of mice, and 4 t1 or 67NR supernatants (metastatic and nonmetastatic murine mammary cancer cell lines, respectively) were injected into the pouches. Twenty hours later cell infiltrates were harvested from the pouches and stimulated with phorbol myristate acetate (PMA) plus ionomycin and brefeldin A, and intracellular levels of interleukin-17 were determined by flow cytometry analysis. From this experiment it was concluded that interleukin-17 producing $CD3^+$ cells were significantly higher in the group treated with the 4 T1 metastatic cancer cell line supernatant [58].

2018: Our group adapted the air pouch model for the screening of antitumor properties of the bio-compound IMMUNEPOTENT CRP. We induced the air pouch model, inoculated L5178Y-R cancer cells, and determined if our compound interfered with tumor implantation. We concluded that our bio-compound has antitumor properties [59].

In our opinion the air pouch model is a valuable technique for the evaluation of compounds with antitumor and tumor-preventive and/or tumor-chemopreventive properties. In **Table 1**, we summed the main advantages and disadvantages of the air pouch model for cancer as compared to in vitro cell culture.

	In vitro	*In vivo* (air-pouch model)
Advantages	Simple Highly reproducible Human origin cells Fast result production There are no concerns regarding animal ethics	Simple Highly reproducible A large number of compounds can be assessed with minimum amount of animals Direct tumor-drug interaction Maintains tumor heterogeneity and complex interactions with immune system and stroma cells Complex tumor interactions (e.g immune cell infiltrates) and systemic parameters (e.g metastasis and pharmacokinetics) can be also be evaluated Easy tumor detachment
Disadvantages	Low correlation with *in vivo* results Artificial culture conditions (nutrient, oxygen levels, drastic pH changes, among others) No tumor cell heterogeneity No interactions with immune or stromal cells Excessive plastic use (non-ecological) Do not assesses metastasis, pharmacokinetics, and other crucial variables	There are concerns regarding animal ethics Specialized installations for animal breeding are needed Not all human targets have a murine homolog target

Table 1. Comparison between the *in vitro* and the *in vivo* air pouch model cell line culturing and drug discovery.

3. Air pouch methodology

Air pouch inflammation model.

In general, a common process to induce air pouches is performed as follows:

1. Sterile air is obtained in a laminar flow station by filtration through a Millipore filter (0.22 μm) directly into a 10 mL syringe (**Figure 2**).
2. Five milliliters of sterile air is injected subcutaneously into the shaved skin site on the back of each mouse (**Figure 3**).
3. The pouches are allowed to settle for 3 days to permit the healing of the wound. The pouch is then reinflated with 5 mL of sterile air and left for 3 more days before treatments.
4. On day 8, the pouches of the experimental groups are filled with necessary doses of the compound to test, according to our requirements.

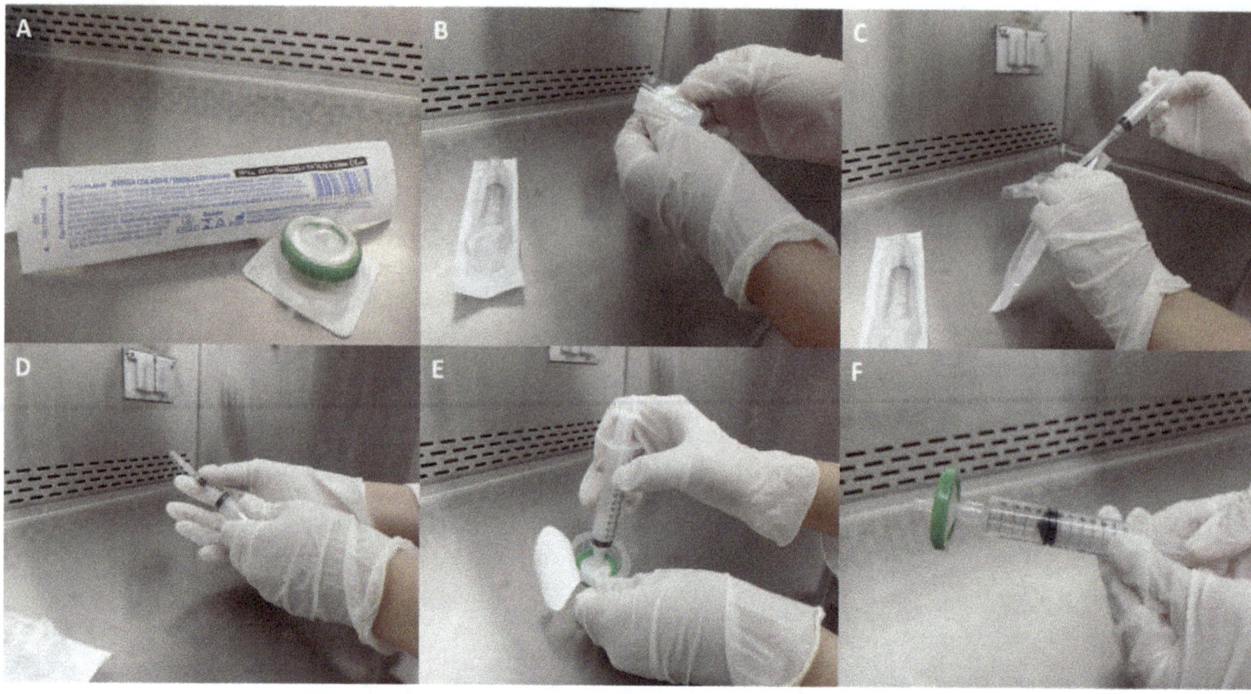

Figure 2. Step-by-step sterile air filtration in laminar flow hood. A sterile syringe that can contain 5 mL of air and a 0.2 μm filter (A) packages is opened inside a laminar flow hood (B), the syringe is carefully removed (C), the needle is detached (D), the syringe is attached to the filter (E), and 5 mL of air is loaded into the syringe (F).

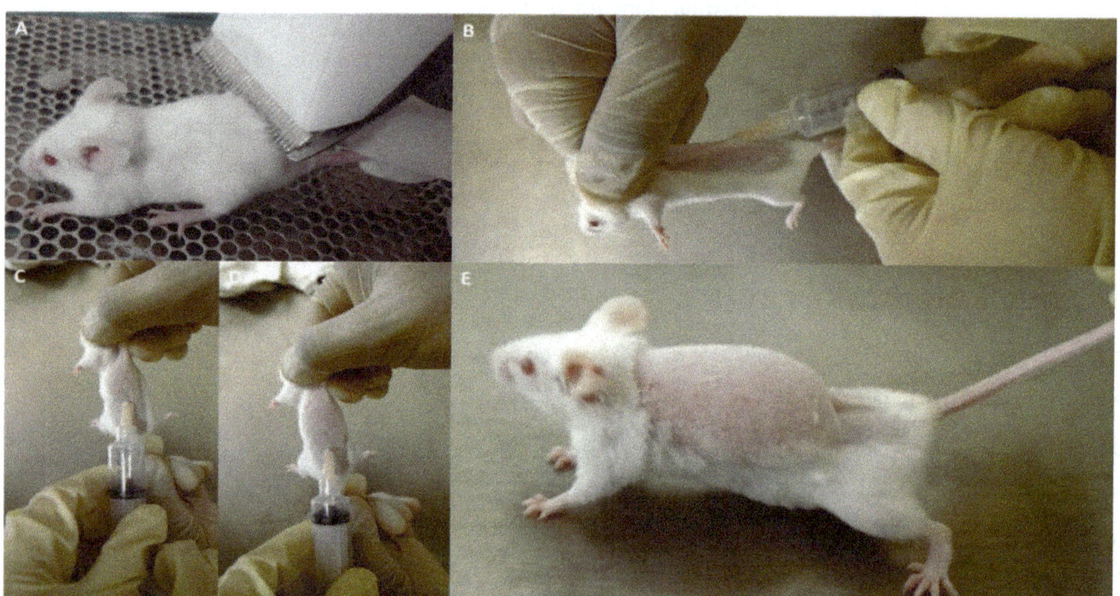

Figure 3. Air pouch formation. The rodent back is shaved (A); the animal is held, and air is injected subcutaneously into the back (B); and air pouch inflation is observed (C and D). The air pouch completely formed (E).

Note: Inflammatory agents commonly used to induce the inflammation air pouch model are ultraviolet radiation, 12-o-tetracanoilphorbol-13-acetate, oxazolone in acetone, turpentine, carrageenan, brewer's yeast, formaldehyde, dextran, egg albumin, kaolin, aerosil, implantation of pellets of compressed cotton, collagen, incomplete Freund's adjuvant, and papaya latex [60].

Air pouch modified for tumor induction

1. Inside the laminar flow hood, 5 mL of air is charged into a sterile syringe through a 0.2 μm filter (to avoid any contaminant particles) (**Figure 2**).

2. The animal back is shaved and sprayed with alcohol (70%). The animal is hold by the scruff and tail (as shown in **Figure 2B**), and the sterile air is injected subcutaneously using a 21G × 1 ¼ caliber needle. The formation of the air pouch is immediately observed (**Figure 3**).

3. After 72 hours the procedure is repeated once again to avoid disinflation. Tumor cells are inoculated 48 hours after the air pouch induction (**Figure 4A**).

Protocol 1: Air pouch model for tumor implantation

1. Viable tumor cells in 100 μL of phosphate-buffered saline are injected subcutaneously into the air pouch using a 30G × 1/2 caliber needle. Cell number varies depending on the cell line.

2. Treatment administration begins 24 hours later. Treatment administration scheme depends on the tested drug. We recommend adding a group treated with placebo (injectable solution) and with the reported treatment (e.g., chemotherapy), as positive and negative tumor growth controls, respectively.

3. For therapy evaluation, mice are sacrificed after 9 days of treatment, and tumor is resected (**Figure 4B**) and can be stained with hematoxylin and eosin (**Figure 4C**).

Protocol 2: Air pouch model for tumor progression

1. Viable tumor cells in 100 μL of phosphate-buffered saline are injected subcutaneously into the air pouch using a 30G × 1/2 caliber needle. Cell number varies depending on the cell line.

2. Treatment administration begins 7 days after tumor inoculation (days can vary depending on tumor growth characteristics). Treatment administration scheme depends on the tested drug. We recommend adding a group treated with placebo (injectable solution) and with the reported treatment group (e.g., chemotherapy), as positive and negative tumor growth controls, respectively.

3. For therapy evaluation, mice are sacrificed after 21 days of treatment.

Note: For each experiment design, it is important to take into consideration variables that could alter the results, such as animal age and sex, and treatment administration.

Animal age: A study performed by Jackson et al. revealed that there is a discrepancy on the age of rodents used for neuroscience, immunology, cancer, genetics, physiology, and toxicology research. The age can vary from 2 to 160 weeks, and 8 to 12 weeks old are the most used.

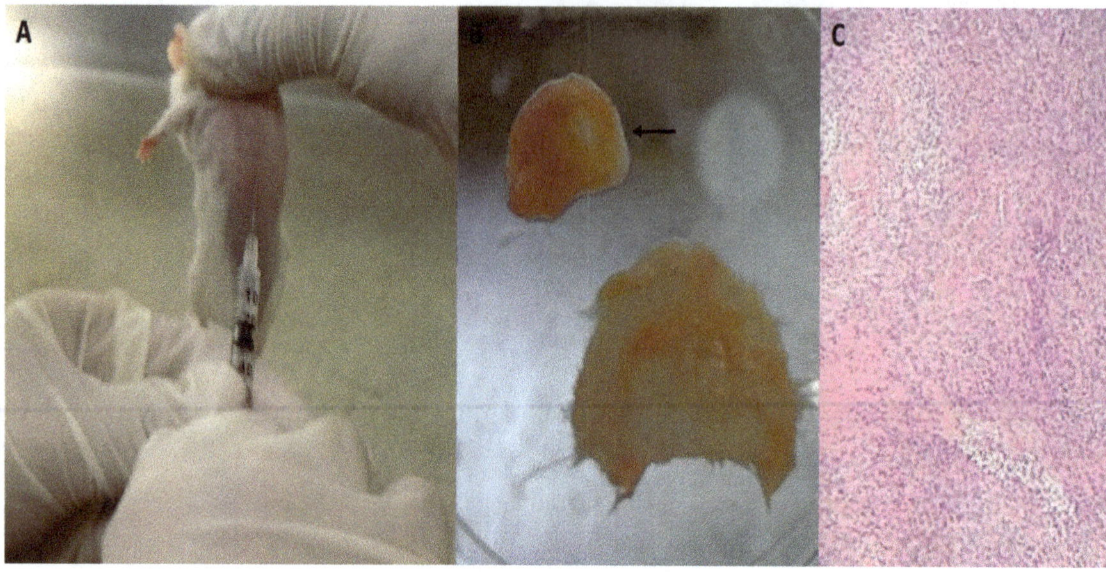

Figure 4. Tumor cell inoculation into the air pouch. 1×10^6 viable 4 T1 cells in 100 μL of PBS were inoculated into an air pouch (A). The mouse was sacrificed 12 days after for tumor evaluation. The skin from the air pouch (B) and tumor (B) were detached (the arrow points to the tumor). Tumor sections were stained with hematoxylin and eosin (C).

The appropriate age range of the rodents depends on the experiment and relevant age for human disease. For example, 3- to 6-month C57BL6/J mouse is comparable with 20–30 human years, 10–14 months compares to 38–47 human years, and 18–24 months compares to 56–69 human years [61].

Animal sex: This can also affect the therapy outcome. This is particularly true for targeted therapies, and it has become a requirement for any applications for the National Institutes of Health (NIH) to report the male and female ratio in preclinical studies [62]. Genes encoded by the Y chromosome include inflammatory pathway genes, while X chromosome encodes genes for Toll-like receptors, cytokine receptors, and transcriptional and translational regulator genes. Although X chromosome is present in both sexes, female X polymorphism allows for mosaicism and randomly silenced alleles. It has been observed that male sex leads to data alteration in pharmacology and neuroscience, and female sex affects immunology results [62, 63]. The National Institutes of Health (NIH) announced in May 2014 that they "require applicants to report their plans for the balance of male and female cells and animals in preclinical studies in all future applications" [62].

Therapy administration: Carcinogenesis is described as a three-stage process, initiation, promotion, and progression. Compounds that prevent or delay any stage of carcinogenesis are called chemopreventive agents. Chemoprevention is defined as the use of compounds to reduce the risk or delay the development of cancer or to avoid its recurrence. Bio-compounds such as food-derived polyphenols, usually inhibit initiation and promotion, of induced cancer, and therefore are considered chemopreventive agents [64]. When testing antitumor activity of natural or synthetic compounds, highly variable results can be obtained based on the type of tumor, the administration dose and frequency, and the evaluation parameters: tumor incidence, growth, or metastasis. Therefore, the purpose and parameters of the preclinical testing must be clearly defined [65].

4. Suggested evaluation parameters

Angiogenesis: Angiogenesis is a physiological process that refers to new blood vessel formation. In the tumor microenvironment, angiogenesis is over-induced to sustain tumor growth and metastasis. The vascular endothelial factor is one of the parameters used to evaluate angiogenesis. This factor induces angiogenesis, and it is increased in tumor and tumor-adjacent stroma tissue and correlates with tumor aggressiveness and with the patient prognosis [66]. Another parameter for angiogenesis evaluation is the tissue hemoglobin content [56, 57]. Angiogenesis can be measured in the tumor and adjacent tissue (skin from the air pouch), with the following techniques: flow cytometry, fluorescence microscopy, Western blot, immunohistochemistry, ELISA, and RT-PCR.

Cancer-associated fibroblasts (CAFs): Fibroblasts are the most abundant cells of the connective tissue; they produce collagen for the extracellular matrix and are involved in wound healing. In the tumor microenvironment (TME), cancer cells and other stroma cells induce fibroblasts to produce tumor-promoting substances such as epidermal and hepatocyte growth factors; chemokines CXCL12, CXCL14, and CCL5 [that attract immature and suppressor immune cells]; vascular endothelial growth factor; and interleukin-6, among others. This type of fibroblasts is termed activated or cancer-associated fibroblasts (CAFs), and they correlate with tumor growth, progression, metastasis, and chemoresistance [67].

Several cell markers are used to detect CAFs, including fibroblast activation protein α, podoplanin-a, S100A4, vimentin, fibroblast-specific protein-1, platelet-derived growth factor receptors α and β, and insulin-like growth factor-binding protein [67]. CAFs can be measured in the tumor tissue with the following techniques: flow cytometry, fluorescence microscopy, Western blot, immunohistochemistry, ELISA, and RT-PCR.

Indoleamine 2,3-dioxygenase 1 (IDO1): It is a cytosolic enzyme that catabolizes tryptophan into kynurenine, a metabolite with immunosuppressive properties. IDO1 is overexpressed in more than 50% of all tumors. Increased levels of IDO1 correlate with the decrease of natural killers and specific effector T cells and increase of regulatory T cells, tolerogenic dendritic cells, and myeloid-derived suppressor cells. IDO1 also correlates with tumor progression and multidrug resistance. It is therefore considered a tumor progression biomarker and a promising therapeutic target [68]. IDO levels can be measured in the tumor tissue with the following techniques: flow cytometry, fluorescence microscopy, Western blot, immunohistochemistry, ELISA, and RT-PCR.

Interferon gamma (IFN-γ): It is a cytokine produced by natural killer cells, natural killer T cells, antigen-specific CD4 Th1 and CD8 cytotoxic effector lymphocytes, non-cytotoxic innate lymphoid cells, and mucosal epithelial cells. IFN-γ induces class I and II major histocompatibility complex expression on antigen-presenting cells, promotes natural killer activity, increases antigen presentation and lysosome activity of macrophages, activates inducible nitric oxide synthase, induces production of IgG by plasma cells, promotes adhesion required for leukocyte migration, and has direct antiviral effect (by induction of tripartite motif-containing protein 5 and apolipoprotein B-mRNA editing enzyme, among others) [69]. IFN-γ levels correlate with good prognosis of patients with different types of cancer [69, 70]. IFN-γ levels can be measured in the tumor tissue or tumor tissue supernatant with the following techniques: flow cytometry, ELISA, and RT-PCR.

Lipid rafts: Cholesterol and sphingolipids form specific domains termed lipid rafts that regulate receptor-ligand interactions. In cancer cells, signaling protein and pro-oncogenic receptor activation correlates with their location inside the lipid rafts; disruption of lipid rafts induces apoptosis in cancer cell lines. Lipid rafts are characterized by the presence of glycosylphosphatidylinositol (GPI)-anchored proteins [71]. Also, acetyl-CoA carboxylase (ACC), fatty acid synthase (FASN), ATP citrate lyase (ACLY), and other lipogenic enzymes that promote cholesterol synthesis are altered in most tumors [72]. Lipid rafts can be measured in the tumor and adjacent tissue (skin from the air pouch), with the following techniques: flow cytometry, fluorescence microscopy, Western blot, immunohistochemistry, and ELISA.

Liver toxicity: Synthetic and biological compounds are often metabolized and excreted by the liver. Certain drugs can be metabolized to reactive compounds that bind to intracellular proteins inducing oxidative stress and cell death. Liver toxicity must be evaluated in early phase or preclinical studies of any drug and can be assessed by transcriptomics, cellular respiration, ATP (adenosine triphosphate), ROS (reactive oxygen species), covalent binding, apoptosis or necrosis, and bile salt export pump inhibition tests [73]. Liver toxicity can be evaluated in the liver tissue by immunohistochemistry or Western blot or in peripheral blood by flow cytometry or ELISA. Furthermore, blood tests can be included (glutamic oxaloacetic transaminase, glutamic pyruvic transaminase, alkaline phosphatase).

Mesenchymal stem cells (MSCs): Mesenchymal stem cells are multipotent stem cells characterized by CD73, CD105, and CD90 surface markers. MSCs have the potential to differentiate into osteoblasts, chondrocytes, and adipocytes and are recruited to injured tissues for healing. However, under different stimuli in the TME, MSCs secrete PGE_2, IL-6, IL-10, IL-17b, EGF, and CCL5, therefore promoting cancer stemness and metastasis; furthermore, they have been shown to regulate cancer cell metabolism by exosome secretion [74]. MSCs can be evaluated in the tumor tissue with the following techniques: flow cytometry, fluorescence microscopy, Western blot, and immunohistochemistry.

Tumor-associated macrophages: Macrophages are specialized phagocytic cells of the immune system. In response to various stimuli, macrophages shift their phenotype to M1 or M2. M1 macrophages are associated to an inflammatory response with antitumor properties; on the other hand, M2 macrophages promote tumor growth and have anti-inflammatory properties. Tumor-associated macrophages (TAMs) found in the TME resemble M2 macrophages and correlate with poor prognosis. TAMs produce IL-23, IL-17, IL-6, PGE2, IL-10, and indoleamine 2,3-dioxygenase and CCL17, CCL18, and CCL22, which are chemotactic factors for regulatory T cells. TAMs are characterized by CD163, CD204, or CD206 surface markers [74, 75]. TAMs can be evaluated in the tumor tissue (skin from the air pouch), with the following techniques: flow cytometry, fluorescence microscopy, Western blot, and immunohistochemistry.

Tumor growth rate: Tumor size is defined by the Response Evaluation Criteria in Solid Tumors (RECIST) as the sum of the longest diameters of the tumor mass; to report the tumor growth rate, tumor size can be measured during relevant therapy time points, for example, before treatment, after the first treatment cycle, after the last cycle of treatment, and after discontinuation of the treatment [76]. Also, the Ki67 levels, a protein in all phases of the cell cycle (G1, S, G2, and mitosis), except for the resting phase (G0), often correlate with tumor growth and progression [77]. Tumor growth can be measured with a caliper (volume), or weighted after removal, or determination of Ki67 levels in the tumor tissue by flow cytometry, fluorescence microscopy, Western blot, and immunohistochemistry.

5. Conclusion

To be clinically successful, an anticancer drug must have effect over cancer cells in the tumor microenvironment context; however, **in vitro** models do not include all of its components. Therefore, the improvement of cancer models is considered a priority for the current drug development [55].

In the current chapter, we propose a modified air pouch model as an alternative or complement to **in vitro** studies. As previously described, the air pouch model has been extensively used to evaluate inflammation process, anti-inflammatory compounds, immune response, biomaterial compatibility, and of course cancer development and treatment.

The air pouch model allows the administration of higher volumes of chemotherapy alone or combined with other treatment modalities, including targeted therapy, immunotherapy, and biological compounds, to determine single or cumulative antitumor effects, simulating the true clinical condition and treatment.

Conflict of interest

The authors declare that there is no conflict of interest.

Notes/thanks/other declarations

We are grateful for the support to the Laboratory of Immunology and Virology of the Faculty of Biological Sciences of the Autonomous University of Nuevo León.

List of abbreviations

poly(DTBzl carbonate)	poly(desaminotyrosyl-tyrosine benzyl carbonate)
poly(DTE carbonate)	poly(desaminotyrosyl-tyrosine ethyl carbonate)

ACC	acetyl-CoA carboxylase
ACLY	ATP citrate lyase
ATP	adenosine triphosphate
BrdU	5-Bromo-2′-deoxyuridine
CAFs	cancer-associated fibroblasts
CP	cyclophosphamide
DMBA	Dalton's lymphoma-associated antigen
ELISA	enzyme-linked immunosorbent assay
ePTFE	polytetrafluoroethylene
FASN	fatty acid synthase
GPI	glycosylphosphatidylinositol
LDPE	low-density polyethylene
MMC	mitomycin C
PLLA	poly(L-lactic acid)
PMA	phorbol myristate acetate
RECIST	response evaluation criteria in solid tumors
ROS	reactive oxygen species
RT-PCR	real-time polymerase chain reaction
TAMs	tumor-associated macrophages
TME	tumor microenvironment

Author details

Moisés Armides Franco-Molina*, Silvia Elena Santana-Krímskaya and Cristina Rodríguez-Padilla

*Address all correspondence to: moyfranco@gmail.com

Laboratory of Immunology and Virology of the Faculty of Biological Sciences of the Autonomous University of Nuevo León, San Nicolás de los Garza, México

References

[1] Breslin S, O'Driscoll L. Three-dimensional cell culture: The missing link in drug discovery. Drug Discovery Today. 2013;**18**:240-249. DOI: 10.1016/j.drudis.2012.10.003

[2] Hait W. Anticancer drug development: The grand challenges. Nature Reviews. Drug Discovery. 2010;**9**(4):253-254. DOI: 10.1038/nrd3144

[3] Hartung T, Daston G. Are *in vitro* tests suitable for regulatory use? Toxicological Sciences. 2009;**111**(2):233-237. DOI: 10.1093/toxsci/kfp149

[4] Day C-P, Merlino G, Van Dyke T. Preclinical mouse cancer models: A maze of opportunities and challenges. Cell. 2015;**163**(1):39-53. DOI: 10.1016/j.cell.2015.08.068

[5] Talmadge JE, Singh RK, Fidler IJ, Raz A. Murine models to evaluate novel and conventional therapeutic strategies for cancer. The American Journal of Pathology. 2007;**170**(3):793-804. DOI: 10.2353/ajpath.2007.060929

[6] Selye H. On the mechanism through which hydrocortisone affects the resistance of tissues to injury: An experimental study with the granuloma pouch technique. Journal of the American Medical Association. 1953 July 25;**152**(13):1207-1213

[7] Kleinfeld G, Habif DV. Effect of trypsin and chymotrypsin on the granuloma pouch. Proceedings of the Society for Experimental Biology and Medicine. 1954;**87**(3):585-586

[8] Hewitt HB. The quantitative transplantation of sarcoma 37 into subcutaneous air pouches in mice. British Journal of Cancer. 1956;**3**:564-569

[9] Selye H. Effect of inflammation upon the growth of transplantable neoplasms as demonstrated by the "double granuloma-pouch" technique. British Journal of Cancer. 1957;**11**(4):550-553. DOI: 10.1038/bjc.1957.67

[10] Robert A, Nezamis JE. The granuloma pouch as a routine assay for antiphlogistic compounds. European Journal of Endocrinology. 1957;**25**(1):105-112

[11] Richer CL. Inhibition of Walker tumor by autolyzed lymphosarcoma cells. Experimental Biology and Medicine. 1957;**96**(2):548-549

[12] Selye H. Topical effects of cortisol upon Walker tumors. Oncologia. 1958;**11**:193-198

[13] Ship AG, Eck RV, Smith RR. Local chemotherapy of experimentally tumor-seeded wounds. Cancer. 1958;**11**(4):687-695

[14] Irino S. Granuloma pouch and skin histamine of the rat. Acta Medica Okayama. 1958;**12**(2): 112-125

[15] Beckfield WJ, Hirata M. A new approach to the study of the lymphocyte. Proceedings of the Society for Experimental Biology and Medicine. 1961;**106**:263-266

[16] Chang JP, Gibley CW, Ichinoe K. Establishment of transplantable hepatomas induced by 3′-methyl-4- dimethylaminoazobenzene with special reference to the histologie features of the transplants of early passages. Journal of Cancer Research. 1967;**27**:2065-2071

[17] Lindhe J, Sonesson B. The effect of sex hormones on inflammation II. Progestogen, oestrogen, and chorionic gonadotropin. Journal of Periodontal Research. 1967;**2**(1):7-12. DOI: 10.1111/j.1600-0765.1967.tb01990.x

[18] Machida T. Effect of progesterone injections on subcutaneous ovarian grafts in orchidectomized rats under stressful conditions. The Proceedings of the Japan Academy. 1968;**44**(9):954-958

[19] Toto PD, Black EE, Sawinski VJ. *In vivo* growth stimulation with tissue culture medium. Oral Surgery, Oral Medicine, and Oral Pathology. 1968;**25**(6):839-843

[20] Toto PD, Santangelo MV. Reutilization of PMN nuclei in repair. Oral Surgery, Oral Medicine, and Oral Pathology. 1969 Jan;**27**(1):68-71

[21] Willis AL. Parallel assay of prostaglandin-like activity in rat inflammatory exudate by means of cascade superfusion. The Journal of Pharmacy and Pharmacology. 1969;**21**(2):126-128

[22] DiPasquale G, Rassaert CL, McDougall E. Modified granuloma pouch procedure for the evaluation of topically applied anti-inflammatory steroids. Journal of Pharmaceutical Sciences. 1970;**59**(2):267-270

[23] Smith DF, Walborg EF, Chang JP. Establishment of a transplantable ascites variant of a rat hepatoma induced by 3′-methyl-4-dimethylaminoazobenzene. Cancer Research. 1970;**30**(9):2306-2309

[24] Reinhold HS, Buisman GH. Radiosensitivity of capillary endothelium. The British Journal of Radiology. 1973;**46**(541):54-57

[25] Hay ED, Papaconstantinou J, King TJ, editors. Macromolecules Regulating Growth and Development. New York: Elsevier; 1974. DOI: 10.1016/B978-0-126-12973-1.X5001-3

[26] Bowden GT, Shapas BG, Boutwell RK. The binding of 7,12-dimethylbenz[a]anthracene to replicating and non-replicating DNA in mouse skin. Chemico-Biological Interactions. 1974;**8**(6):379-394

[27] Folkman J. Proceedings: Tumor angiogenesis factor. Cancer Research. 1974;**34**(8):2109-2113

[28] Roberts DD, Leighton J, Abaza NA, Troll W. Heterotopic urinary bladders in rats produced by an isograft inoculum of bladder fragments and air. Cancer Research. 1974;**34**(10):2773-2778

[29] van den Brenk H, Sharpington C, Orton C, Stone M. Effects of X-radiation on growth and function of the repair blastema (granulation tissue). II. Measurements of angiogenesis in the Selye pouch in the rat. International Journal of Radiation Biology and Related Studies in Physics, Chemistry and Medicine. 1974;**25**(3):277-289

[30] Chang WC, Tsurufuji S. Differences in the mode of exudative reaction between early phase and late phase of carrageenin-induced inflammation in rats. European Journal of Pharmacology. 1976;**36**(1):7-14

[31] Justus DE, Adams DA. Evaluation of tobacco hypersensitivity responses in the mouse. International Archives of Allergy and Immunology. 1976;**51**(6):687-695

[32] Adams SS, Burrows CA, Skeldon N, Yates DB. Inhibition of prostaglandin synthesis and leucocyte migration by flurbiprofen. Current Medical Research and Opinion. 1977;**5**(1):11-16

[33] Reinhold HS, Blachiewicz B, Berg-Blok A. Reoxygenation of tumours in "sandwich" chambers. European Journal of Cancer (1965). 1979;**15**(4):481-489

[34] Maier P, Manser P, Zbinden G. Granuloma pouch assay. Mutation Research/Genetic Toxicology. 1980;**77**(2):165-173

[35] van den Boocaard L. Enhanced tumor cure rates by combination of heat and irradiation with a one-day interval. An experimental study. European Journal of Cancer (1965) 1980;**16**(1):93-97

[36] Maier P, Frei K, Weibel B, Zbinden G. Granuloma pouch assay IV. Induction of sister-chromatid exchanges in vivo. Mutation Research: Environmental Mutagenesis & Relat ed Subjects. 1982;**97**(5):349-357

[37] Zbinden G, Maier P. Single dose carcinogenicity of procarbazine in rats. Cancer Letters. 1983;**21**(2):155-161

[38] Arun B, Udayachander M, Meenakshi A. 7,12-dimethylbenzanthracene induced mammary tumours in Wistar rats by air pouch technique–A new approach. Cancer Letters. 1984;**25**(2):187-194

[39] Sedgwick AD, Lees P. A comparison of air pouch, sponge and pleurisy models of acute carrageenan inflammation in the rat. Agents and Actions. 1986;**18**:439-446

[40] Suarez CR, Pickett WC, Bell DH, McClintock DK, Oronsky AL, Kerwar SS. Effect of low dose methotrexate on neutrophil chemotaxis induced by leukotriene B4 and complement C5a. The Journal of Rheumatology. 1987;**14**(1):9-11

[41] Punnia-Moorthy A. Evaluation of pH changes in inflammation of the subcutaneous air pouch lining in the rat, induced by carrageenan, dextran and *Staphylococcus aureus*. Journal of Oral Pathology & Medicine. 1987;**16**(1):36-44

[42] Yoshino S, Ellis EF. Effect of a fish-oil-supplemented diet on inflammation and immunological processes in rats. International Archives of Allergy and Immunology. 1987; **84**(3):233-240

[43] Udayachander M, Meenakshi A, Muthiah R, Sivanandham M. Tumor targeting potential of liposomes encapsulating GA-67 and antibody to Dalton's lymphoma associated antigen (anti-DLAA). International Journal of Radiation Oncology. 1987;**13**(11):1713-1719

[44] Bottomley KMK, Griffiths RJ, Rising TJ, Steward A. A modified mouse air pouch model for evaluating the effects of compounds on granuloma induced cartilage degradation. British Journal of Pharmacology. 1988;**93**(3):627-635

[45] Alpar HO, Field WN, Hyde R, Lewis DA. The transport of microspheres from the gastro-intestinal tract to inflammatory air pouches in the rat. The Journal of Pharmacy and Pharmacology. 1989;**41**(3):194-196

[46] Appleton I, Brown NJ, Willis D, Colville-Nash PR, Alam C, Brown JR, et al. The role of vascular endothelial growth factor in a murine chronic granulomatous tissue air pouch model of angiogenesis. The Journal of Pathology. 1996;**180**(1):90-94

[47] Fischer SM, Lo H-H, Gordon GB, Seibert K, Kelloff G, Lubet RA, et al. Chemopreventive activity of celecoxib, a specific cyclooxygenase-2 inhibitor, and indomethacin against ultraviolet light-induced skin carcinogenesis. Molecular Carcinogenesis. 1999;**25**(4): 231-240

[48] Ellis L. Activation of the transcription factor NF-kappa B in the rat air pouch model of inflammation. Annals of the Rheumatic Diseases. 2000;**59**(4):303-307

[49] Hooper KA, Nickolas TL, Yurkow EJ, Kohn J, Laskin DL. Characterization of the inflammatory response to biomaterials using a rodent air pouch model. Journal of Biomedical Materials Research. 2000;**50**(3):365-374

[50] Sud S, Yang SY, Evans CH, Robbins PD, Wooley PH. Effects of cytokine gene therapy on particulate-induced inflammation in the murine air pouch. Inflammation. 2001; **25**(6):361-372

[51] Terkeltaub R, Baird S, Sears P, Santiago R, Boisvert W. The murine homolog of the interleukin-8 receptor CXCR-2 is essential for the occurrence of neutrophilic inflammation in the air pouch model of acute urate crystal-induced gouty synovitis. Arthritis and Rheumatism. 1998;**41**(5):900-909

[52] Chihara Y, Fujimoto K, Kondo H, Moriwaka Y, Sasahira T, Hirao Y, et al. Anti-tumor effects of liposome-encapsulated titanium dioxide in nude mice. Pathobiology. 2007;**74**(6): 353-358

[53] Maione F, Paschalidis N, Mascolo N, Dufton N, Perretti M, D'Acquisto F. Interleukin 17 sustains rather than induces inflammation. Biochemical Pharmacology. 2009;**77**(5): 878-887

[54] Kourelis A, Zinonos I, Kakagianni M, Christidou A, Christoglou N, Yiannaki E, et al. Validation of the dorsal air pouch model to predict and examine immunostimulatory responses in the gut: Probiotic injection in mouse air pouch. Journal of Applied Microbiology. 2010;**108**(1):274-284

[55] Vandooren J, Berghmans N, Dillen C, Van Aelst I, Ronsse I, Israel LL, et al. Intradermal air pouch leukocytosis as an *in vivo* test for nanoparticles. International Journal of Nanomedicine. 2013;**8**(1):4745-4756. DOI: 10.2147/IJN.S51628

[56] Eteraf-Oskouei T, Najafi M, Gharehbagheri A. Natural honey: A new and potent anti-angiogenic agent in the air-pouch model of inflammation. Drug Research. 2013; **64**(10):530-536. DOI: 10.1055/s-0033-1363229

[57] Eteraf-Oskouei T, Allahyari S, Akbarzadeh-Atashkhosrow A, Delazar A, Pashaii M, Gan SH, et al. Methanolic extract of *Ficus carica Linn*. leaves exerts antiangiogenesis effects based on the rat air pouch model of inflammation. Evidence-based Complementary and Alternative Medicine. 2015;**2015**:1-9

[58] Benevides L, da Fonseca DM, Donate PB, Tiezzi DG, De Carvalho DD, de Andrade JM, et al. IL17 promotes mammary tumor progression by changing the behavior of tumor cells and eliciting tumorigenic neutrophils recruitment. Cancer Research. 2015;**75**(18):3788-3799. DOI: 10.1155/2015/760405

[59] Franco-Molina MA, Santana-Krímskaya SE, Coronado-Cerda EE, Hernández-Luna CE, Zarate-Triviño DG, Zapata-Benavides P, et al. Increase of the antitumour efficacy of the biocompound IMMUNEPOTENT CRP by enzymatic treatment. Biotechnology and Biotechnological Equipment. 2018;**32**(1):1-8. DOI: 10.1080/13102818.2018.1460622

[60] Sindhu RK, Sood N, Puri V, Arora S. Various animal models for preclinical testing of anti-inflammatory agents. International Journal of Pharmaceutical Sciences and Research. 2017;**8**(4):1550-1557

[61] Jackson SJ, Andrews N, Ball D, Bellantuono I, Gray J, Hachoumi L, et al. Does age matter? The impact of rodent age on study outcomes. Laboratory Animals. 2017;**51**(2):160-169. DOI: 10.1177/0023677216653984

[62] Klein SL, Schiebinger L, Stefanick ML, Cahill L, Danska J, de Vries GJ, et al. Opinion: Sex inclusion in basic research drives discovery. Proceedings of the National Academy of Sciences 2015;**112**(17):5257-5258. DOI: 10.1073/pnas.1502843112

[63] Giefing-Kröll C, Berger P, Lepperdinger G, Grubeck-Loebenstein B. How sex and age affect immune responses, susceptibility to infections, and response to vaccination. Aging Cell. 2015;**14**(3):309-321. DOI: 10.1111/acel.12326

[64] Siddiqui IA, Sanna V, Ahmad N, Sechi M, Mukhtar H. Resveratrol nanoformulation for cancer prevention and therapy: Resveratrol nanoformulations for cancer. Annals of the New York Academy of Sciences. 2015;**1348**(1):20-31. DOI: 10.1111/nyas.12811

[65] Bugelski PJ, Volk A, Walker MR, Krayer JH, Martin P, Descotes J. Critical review of preclinical approaches to evaluate the potential of immunosuppressive drugs to influence human neoplasia. International Journal of Toxicology. 2010;**29**(5):435-466. DOI: 10.1177/1091581810374654

[66] Nishida N, Yano H, Nishida T, Kamura T, Kojiro M. Angiogenesis in cancer. Vascular Health and Risk Management. 2006;**2**(3):213-219

[67] Tao L, Huang G, Song H, Chen Y, Chen L. Cancer associated fibroblasts: An essential role in the tumor microenvironment. Oncology Letters. 2017;**14**(3):2611-2620. DOI: 10.3892/ol.2017.6497

[68] Hornyák L, Dobos N, Koncz G, Karányi Z, Páll D, Szabó Z, et al. The role of indoleamine-2,3-dioxygenase in cancer development, diagnostics, and therapy. Frontiers in Immunology. 2018;**9**:151. DOI: 10.3389/mmu.2018.00151

[69] Kosmidis C, Sapalidis K, Koletsa T, Kosmidou M, Efthimiadis C, Anthimidis G, et al. Interferon-γ and colorectal cancer: An up-to date. Journal of Cancer. 2018;**9**(2):232-238. DOI: 10.7150/jca.22962

[70] Lee IC, Huang YH, Chau GY, Huo TI, Su CW, Wu JC, et al. Serum interferon gamma level predicts recurrence in hepatocellular carcinoma patients after curative treatments. International Journal of Cancer. 2013;**133**(12):2895-2902

[71] Badana AK, Chintala M, Gavara MM, Naik S, Kumari S, Kappala VR, et al. Lipid rafts disruption induces apoptosis by attenuating expression of LRP6 and survivin in triple negative breast cancer. Biomedicine & Pharmacotherapy. 2018;**97**:359-368. DOI: 10.1016/j.biopha.2017.10.045

[72] Beloribi-Djefaflia S, Vasseur S, Guillaumond F. Lipid metabolic reprogramming in cancer cells. Oncogene. 2016;**5**(1):1-10. DOI: 10.1038/oncsis.2015.49

[73] Noureddin N, Kaplowitz N. Overview of mechanisms of drug-induced liver injury (DILI) and key challenges in DILI research. New York. 2018;3-18. DOI: 10.1007/978-1-4939-7677-5_1

[74] Papaccio F, Paino F, Regad T, Papaccio G, Desiderio V, Tirino V. Concise review: Cancer cells, cancer stem cells, and mesenchymal stem cells: Influence in cancer development. Translational Medicine. 2017;**6**(12):2115-2125. DOI: 10.1002/sctm.17-0138

[75] Yang L, Zhang Y. Tumor-associated macrophages: From basic research to clinical application. Journal of Hematology & Oncology. 2017;**10**(1):58. DOI: 10.1186/s13045-017-0430-2

[76] Ferté C, Koscielny S, Albiges L, Rocher L, Soria J-C, Iacovelli R, et al. Tumor growth rate provides useful information to evaluate sorafenib and everolimus treatment in metastatic renal cell carcinoma patients: An integrated analysis of the target and record phase 3 trial data. European Urology. 2014;**65**(4):713-720. DOI: 10.1016/j.eururo.2013.08.010

[77] Scholzen T, Gerdes J. The Ki-67 protein: From the known and the unknown. Journal of Cellular Physiology. 2000 Mar;**182**(3):311-322

8

Culturing Adult Stem Cells for Cell-Based Therapeutics: Neuroimmune Applications

Victoria Moreno-Manzano and Elisa Oltra García

Abstract

Pluripotent stem cells can be successfully isolated from a variety of tissues from adult organisms. This fact opens the exciting possibility of cell-based therapies for a large number of clinical treatments. However, the development of optimized protocols to obtain, grow, and cryopreserve cells, as well as that of effective clinical treatment procedures, is no easy task. The therapeutic potential of cells expanded *in vitro* depends on a multitude of factors including isolation procedures, donor and tissue types, expansion and preservation methods, etc. Researchers are investing great efforts to determine which of these many variables significantly impact downstream performance of *in vitro* expanded stem cells by studying associated changes in molecular profiles and their effect on the host immune system. This chapter reviews the current status of stem cell production and its derivatives, which are paving the way to different treatments in the clinic. Due to the research interests of our labs, particular emphasis is placed on the potential benefits of stem cell-based therapeutics for the treatment of spinal cord injuries and the neuroimmune disease myalgic encephalomyelitis/chronic fatigue syndrome (ME/CFS) not only derived from differentiation and cell engraftment mechanisms but also due to the anti-inflammatory and immunoregulatory capacities of these cells.

Keywords: mesenchymal stem cell (MSC), induced pluripotent stem cell (iPSC), spinal cord injury (SCI), myalgic encephalomyelitis/chronic fatigue syndrome (ME/CFS), extracellular vesicle (EV)

1. Introduction

Stem cells present particular characteristics that make them different from other cell types. Firstly, they are unspecialized self-renewing tissue resident cells, and secondly they can be induced to differentiate into a milieu of specialized cell types, thus holding promise for regenerative medicine. When these cells are isolated from adult fully differentiated tissues, they receive the attribute of adult stem cells, even though they are also present in infants and fetus. Therefore, it would be more appropriate to refer to them as tissue stem cells or mesenchymal stem cells to differentiate them from resident progenitors with limited differentiation capacity. MSCs can be isolated from a large number of tissues, such as bone marrow, adipose tissue, dental pulp, hair follicles, amniotic fluid, Wharton's jelly in the umbilical cord, and even from nervous or cardiac tissue. MSCs are multipotent and can be differentiated into chondrocytes, adipocytes, and osteoblasts under proper conditions [1, 2]. MSCs can be cloned and expanded *in vitro* more than a million fold without losing their differentiation potential [3] constituting, theoretically, a rich resource for tissue repair. However, their sensitivity to environmental cues and genetic factors together with a lack of standardized good manufacturing procedures (GMPs) using defined components has hampered their true therapeutic potential. Since the finding by Bartholomew et al. that MSCs inhibit mixed lymphocyte reactions and prevent the rejection of allogeneic skin grafts [4], a large number of reports have evidenced that MSCs are immunosuppressive and immunoregulatory, properties that can be harnessed therapeutically. However, challenges to fully understand and control MSC regenerative potential remain.

In addition to MSC, the reprograming of terminally differentiated cells or induction of de-differentiation by the introduction of particular sets of transcription factors [5–7] opened an additional avenue of opportunities in the field of regenerative medicine. iPScs or induced pluripotent stem cells facilitate the production of patient-specific cells overcoming immune rejection and also ethical concerns. Although they have shown their value in the generation of *in vitro* models of human disease [8, 9], the low efficiency of reprogramming events and the safety concerns associated with the process of reprogramming has prevented their use in the clinic [6, 10].

Based on the research interests of our labs, this chapter, while reviewing the advances to generate clinical-grade stem cells or their by-products, highlights the potential benefits of stem cell-based therapeutics for the treatment of spinal cord injuries (SCI) and the neuroimmune disease myalgic encephalomyelitis/chronic fatigue syndrome (ME/CFS).

2. Stem cell therapy for spinal cord repair

2.1. Spinal cord injury pathological events and timing sequence

Spinal cord injury (SCI) is often mentioned among the first conditions for which stem cells may provide a new therapy. While recent decades have brought significant improvements in rescuing neuronal activity after SCI at preclinical phases testing several individual approaches, translation to the clinic still remains inefficiently explored. Management for SCI efficient treatment is a difficult task by the intrinsic nature of the pathological cascade of events that makes the SCI a dynamic and progressive disorder. The sentence "Time is spine" defines the crucial importance of timing to rapidly diagnose patients and implement neuroprotective interventions during the acute injury phase (≤2 h) in order to diminish the devastating effects of the secondary phase of the injury (≥2–48 h) which are known to be key determinants of the final extent of neurological deficits. The secondary injury leads to necrosis and/or apoptosis of neurons and glial cells, such as oligodendrocytes, which can lead to demyelination and the loss of neural circuits. Later, in a subacute phase (2–4 days after injury), further ischemia occurs owing to ongoing edema, vessel thrombosis, and vasospasm. Persistent inflammatory cell infiltration causes further cell death and formation of very toxic cystic microcavities over time. Astrocytes, fibroblast, and pericytes proliferate and deposit extracellular matrix molecules into the perilesional area in the already intermediate and chronic phases, few weeks after SCI, when axons continue degenerating (**Figure 1a**) [11].

Spontaneous regeneration during and after having reached the chronic stage occurs due to the neuroplasticity capacity of the central nervous system (CNS); however, very limited gain of function is obtained decreasing advancing age, attributed to both extrinsic and intrinsic factors that modulate further onset, severity, and progression of the injury [12]. The cumulative myelin-associated protein anchorage to myelin sheet debris, in and around the epicenter of the injury, has a strong inhibitory nature. Nogo-A (reticulon-4 isoform A) and myelin-associated glycoprotein (MAG), among other myelin-associated proteins, bind to NOGO receptors to activate the GTPase Rho A, which activates Rho-associated protein kinase (ROCK), a regulator of further downstream effectors, leading to apoptosis and growth-cone collapse of regenerating axons involving neurite retraction [13–17]. Additional external barriers are potently adding to the inhibition of regeneration like the hypertrophic astrocytes and the reactive chemical scar with a number of axonal growth inhibitory chondroitin sulfate proteoglycans (CSPGs) [18].

The chronic SCI repair demands an intensive effort to overcome the impediments and enhance the intrinsic axon regeneration involving an efficient anatomical reorganization [19, 20]. Fortunately, although long distances for axonal reconnection or spared degenerated tracts are normally required, involving a long-term process (a rate of 1 mm/month for axon growth is estimated), it has been shown that as little as 10% of particular tracts can subserve substantial function [19, 20]. This in fact allows hypothesizing for a real recovery, mediated by bridging and partially reconnecting the spared axons allowing subsequent plasticity. Additionally, both, humans and rats, can regain a degree of function after incomplete injury, thought to be mostly due to local structural rearrangements, such as collateral sprouting from remaining axons in the gray matter, rather than by long-distance regeneration of axons in the white matter (**Figure 1b**) [21].

2.2. Stem cell therapy for SCI repair

Cell transplantation methods constitute a very promising strategy for SCI repair. Numerous studies with a diversity of cell types have clearly showed benefits to different extents, for

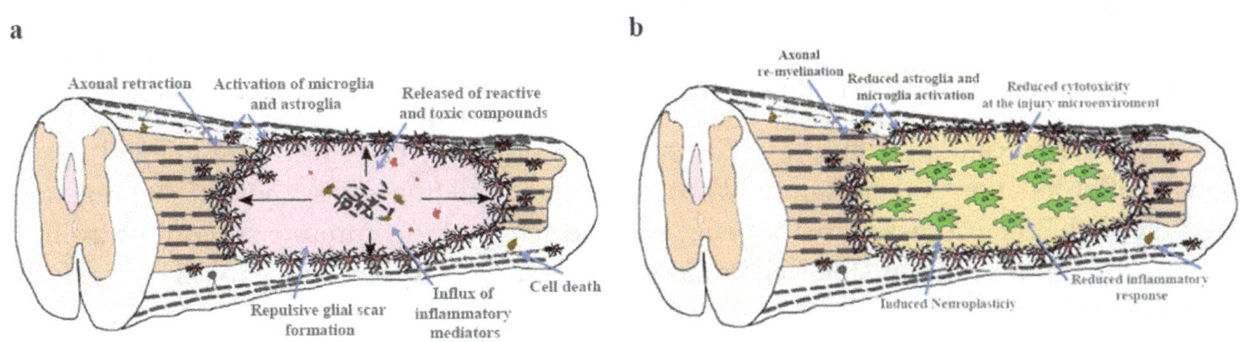

Figure 1. Summary of physiopathological events after SCI (a) and stem cell transplantation (b).

instance, therapeutic effects in sensorimotor function recovery in preclinical models [22, 23] and in several ongoing clinical trials [24–26].

Among these cell types used, mesenchymal stem cells (MSCs) from the adipose tissue, bone marrow [27], umbilical cord, [28] or dental pulp [29], in addition to olfactory ensheathing glial (OEG) cells [30], Schwann cells [31], or neural precursor cells [32], have been used.

Preclinical studies have shown that MSCs have potent anti-inflammatory, anti-apoptotic, immunomodulatory, and angiogenic effects post-SCI [33]. MSC transplantation, overall, results in substantially improved locomotor recovery among animal models of SCI [34]. There have also been several clinical trials using autologous bone marrow-derived MSCs. These early studies confirm the safety of different administration protocols using MSCs post-SCI. It seems that sufficient quantities of transplanted allogeneic MSCs combined with immunosuppression prolong the survival of engrafted cells and improve functional and morphological outcomes after SCI [35].

Transplantation of neural stem/progenitor cells (NSPCs) has shown promising results in the repair and regeneration of lost neural tissues and the associated restoration of neurological deficits [36] with particular benefits among other cell types. The engrafted transplanted NSPCs generate a favorable non-inhibitory environment for functional recovery creating additional paracrine activity modulating the post-SCI inflammatory response, feeding the injured area with growth factors, and rendering additional neurotrophic support by releasing, among others, GDNF. NSPCs include multipotent stem cells present in the ependymal region lining the central canal of the spinal cord (epSPC) [37]. epSPC represents an ideal candidate for stem cell therapy based on noted functional improvements after transplantation and the absence of malignant transformation, offering a safe and relevant cell type for clinical applications. The rationale for the therapeutic application of epSPC for SCI includes the replacement of damaged neurons and glial cells, secretion of trophic factors, regulation of gliosis and scar formation, prevention of cyst formation, and enhancement of axon elongation. After SCI, epSPCs proliferate and migrate to the injured area and produce new oligodendrocyte precursor cells (OPCs) [37]. Acute [38] and chronic [39] transplantation of undifferentiated epSPCs from SCI donors or *in vitro* differentiated OPCs into a rat model of severe spinal cord contusion produced significant locomotion recovery 1 week after injury. Transplantation of epSPCs provides trophic support and positively modulates the local immune response. It reduces purinergic receptor expression associated with neurodegenerative and neuropathic pain, thereby inducing signals promoting neuronal protection and survival with axonal outgrowth [40]. Interestingly, an immortalized human fetal epSPC line (HuCNS-SC) has been the main focus of cell therapies developed by the company Neuralstem, Inc. (USA), and therapies based on this product have been applied to human subjects in a phase II clinical trial. Phase I/II trial has declared no adverse effects of treatment, with modest functional improvements [41].

Stem cell therapy can contribute to SCI repair not only by restoring the damaged tissue through differentiation and engraftment but also by potentiating endogenous tissue regenerative potential. Both processes are influenced by the action of the immune system controlling local inflammation, and thus stem cell paracrine factor role in this context should be carefully evaluated.

3. Stem cell therapy for other neuroimmune-related health problems: potential benefits for the treatment of myalgic encephalomyelitis/chronic fatigue syndrome (ME/CFS)

Mesenchymal stromal cells (MSCs) have been used in clinical trials (CTs) for a broad range of immune-related health problems such as acute and chronic inflammatory disorders, autoimmune diseases, and transplant rejection by their potent immunosuppressive and anti-inflammatory properties [42–45]. As reviewed by Wang et al., as of April 2016, over 500 MSC-related clinical trials were registered on the NIH clinical trial database (https://clinicaltrials.gov/). Although the immunomodulatory properties of MSCs have more recently been identified, almost half of the registered CTs (230 or 42% of them) have or are being conducted for immune- or inflammation-mediated diseases (see **Figure 2**) [45].

Multiple sclerosis (MS) and its animal model (experimental autoimmune encephalomyelitis or EAE) associate with CNS inflammation, gliosis, demyelination, and axonal loss. MSCs' pleiotropic properties, including immunomodulation, immunosuppression, neurotrophy, and repair-promotion, make them attractive candidates for the treatment of neurodegenerative diseases, including MS [42–46]. The remyelination benefits reported in MS are largely attributed to paracrine signals and secreted soluble molecules such as tumor growth factor (TGF-β1), interferon (INF)-γ, indoleamine 2,3-dioxygenase (IDO), and prostaglandin E2 (PGE2) [46–48]. On another side, neural precursors obtained from induced pluripotent stem cells (iPSCs) promote the viability of endogenous OPCs facilitating remyelination through the secretion of leukemia inhibitory factor (LIF) in EAE [46, 49–51]. LIF, a member of the IL-6 cytokine family implicated in the pathophysiology of MS, has shown to offer neuroprotection and axonal regeneration as well as prevention of demyelination [49–51].

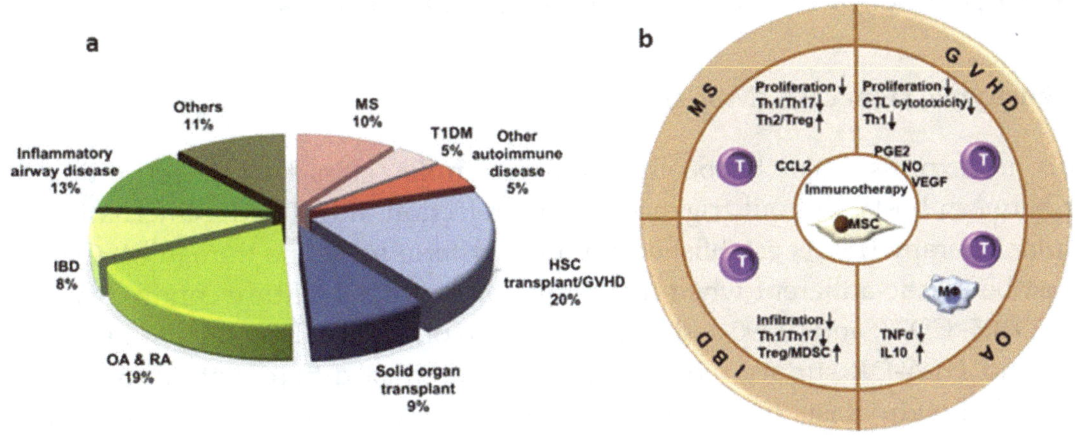

Figure 2. Summary of the number of clinical trials using MSC therapy in immune- or inflammation-mediated diseases, as registered on the website https://clinicaltrials.gov (accessed April 2016). MS, multiple sclerosis; T1DM, type 1 diabetes mellitus; GVHD, graft-versus-host disease; OA, osteoarthritis; IBD, inflammatory bowel disease (a). MSC-derived paracrine factors mediating immunomodulatory functions, particularly toward T lymphocytes, in preclinical animal studies of various immune- and inflammation-mediated diseases (b). Source: Wang et al. [45].

Myalgic encephalomyelitis/chronic fatigue syndrome (ME/CFS) is a complex, multiorgan system disease, often devastating, for which no single diagnostic test yet exists. The diagnosis of ME/CFS is based on exclusion, meaning other medical conditions, including psychiatric disorders, must be first ruled out. The disease is characterized by profound fatigue and disability lasting for at least 6 months, episodes of cognitive dysfunction, sleep disturbance, autonomic abnormalities, chronic or intermittent pain syndromes, microbiome abnormalities [52], cerebral cytokine dysregulation [53, 54], natural killer cell dysfunction [55], and other symptoms that are made worse by exertion of any kind [56, 57]. The Institute of Medicine (IOM) recently published an update of the diagnostic criteria recommended for CFS [56, 57]. The estimated worldwide prevalence of ME/CFS is 0.4–1%. The disease predominantly affects young adults, with a peak age of onset of between 20 and 40 years, and women, with a female-to-male ratio of 6:1 [58]. Although the etiological agent of ME/CFS remains unknown, the many hypotheses raised based on patient testimonies and clinical observations seem to lead to pathological immune system malfunctioning as one major factor. Autoimmune features on one side [59] and latent infection of unknown microorganisms, with a chronically activated immune system leading to inflammatory type situations, on another [60] have led our group to propose that stem cell-based therapeutics, as evidenced for MS, might be of benefit to these patients as well. The World Health Organization (WHO) has classified ME/CFS as a neurological disorder (International Classification of Diseases, Tenth Revision, Clinical Modification or ICD-10-CM R53.82; G93.3 if post-viral) based on the cognitive and other neurologic associated symptoms these patients suffer from. The neurological symptoms, however, could be explained by microglial activation and the lower-than-normal production of cortisol and adrenocorticotropic hormone (ACTH) these patients show, causing serotonin and corticotropin (CRH) deregulation [61]. A decrease in cortisol production by adrenal glands in turn can influence immune system activity [62]. MSC therapeutics could, at least partially, restore normal immune and, perhaps, neural functioning. Preclinical safety studies, however, should precede CT in ME/CFS.

4. Current protocols for stem cell-based therapeutics

4.1. Mesenchymal stem cells (MSCs)

MSCs do not express major histocompatibility complex class I or II, permitting adoptive transfer between hosts without triggering acute rejection. In 2006, the International Society for Cellular Therapy (ISCT) established minimal criteria to define human MSCs as follows: MSC must be plastic-adherent when maintained in standard culture conditions; MSC must express CD105, CD73, and CD90 and lack expression of CD45, CD34, CD14 or CD11b, CD79α or CD19, and HLA-DR surface molecules; and MSC must differentiate to osteoblasts, adipocytes, and chondroblasts *in vitro* [63, 64]. The latter position paper adds to the original characterization criteria viability and proliferation features [64]. Cells that fulfill these criteria can be isolated from different sources (such as fat, bone marrow, umbilical cord blood, dental pulp, etc.), but tissue source, donor's age, extent, and conditions of *in vitro* expansion, among others, are known to influence the regenerative potential based on engraftment, paracrine effects, and differentiation capacity of these cells [65]. Freeze-thawing effects on the whole genome expression profile of MSC have been observed, although they did not exceed inter-donor differences [66]. This high inherent heterogeneity of MSCs remains a challenge for data harmonization, particularly across lab comparisons. Despite this limitation, consensus good manufacturing procedures (cGMP) for large-scale clinical-grade MSC have been developed [67–69], based on original low-scale lab preparation methods consisting of tissue trimming, enzyme-based (collagenase) dissociation, cell filtration, and cell-type selection through adherence to plastic and extended survival *in vitro* [70, 71].

Although the numbers of CT with MSCs are already considerable and increasing, only 13 human MSC-based products count with marketing authorization. As shown in **Table 1**, nine are developed for allogeneic therapies and only four for autologous. The main source for MSC manufacturing is the bone marrow, followed by adipose tissue, although others such as umbilical cord, cord blood, placental tissue, and Wharton's jelly are being explored. However, as ASCs (adipose stromal stem cells) possess similar therapeutic potential other than bone marrow MSCs as described by the ISCT and the International Federation of Adipose Therapeutics and Science (IFATS), and since they are obtained by minimally invasive procedures from a generally undesired tissue, the fat, they may shortly become the main choice of adult stem cells for clinical applications. In fact, as reported by Nordberg and Loboa, clinical trials using ASC raised from 18 to 152 in less than 5 years (2010 to the first quarter of 2015) [72]. Standard procedures based on single-use bioreactors yield superior quantities and quality of cells when compared to traditional planar multilayer cultivation systems, such as CELLstack, HYPERStack, and CellFactories (Corning, Nalge) [67].

Efficient manufacture of MSC-based products also takes costs into account. Either allogeneic or autologous therapies involve cGMP upstream processing (USP) through master and working cell banks (MCB and WCB, respectively) and downstream processing (DSP) events, a summary of which are shown in **Figure 3**.

These manufacturing processes are tightly regulated by the Advanced Therapeutic Medicinal Product (ATMP) path [73], the European Medicines Agency (EMA) in Europe, the Center for Biologics Evaluation and Research/Food and Drug Administration (FDA) in the USA, and the Central Drugs Standard Control Organization in Asia (readers are directed to selected reviews for further legal regulatory details) [67, 73, 74].

A potential formulation to standardize cell source has been proposed by Yi et al. who using GMPs could expand clonal MSCs from a single colony-forming unit (CFU)-derived colonies derived from a small amount of bone marrow to treat a number of patients [75].

Typically, the conventional media used for clinical production of MSCs are the common, defined Dulbecco's Modified Eagle Medium (DMEM) and Minimum Essential Medium (MEM) basal media supplemented with 10–20% fetal bovine serum (FBS), due to limitations of human alternatives and to cost reasons, although FBS is not cGMP compliant. FBS is prone to batch-to-batch variation and to contamination with prions, viral and zoonotic agents [76]. Thus, most clinical trials (phases I to III) used ASCs or other MSCs produced in the presence of FBS, some of them reporting immunogenic effects in patients, elicited by components of FBS (antibodies against components of FBS, Arthus, and anaphylactic reactions) [77–79]. In addition, the immune responses elicited by FBS could turn into rejection of the transplanted cells in cell-based therapies restricting their therapeutic efficacy. FBS-free alternatives can be basically grouped into serum-free (SF) medium containing animal-derived or human serum albumin and growth factors (GFs). Among human alternatives, the use of autologous products obviates the need for infectious or other pathological agent testing but limits the

Medicinal product	Company	hMSC type	Indication	Marketing authorization
Allostem	AlloSource	Allogeneic ASC	Bone regeneration	US medical device
Cartistem	Medipost	Allogeneic UCB-MSC	Osteoarthritis	Korea
Grafix	Osiris Therapeutics	Allogeneic BM-MSC	Soft tissue defects	US medical device
Prochymal	Mesoblast	Allogeneic BM-MSC	Graft-versus-host disease	Canada and New Zealand
OsteoCel	NuVasive	Allogeneic BM-MSC	Spinal bone regeneration	US medical device
OvationOS	Osiris Therapeutics	Allogeneic BM-MSC	Bone regeneration	US medical device
TEMCELL HS	JCR Pharmaceuticals	Allogeneic BM-MSC	Graft-versus-host disease	Japan
Trinity Evolution	Orthofix	Allogeneic BM-MSC	Bone regeneration	US medical device
Trinity Elite	Orthofix	Allogeneic BM-MSC	Bone regeneration	US medical device
Hearticellgram-AMI	Pharmicell	Autologous BM-MSC	Acute myocardial infarction	Korea
Cupistem	Anterogen	Autologous ASC	Crohn's fistula	Korea
QueenCell	Anterogen	Autologous ASC	Regeneration of subcutaneous adipose tissue	Korea
Ossron	RMS	Autologous BM-MSC	Bone regeneration	Korea

Source: Adapted from Jossen et al. [67]. ASC human adipose tissue-derived stromal/stem cells, BM-MSC human bone marrow-derived mesenchymal stem cells, and UCB-MSC umbilical cord-derived mesenchymal stem cells.

Table 1. MSC-based products with marketing authorization for allogeneic and autologous therapies

production to few doses. The allogeneic alternative permits larger cell production by pooling samples from different donors but requires pathological agent screenings. Human derivatives show improved proliferation when compared to FBS-supplemented media reducing the time for cell expansion and lowering threats of senescence and transformation; however, human serum seems to limit osteogenic differentiation [63, 64, 80], and human platelet-poor plasma limits chondrogenesis [81, 82], while human platelet-rich plasma or platelet lysate preserves trilineage differentiation [83–85]. In addition, human platelet lysate, obtained by temperature-shock protocols (freezing platelets from −30 to −80°C during 24 h followed by a thawing a centrifugation step), can be prepared from banked blood with 4 or 5 days passed expiration date [86], making platelet lysate a preferable choice. To avoid MSC senescence, forced expression of telomerase reverse transcriptase (TERT) has been tried [86, 87]; however, nongenetic manipulations will be more suitable for clinical translation. On another side, the serine/threonine kinase AKT activation by plasma rich in growth factors leads to enhanced

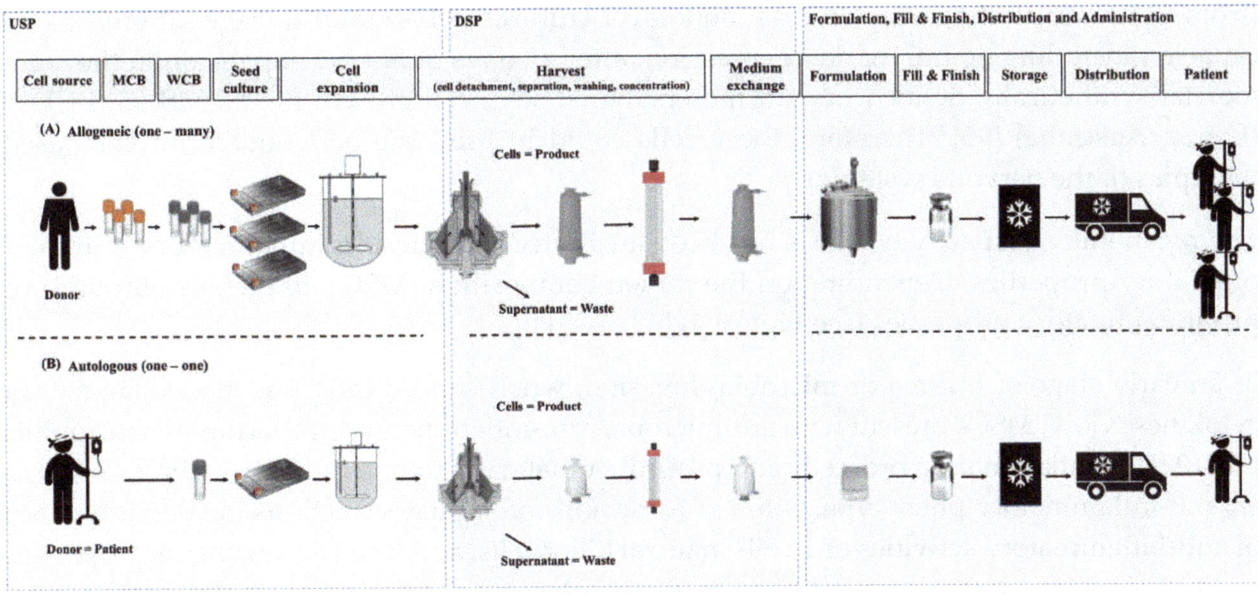

Figure 3. Main operations required to manufacture clinical-grade human MSCs for allogeneic (A) or autologous (B) therapies. USP operations typically include manufacturing of the MCB and WCB, seed cell production, and expansion at a large scale. DSP steps include cell harvest, cell detachment and separation, washing, concentration procedures, and medium exchange. Formulation, fill and finish, and storage and distribution will complete processes prior to clinical administration. Source: Jossen et al. [67].

survival and regenerative potential of MSCs and also confers resistance to hostile environments associated with inflammation which induces cell death by oxidative stress [88–90], evidencing the presence of GFs as a conditioning advantage for stem cell production.

Chemically defined, xeno-free medium does not involve donor or batch-to-batch variation; neither requires pathogenic agent screenings and presents minimal immunogenicity [91]. Drawbacks are the high cost of commercial versions and the fact that cells grown under these conditions lose their ability to adhere to plastic, requiring additional coating agent steps [92, 93]. Cells grown in xeno-free protocols also show improved proliferation potential when compared with FBS-based growing media with the previously mentioned consequent advantages [76, 92–94]. In addition to the choice of source of proteins and growth factors, basal media, seeding density, oxygen tension, confluency, and dissociation protocols may also influence outcomes. Further studies that carefully control growing conditions and at the time explore aspects such as senescence, genetic stability, immunogenicity and cytokine production, and transcriptome and proteome are needed.

In vivo ASCs reside under low oxygen tension (physioxia); Chen et al. have recently shown that the simulation of physioxic conditions (2% O_2) instead of the typical atmospheric oxygen concentration used in cell culture (20–21%) leads to increased proliferation, migration, and angiogenesis plus decreased senescence and apoptosis suggesting that the maintenance of native bioactivities may translate into production of superior cell products [95].

It is clear that tissue source and optimal growing conditions for MSC manufacturing will depend on the needs of downstream applications. In this sense it is important to mention that Okolicsanyi et al. have shown that MSCs isolated from the bone marrow of normal donors (from Lonza, Australia), expanded as monolayer cultures, retain multilineage differentiation capacity, including neural marker expression, after 43 days of *in vitro* expansion in the commercial synthetically defined human mesenchymal stem cell-growth media MSCGM-CD™ (Lonza, Australia) [96]. Therefore, these cells could in principle be a source for cell-based therapies of the nervous system.

The regenerative capacity of MSCs has been attributed to their anti-inflammatory immunoregulatory properties. Depending on the milieu composition, MSCs, in fact, exhibit anti- or pro-inflammatory properties (see **Figure 4**) [42, 97–100].

In an early stage of trauma or microbial invasion, when concentration of pro-inflammatory cytokines is low, MSCs present with antimicrobial pro-inflammatory properties of neutrophils [97–102]. As inflammation proceeds and pro-inflammatory cytokines build up, MSCs switch to an anti-inflammatory phenotype. Some of these anti-inflammatory actions include inhibition of anti-inflammatory activities of T cells, natural killer cells, and B cells; skewing macrophages to an M2 immunosuppressive state and monocyte-derived dendritic cells to a regulatory phenotype; and increasing their phagocytic capacity and inhibit mast cell degranulation [103, 104]. Increased immunomodulatory capacity of MSCs correlates with high levels of activated complement C3 [105, 106]. As MSCs express the complement factor H and the complement regulatory protein CD59, MSCs are protected from lysis. All this endows MSCs with the potential to suppress uncontrolled immune responses making them a suitable candidate for inflammation and immune dysfunction therapeutics by themselves or in combination with other cell types.

4.2. Induced pluripotent stem cells (iPSCs)

NSPCs may differentiate into neural cells after transplantation into an injured spinal cord, replacing lost or damaged cells, providing trophic support, restoring connectivity, and facilitating regeneration as a large number of studies have reported [107]. NSPC has produced some degree of functional recovery. The fetal, adult brain and adult spinal cord are the main sources for NSPCs resulting in advantageous cells for transplantation because they can be expanded and self-renewed in culture. Fetal NSPCs can be expanded for long periods by *in vitro* conditions, while adult NSPCs have more limited capabilities.

Despite a large number of studies using NSPCs, reviewed by Mothe et al. [108], some important issues such as isolation from their natural niche and their purification and expansion have to be taken in consideration [109]. Also, NSPCs have been reported to promote neuropathic pain, a concerning adverse effect. Most experimental SCI studies with NSPC transplants have involved rodent NSPCs because human NSPCs were either not available or difficult to grow. Human NSPCs have been isolated from the fetal brain and spinal cord of aborted fetuses [110] and postmortem tissue, but actually NSPCs can also be derived from human iPSCs [111].

Human iPSC-derived NSPCs have been transplanted into SCI models [112–114]. In these studies, nonobese diabetic (NOD)-severe combined immunodeficient (SCID) mice were used for SCI. The studies revealed an improved functional recovery with expression of neurotrophic factors from the grafted cells, axonal growth and stimulation of angiogenesis, increased myelination, and new forming synaptic connections between grafted cells and host neurons. In addition these studies showed the safety of human iPSC-derived NSPCs. All studies were performed in the subacute stage with just epicenter transplants.

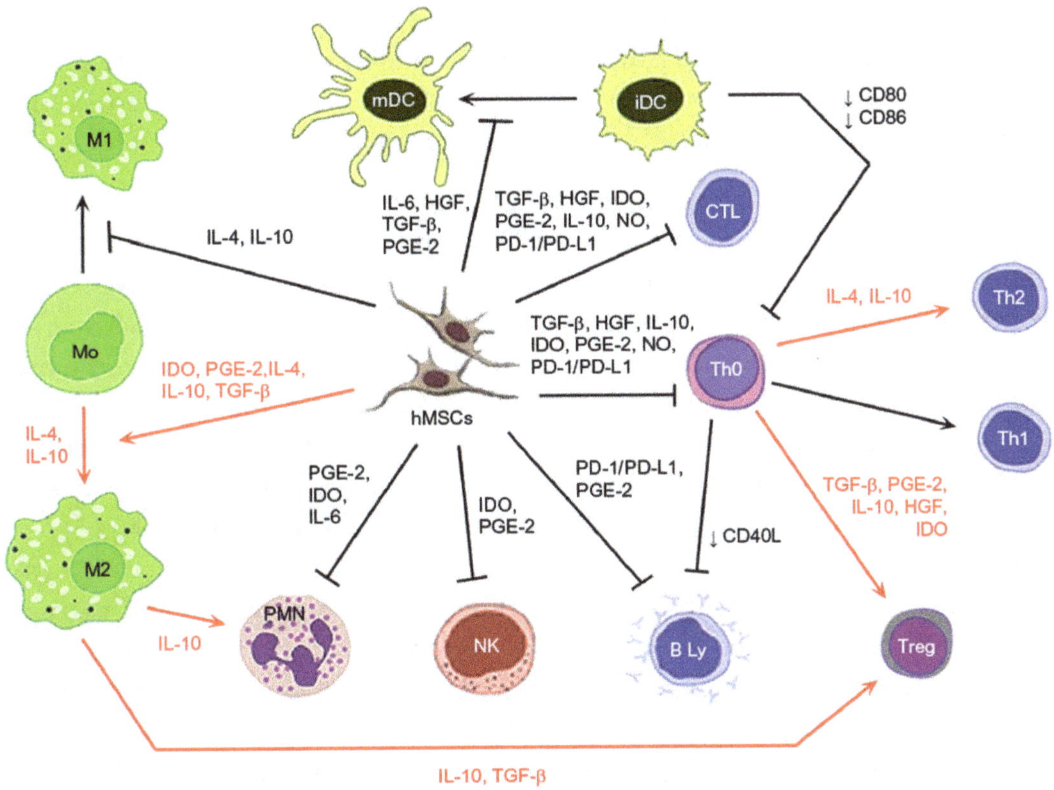

Figure 4. Immunomodulatory action of activated MSCs. Notes: Red arrow, stimulation; black arrow, suppression; blunt-ended arrow, direct inhibition. Abbreviations: iDC, immature dendritic cell; IL, interleukin; HGF, hepatocyte growth factor; TGF-β, transforming growth factor-β; PGE-2, prostaglandin E2; IDO, indoleamine 2,3-dioxygenase; NO, nitric oxide; PD-L1, programmed death-ligand 1; hMSC, human mesenchymal stem cell; Treg, T regulatory; Th, T helper; CTL, cytotoxic T cell; mDC, mature dendritic cell; PD-1, programmed cell death protein 1; PMN, polymorphonuclear leukocyte; NK, NK cell. Source: Zachar et al. [97].

All the previous reports support the potential use of iPSC-derived NSPCs in SCI. They have significant advantages, such as the lack of ethical controversy regarding their source and the potential for providing autologous transplants, thus avoiding the risk of rejection or side effects associated with immunosuppression. Recent data demonstrated the effect of the microenvironment of the injured spinal cord in the grafted iPSC-derived NSPCs. This pro-inflammatory environment induced proliferation of grafted cells [115]. Therefore, new approaches are needed to promote and guide cell differentiation, as well as to reduce tumorigenicity. Protocols for NSPC reprogrammed cells are actually improved to avoid rejection [116].

The most current iPSC protocols for neural differentiation require GFs or embryoid body formation, decreasing yields and limiting medical applications. Our lab recently developed a simple animal-free medium formula based on the inclusion of insulin and human extracellular matrix components leading to direct conversion of >98% of iPSCs into expandable and functional neural progenitors with neural rosette characteristics [111]. Further differentiation into dopaminergic and spinal motoneurons as well as oligodendrocytes and astrocytes supports the proposal that these neural progenitors retain responsiveness to environmental cues supporting applicability of the protocol for the treatment of neurodegenerative diseases. The fact that this protocol avoids embryoid body formation makes it suitable for the clinic [111].

Formerly, a feeder-free, single-step, and quick (less than 40 days) generation of mature neurons from iPSC strategy using the chemically defined medium mTeSR from STEMCELL

Technologies, overcoming the need for embryoid body formation and neuronal rosette isolation, was developed by Badja et al. Authors show that after induction the cells express voltage-gated and ionotrophic receptors for GABA, glycine, and acetylcholine (ACh) receptors and recommend the method to model human pathologies [117].

Apart from efficient differentiation methods, iPSC generation presents with the limitations of low reprogramming efficiencies (below 0.02%) and genetic modification requirements, as described by Yamanaka et al. [5, 6] and concurrently James Thompson's group [7]; thus, chromosomal instability and tumorigenic potential derived from oncogene overexpression concerns arise for their use in the clinic. An advance for safety is provided by the use of polycistronic plasmids to lead ectopic expression of the transcription factors OCT4, SOX2, KLF4, and C-MYC [118]. Other improvements based on the choice of somatic cell source, choice of reprogramming factors, culture procedures, and delivery methods have been described. For example, reprogramming kinetics and efficiencies vary between somatic cell types, in particular, keratinocytes reprogramed 2 times faster and 100 times more efficient than skin fibroblasts [10], and in general, immature cells are more readily reprogrammed than terminally differentiated cells [119]. The requirement of reprogramming factors also varies according to the cell type, so neural stem cells need only the introduction of OCT4 to be reprogrammed [120]. Reprogramming efficiency can be increased by different methods including adjustment of expression levels of noncoding RNAs, such as microRNAs or lincRNAs [111, 121]. ncRNAs can reduce the amount of reprogramming factors as they specifically target multiple pathways. Traditionally, lentivirus has been the reprogramming vector of choice; other viral vectors such as Sendai and adenovirus to lower transformation risks have been used with lower efficacy [122–124]. Excellent reviews describing more details for reprogramming protocol improvements are available [125–127].

In addition to iPSC-derived NPCs, other neural types such as neurons or astrocytes have shown some potential for SCI recovery either to improve synaptic connections or reduce neuropathic pain for the first or to protect the lesion epicenter from infiltrating peripheral inflammatory cells for the second [122, 128]. Peripheral inflammatory cell infiltration can be reduced by the immunoregulatory functions of central nervous system perivascular stromal cells (PSCs). Their low abundance, inaccessibility, and limited proliferation capacity hampers its clinical use. However, PSCs can be successfully generated from iPSCs [129] and expanded *in vitro* without senescing. Thus, SCI stem cell-based therapeutics may benefit including PSCs as part of combinatorial treatments.

4.3. Preconditioning of stem cells

It may result advantageously to precondition MSCs or iPSCs before transplantation according to the particular application or therapy, so that the cells are committed to a particular desired phenotype. Okolicsanyi et al. have more recently shown that heparan sulfate proteoglycans (HSPGs) act as drivers of neural progenitors in expanded bone marrow-derived MSCs [96]. Treatment of MSCs with heparin increased proliferation in addition to the expression of neural markers although these changes were not uniform across growth phases indicating that the direct lineage specifications and functionality of expanded cells might need fine-tuning. Nevertheless, the data sustains that MSCs may provide an abundant source that can be manipulated for purposes of neural repair and regeneration. Biomimetic approaches to exploit the role of HSPGs in neurogenesis, such as synthetic glycopolymers or heparin conjugates, are being developed so that neural differentiation into specific lineages can be controlled and tailored [96, 130].

Stem cell transplantation has shown an important limitation due to its poor survival and engraftment at the injured spinal cord, where cells are exposed to hypoxic conditions, nutritional deficiency, or oxidative stress among others. We recently showed that FM19G11, a small chemical, first described as a HIFα protein inhibitor, is able to allow progenitor cells to differentiate into more mature oligodendrocytes under hypoxia without cytotoxic effects at nanomolar doses [IC50 (80 nM)]. Moreover, FM19G11 induces self-renewal by inducing insulin-like signaling pathway and inducing ATP accumulation, activated glucose metabolism with glucose uptake by upregulation of the GLUT4 transporter. The over-induction of AKT/mTOR signaling was directly correlated to the FM19G11-dependent induction of the self-renewal-related markers Sox2, Oct4, Nanog, and Notch1 [130]. Interestingly, the use of a combination of FM19G11 treatment and epSPC transplantation for SCI therapy reduced the glial scar extension and increased the number of neuronal fibers at the epicenter of the lesion. It also increased expression markers for neuronal plasticity and induced oligodendrocyte turnover for potential remyelination [131, 132].

In addition to preconditioning toward enhancing cell survival and proliferation or inducing differentiation into particular cell types, various treatments have shown to impact MSC secretome which could be advantageous for particular therapies. For example, the treatment of MSCs with IL-1β increases the expression levels of a number of cytokines and chemokines as well as induces the expression of cell adhesion molecules improving the migration ability of preconditioned cells to the site of inflammation *in vivo* [133]. Preconditioning protocols typically include physical treatments such as different degrees of hypoxia, mechanical stretching, application of electromagnetic fields or mimicking of three-dimensional environments on one side, and chemical or pharmacological treatments, including herbal medicines or natural extracts on another. For a recent quite complete review of preconditioning treatments of MSCs and their effects, readers are directed to the review by Hu and Li [134]. It is interesting to note that preconditioning of MSCs with low-dose lipopolysaccharide (LPS), a major component of Gram-negative bacteria, preserves mitochondrial membrane potential inhibiting cytochrome c release in hypoxia serum-deprived cultured cells [135], suggesting that a mild local infection could in fact potentiate stem cell treatment. Therefore, it should be taken into account that patients undergoing stem cell therapies are often subjected to additional pharmacological treatments and exposed to particular environmental factors which may impact the performance of the introduced stem cells at the post-implant level. To circumvent the uncertainty associated with these hard-to-control variables, genetic modification of MSCs toward the production of defined immunoregulatory effects or homing molecules is starting to be explored. For example, EAE was shown to be consistently attenuated by using engineered MSCs with CNS-homing ligand genes along with overexpression of IL-10 [136].

4.4. Extracellular vesicle-based therapeutics

Although autologous MSCs constitute a safer choice in terms of avoiding unwanted immune responses, donor comorbidities may hamper the use of their own stem cells. Expanded allogeneic MSCs were initially believed to be immune privileged due to their low expression of major histocompatibility complex (MHC) and costimulatory molecules and the fact that they can suppress the activity of numerous immune cell populations [42–45]. Despite the overall safety reported by a large number of CTs [25, 41, 45], substantial evidence now supports both cell-mediated and humoral immune responses against donor antigens following administration of these cells highlighting that MSCs can be recognized by the host immune system (reviewed by Berglund et al. and Lohan et al.) [137, 138]. On another end, iPSCs are envisioned as a source to eliminate immune rejection; however, this remains theoretical, as therapeutic human trials have yet to be conducted. It will be important to monitor DNA methylation status and gene expression changes that could evoke immune responses in transplanted hosts even if iPSCs are autologously derived. Therefore, the possibility of a therapeutic cell-free product could be highly relevant on safety terms.

GFs and cytokines packed and secreted by MSCs (secretome) are thought to play a significant role in SCI repair, mainly by lowering pro-inflammatory cytokines (i.e., IL-2 or IL-6 and TNF-α) [139]. In fact, Cizkova et al. attributed motor function recovery, attenuated inflammatory response, and spared spinal cord tissue to a molecular cocktail found in the MSCs after transplantation [140]. MSC paracrine secretion or secretome was first described by Haynesworth et al. in 1996 [141]; since then multiple actions are endowed to MSC secretome rather than to their engraftment. Such actions include increased angiogenesis, decreased apoptosis and fibrosis, enhanced neuronal survival and differentiation, restriction of local inflammation, and adjustment of immune responses, effects that translate into induction of regeneration of damaged tissues [142]. Therefore, the therapeutic value of stem cells may mainly derive from the released factors or secretome including soluble and vesicle-packed factors. This latter fraction, termed extracellular vesicles (EVs), is a heterogeneous mix of vesicles including exosomes, a subset of double-membrane vesicles characterized by the expression of a set of markers, including tetraspanins CD9, CD63, and CD81 with attributed intercellular communication role including the transfer of their cargo (DNA, RNA, and proteins) [143].

The first documented clinical administration of EVs was performed in 2011, by administration of EVs intravenously infused at intervals of 2 or 3 days during a period of 2 weeks to a steroid-refractory GvHD patient who showed declined symptoms and stability for over 4 months [144]. Many preclinical models have shown the benefit of EV-based therapy including long-term neuroprotection. Treatment with MSC-derived EVs promoted long-lasting recovery of cognitive functions in inflammation-induced preterm brain injury [145]. EV-based therapy of SCI in rats showed a reduction of inflammatory response with apparent astrocyte and microglia disorganization in cord tissue up to 10 mm caudal to the injury site as well as locomotor recovery [146]. This illustrates the multiple potential benefits of EV-based therapies to treat neuroimmune defects. EV superiority with respect to cell-based therapeutics resides in its ready availability, ease of storage and distribution, reduced immunoantigenicity, scalability, and possibility of multiple routes of administration. EVs can also be used as delivery particles by directionally packaging molecules of interest from genetically modified cells while avoiding the risk of transfer of transformed live cells and could be obtained from iPSCs as well. Guidelines and recommendations for production, quality assurance, and application of EV-based therapeutics have been provided in an International Society for Extracellular Vesicles (ISEV) and European Network on Microvesicles and Exosomes in Health and

Disease (ME-HaD) position paper [147]. Also, the International Council for Harmonization of Technical Requirements for Pharmaceuticals of Human Use (ICH) mission guidelines to ensure the production of safe and effective high-quality medicines can be accessed on the following link: http://www.ich.org/products/guidelines.

Although clinical trials using EVs are still seldom, several companies have already engaged in EV or secretome production. A list of these companies with the corresponding links to their web pages can be found in the recent review by Gimona et al. [148]. This review includes a further in-depth review of clinical-grade EV production current status and remaining challenges. Involvement of biobank networks with pharmaceuticals may be relevant for granting standardized GMP production consistency of EVs [149].

Lastly, it should be mentioned that EVs can also be used as a sensor of stem cell plasticity or other cell features as they reflect characteristics of the cell of origin [150] constituting a helpful tool to develop optimized differentiation and preconditioning protocols.

5. Conclusions

Although CTs have in general evidenced MSC safety, the removal of FBS from clinical-grade stem cell protocols results imperative. The pooling of a large number of donors of cells and human blood fraction-based media through the use of stem cell banks or the use of xeno-free synthetic defined media should translate into allogeneic MSC preparations leading to more homogeneous clinical results. Thus, allowing minimal immune-related safety concerns derived from FBS and unveiling the real therapeutic value of *in vitro* expanded off-the-shelf MSCs.

The iPSC manufacturing technology offers the possibility of developing patient-tailored cell therapies with the consequent safety and immune-related advantages, as genetically identical cells should prevent immune rejection. iPSCs can differentiate into all three germ layers and, by their nature, do not raise bioethical debate. However, safety concerns related to *in vivo* properties of immortal cell types and the use of genetically manipulated cells raise regulation hurdles for their use in the clinic.

Preconditioning of *in vitro* expanded MSCs to ensure cell lineage commitment might result advantageously for improved treatment of particular diseases. Optimizations for the treatment of SCI and other neuroimmune health problems such as ME/CFS remain. Also, EVs and in particular exosome-enriched MSC-derived fractions may eventually become the treatment of choice for cell-based-free therapeutics by themselves or in combination with other clinical treatments once GMP production is optimized.

Acknowledgements

This work was supported by the Universidad Católica de Valencia San Vicente Mártir Research Grant 2018-121-001 and MINECO (MAT2015-66666-C3-2-R FEDER).

Conflict of interest

Authors declare no conflict of interest.

Author details

Victoria Moreno-Manzano[1] and Elisa Oltra García[2,3*]

*Address all correspondence to: elisa.oltra@ucv.es

1 Cellular Reprogramming Laboratory, Centro de Investigación Príncipe Felipe (CIPF), Valencia, Spain

2 School of Medicine, Universidad Católica de Valencia San Vicente Mártir, Valencia, Spain

3 Unidad Mixta CIPF-UCV, Centro de Investigación Príncipe Felipe, Valencia, Spain

References

[1] Uccelli A, Moretta L, Pistoia V. Mesenchymal stem cells in health and disease. Nature Reviews. Immunology. 2008;**8**(9):726-736. DOI: 10.1038/nri2395

[2] Pittenger MF, Mackay AM, Beck SC, Jaiswal RK, Douglas R, Mosca JD, et al. Multilineage potential of adult human mesenchymal stem cells. Science. 1999;**284**(5411):143-147

[3] Huss R. Isolation of primary and immortalized CD34-hematopoietic and mesenchymal stem cells from various sources. Stem Cells. 2000;**18**(1):1-9. Review

[4] Bartholomew A, Sturgeon C, Siatskas M, Ferrer K, McIntosh K, Patil S, et al. Mesenchymal stem cells suppress lymphocyte proliferation *in vitro* and prolong skin graft survival *in vivo*. Experimental Hematology. 2002;**30**(1):42-48

[5] Takahashi K, Yamanaka S. Induction of pluripotent stem cells from mouse embryonic and adult fibroblast cultures by defined factors. Cell. 2006;**126**(4):663-676 [Epub Aug 10, 2006]

[6] Takahashi K, Tanabe K, Ohnuki M, Narita M, Ichisaka T, Tomoda K, et al. Induction of pluripotent stem cells from adult human fibroblasts by defined factors. Cell. 2007;**131**(5): 861-872

[7] Yu J, Vodyanik MA, Smuga-Otto K, Antosiewicz-Bourget J, Frane JL, Tian S, et al. Induced pluripotent stem cell lines derived from human somatic cells. Science. 2007;**318**(5858):1917-1920 [Epub Nov 20, 2007]

[8] McKinney CE. Using induced pluripotent stem cells derived neurons to model brain diseases. Neural Regeneration Research. 2017;**12**(7):1062-1067. DOI: 10.4103/1673-5374.211180

[9] Anderson RH, Francis KR. Modeling rare diseases with induced pluripotent stem cell technology. Molecular and Cell Probes. 2018;**40**:52-59. DOI: 10.1016/j.mcp.2018.01.001

[10] Aasen T, Raya A, Barrero MJ, Garreta E, Consiglio A, Gonzalez F, et al. Efficient and rapid generation of induced pluripotent stem cells from human keratinocytes. Nature Biotechnology. 2008;**26**(11):1276-1284. DOI: 10.1038/nbt.1503

[11] Ahuja CS, Wilson JR, Nori S, Kotter MRN, Druschel C, Curt A, et al. Traumatic spinal cord injury. Nature Reviews Disease Primers. 2017;**3**:17018. DOI: 10.1038/nrdp.2017.18

[12] Kwon BK, Liu J, Messerer C, Kobayashi NR, McGraw J, Oschipok L, et al. Survival and regeneration of rubrospinal neurons 1 year after spinal cord injury. Proceedings of the National Academy of Sciences of the United States of America. 2002;**99**(5):3246-3251 [Epub Feb 26, 2002]

[13] Chen MS, Huber AB, van der Haar ME, et al. Nogo-A is a myelin-associated neurite outgrowth inhibitor and an antigen for monoclonal antibody IN-1. Nature. 2000;**403**(6768): 434-439

[14] Freund P, Schmidlin E, Wannier T, et al. Nogo-A-specific antibody treatment enhances sprouting and functional recovery after cervical lesion in adult primates. Nature Medicine. 2006;**12**(7):790-792

[15] Cafferty WB, Duffy P, Huebner E, Strittmatter SM. MAG and Omgp synergize with Nogo-A to restrict axonal growth and neurological recovery after spinal cord trauma. The Journal of Neuroscience. 2010;**30**(20):6825-6837

[16] DeBellard ME, Tang S, Mukhopadhyay G, Shen YJ, Filbin MT. Myelin-associated glycoprotein inhibits axonal regeneration from a variety of neurons via interaction with a sialoglycoprotein. Molecular and Cellular Neurosciences. 1996;**7**(2):89-101

[17] Barton WA, Liu BP, Tzvetkova D, et al. Structure and axon outgrowth inhibitor binding of the Nogo-66 receptor and related proteins. The EMBO Journal. 2003;**22**(13):3291-3302

[18] McKeon RJ, Schreiber RC, Rudge JS, Silver J. Reduction of neurite outgrowth in a model of glial scarring following CNS injury is correlated with the expression of inhibitory molecules on reactive astrocytes. The Journal of Neuroscience. 1991;**11**(11):33983411

[19] Woolfe F, Waxman SG, Hains BC. In silico modeling of axonal reconnection within a discrete fiber tract after spinal cord injury. Journal of Neurotrauma. 2007;**24**(2):421-432

[20] Dietz V, Curt A. Neurological aspects of spinal-cord repair: Promises and challenges. Lancet Neurology. 2006;**5**(8):688-694

[21] Friedli L, Rosenzweig ES, Barraud Q, Schubert M, Dominici N, Awai L, et al. Pronounced species divergence in corticospinal tract reorganization and functional recovery after lateralized spinal cord injury favors primates. Science Translational Medicine. 2015;**7**(302):302ra134. DOI: 10.1126/scitranslmed.aac5811

[22] Assinck P, Duncan GJ, Hilton BJ, Plemel JR, Tetzlaff W. Cell transplantation therapy for spinal cord injury. Nature Neuroscience. 2017;**20**(5):637-647. DOI: 10.1038/nn.4541

[23] Ahuja CS, Fehlings M. Concise review: Bridging the gap: Novel neuroregenerative and neuroprotective strategies in spinal cord injury. Stem Cells Translational Medicine. 2016;**5**(7):914-924

[24] Vaquero J, Zurita M, Rico MA, Bonilla C, Aguayo C, Fernández C, et al. Repeated subarachnoid administrations of autologous mesenchymal stromal cells supported in autologous plasma improve quality of life in patients suffering incomplete spinal cord injury. Cytotherapy. 2017;**19**(3):349-359

[25] Oh SK, Choi KH, Yoo JY, Kim DY, Kim SJ, Jeon SR. A phase III clinical trial showing limited efficacy of autologous mesenchymal stem cell therapy for spinal cord injury. Neurosurgery. 2016;**78**(3):436-447

[26] Kjell J, Olson L. Rat models of spinal cord injury: From pathology to potential therapies. Disease Models & Mechanisms. 2016;**9**(10):1125-1137

[27] Takahashi A, Nakajima H, Uchida K, Takeura N, Honjoh K, Watanabe S, et al. Comparison of mesenchymal stromal cells isolated from murine adipose tissue and bone marrow in the treatment of spinal cord injury. Cell Transplantation. 2018;**27**(7):1126-1139. DOI: 10.1177/0963689718780309

[28] Wang N, Xiao Z, Zhao Y, Wang B, Li X, Li J, et al. Collagen scaffold combined with human umbilical cord-derived mesenchymal stem cells promote functional recovery after scar resection in rats with chronic spinal cord injury. Journal of Tissue Engineering and Regenerative Medicine. 2018;**12**(2):e1154-e1163

[29] Nicola FDC, Marques MR, Odorcyk F, Arcego DM, Petenuzzo L, Aristimunha D, et al. Neuroprotector effect of stem cells from human exfoliated deciduous teeth transplanted after traumatic spinal cord injury involves inhibition of early neuronal apoptosis. Brain Research. 2017;**1663**:95-105

[30] Ziegler MD, Hsu D, Takeoka A, Zhong H, Ramón-Cueto A, Phelps PE, et al. Further evidence of olfactory ensheathing glia facilitating axonal regeneration after a complete spinal cord transection. Experimental Neurology. 2011;**229**(1):109-119

[31] Bunge MB, Monje PV, Khan A, Wood PM. From transplanting Schwann cells in experimental rat spinal cord injury to their transplantation into human injured spinal cord in clinical trials. Progress in Brain Research. 2017;**231**:107-133

[32] Rosenzweig ES, Brock JH, Lu P, Kumamaru H, Salegio EA, Kadoya K, et al. Restorative effects of human neural stem cell grafts on the primate spinal cord. Nature Medicine. 2018;**24**(4):484-490

[33] Vawda R, Fehlings MG. Mesenchymal cells in the treatment of spinal cord injury: Current & future perspectives. Current Stem Cell Research & Therapy. 2013;**8**(1):25-38

[34] Matyas JJ, Stewart AN, Goldsmith A, Nan Z, Skeel RL, Rossignol J, et al. Effects of bone-marrow-derived MSC transplantation on functional recovery in a rat model of spinal cord injury: Comparisons of transplant locations and cell concentrations. Cell Transplantation. 2017;**26**(8):1472-1482

[35] Torres-Espín A, Redondo-Castro E, Hernandez J, Navarro X. Immunosuppression of allogenic mesenchymal stem cells transplantation after spinal cord injury improves graft survival and beneficial outcomes. Journal of Neurotrauma. 2015;**32**(6):367-380. DOI: 10.1089/neu.2014.356236

[36] Zhu Y, Uezono N, Yasui T, Nakashima K. Neural stem cell therapy aiming at better functional recovery after spinal cord injury. Developmental Dynamics. 2018;**247**(1):75-84. DOI: 10.1002/dvdy.24558. 37

[37] Meletis K, Barnabé-Heider F, Carlén M, Evergren E, Tomilin N, Shupliakov O, et al. Spinal cord injury reveals multilineage differentiation of ependymal cells. PLoS Biology. 2008;**6**:e182

[38] Moreno-Manzano V, Rodríguez-Jiménez FJ, García-Roselló M, Laínez S, Erceg S, Calvo MT, et al. Activated spinal cord ependymal stem cells rescue neurological function. Stem Cells. 2009;**27**:733-743

[39] Requejo-Aguilar R, Alastrue-Agudo A, Cases-Villar M, Lopez-Mocholi E, England R, Vicent MJ, et al. Combined polymer-curcumin conjugate and ependymal progenitor/stem cell treatment enhances spinal cord injury functional recovery. Biomaterials. 2017;**113**:18-30

[40] Gómez-Villafuertes R, Rodríguez-Jiménez FJ, Alastrue-Agudo A, Stojkovic M, Miras-Portugal MT, Moreno-Manzano V. Purinergic receptors in spinal cord-derived ependymal stem/progenitor cells and their potential role in cell-based therapy for spinal cord injury. Cell Transplantation. 2015;**24**:1493-1509

[41] Curtis E, Martin JR, Gabel B, Sidhu N, Rzesiewicz TK, Mandeville R, Van Gorp S, et al. A first-in-human, phase I study of neural stem cell transplantation for chronic spinal cord injury. Cell Stem Cell. 2018;**22**(6):941-950.e6

[42] Shi Y, Su J, Roberts AI, Shou P, Rabson AB, Ren G. How mesenchymal stem cells interact with tissue immune responses. Trends in Immunology. 2012;**33**(3):136-143. DOI: 10.1016/j.it.2011.11.004

[43] Uccelli A, de Rosbo NK. The immunomodulatory function of mesenchymal stem cells: Mode of action and pathways. Annals of the New York Academy of Sciences. 2015;**1351**:114-126. DOI: 10.1111/nyas.12815

[44] Wang M, Yuan Q, Xie L. Mesenchymal stem cell-based immunomodulation: Properties and clinical application. Stem Cells International. 2018;**2018**:3057624. DOI: 10.1155/2018/3057624

[45] Wang LT, Ting CH, Yen ML, Liu KJ, Sytwu HK, Wu KK, et al. Human mesenchymal stem cells (MSCs) for treatment towards immune- and inflammation-mediated diseases: Review of current clinical trials. Journal of Biomedical Science. 2016;**23**(1):76

[46] Xiao J, Yang R, Biswas S, Qin X, Zhang M, Deng W. Mesenchymal stem cells and induced pluripotent stem cells as therapies for multiple sclerosis. International Journal of Molecular Sciences. 2015;**16**(5):9283-9302. DOI: 10.3390/ijms16059283

[47] Freedman MS, Bar-Or A, Atkins HL, Karussis D, Frassoni F, Lazarus H, et al. The therapeutic potential of mesenchymal stem cell transplantation as a treatment for multiple sclerosis: Consensus report of the International MSCT Study Group. Multiple Sclerosis. 2010;**16**(4):503-510. DOI: 10.1177/1352458509359727. 20086020

[48] Scolding NJ, Pasquini M, Reingold SC, Cohen JA. International Conference on Cell-Based Therapies for Multiple Sclerosis. Cell-based therapeutic strategies for multiple sclerosis. Brain. 2017;**140**(11):2776-2796. DOI: 10.1093/brain/awx154

[49] Laterza C, Merlini A, De Feo D, Ruffini F, Menon R, Onorati M, et al. iPSC-derived neural precursors exert a neuroprotective role in immune-mediated demyelination via the secretion of LIF. Nature Communications. 2013;**4**:2597. DOI: 10.1038/ncomms3597

[50] Marriott MP, Emery B, Cate HS, Binder MD, Kemper D, Wu Q, et al. Leukemia inhibitory factor signaling modulates both central nervous system demyelination and myelin repair. Glia. 2008;**56**(6):686-698. DOI: 10.1002/glia.20646

[51] Butzkueven H, Emery B, Cipriani T, Marriott MP, Kilpatrick TJ. Endogenous leukemia inhibitory factor production limits autoimmune demyelination and oligodendrocyte loss. Glia. 2006;**53**(7):696-703

[52] Carruthers BM, van de Sande MI, De Meirleir KL, Klimas NG, Broderick G, Mitchell T, et al. Myalgic encephalomyelitis: International consensus criteria. Journal of Internal Medicine. 2011;**270**(4):327-338. DOI: 10.1111/j.1365-2796.2011.02428.x [Epub Aug 22, 2011]. (Review. Erratum in: Journal of Internal Medicine. 2017;**282**(4):353)

[53] Hornig M, Gottschalk CG, Eddy ML, Che X, Ukaigwe JE, Peterson DL, et al. Immune network analysis of cerebrospinal fluid in myalgic encephalomyelitis/chronic fatigue syndrome with atypical and classical presentations. Translational Psychiatry. 2017; **7**(4):e1080

[54] Broderick G, Fuite J, Kreitz A, Vernon SD, Klimas N, Fletcher MA. A formal analysis of cytokine networks in chronic fatigue syndrome. Brain, Behavior, and Immunity. 2010;**24**(7):1209-1217. DOI: 10.1016/j.bbi.2010.04.012

[55] Brenu EW, Hardcastle SL, Atkinson GM, van Driel ML, Kreijkamp-Kaspers S, Ashton KJ, et al. Natural killer cells in patients with severe chronic fatigue syndrome. Autoimmunity Highlights. 2013;**4**(3):69-80

[56] Institute of Medicine, Committee on the Diagnostic Criteria for Myalgic Encephalomyelitis/Chronic Fatigue Syndrome, Board on the Health of Select Populations. Beyond Myalgic Encephalomyelitis/Chronic Fatigue Syndrome: Redefining an Illness. Washington (DC): National Academies Press; 2015. 304p

[57] Clayton EW. Beyond myalgic encephalomyelitis/chronic fatigue syndrome: An IOM report on redefining an illness. Journal of the American Medical Association. 2015;**313**(11): 1101-1102

[58] Capelli E, Zola R, Lorusso L, Venturini L, Sardi F, Ricevuti G. Chronic fatigue syndrome/myalgic encephalomyelitis: An update. International Journal of Immunopathology and Pharmacology. 2010;**23**(4):981-989

[59] Sotzny F, Blanco J, Capelli E, Castro-Marrero J, Steiner S, Murovska M, et al. Myalgic encephalomyelitis/chronic fatigue syndrome—Evidence for an autoimmune disease. Autoimmunity Reviews. 2018;**17**(6):601-609. DOI: 10.1016/j.autrev.2018.01.009

[60] Gow JW, Simpson K, Behan PO, Chaudhuri A, McKay IC, Behan WM. Antiviral pathway activation in patients with chronic fatigue syndrome and acute infection. Clinical Infectious Diseases. 2001;**33**(12):2080-2081 [Epub Nov 6, 2001]

[61] Roerink ME, Roerink SHPP, Skoluda N, van der Schaaf ME, Hermus ARMM, van der Meer JWM, et al. Hair and salivary cortisol in a cohort of women with chronic fatigue syndrome. Hormones and Behavior. 2018;**103**:1-6. DOI: 10.1016/j.yhbeh.2018.05.016

[62] Sedghamiz H, Morris M, Craddock TJA, Whitley D, Broderick G. High-fidelity discrete modeling of the HPA axis: A study of regulatory plasticity in biology. BMC Systems Biology. 2018;**12**(1):76. DOI: 10.1186/s12918-018-0599-1

[63] Dominici M, Le Blanc K, Mueller I, Slaper-Cortenbach I, Marini F, Krause D, et al. Minimal criteria for defining multipotent mesenchymal stromal cells. The International Society for Cellular Therapy position statement. Cytotherapy. 2006;**8**(4):315-317

[64] Bourin P, Bunnell BA, Casteilla L, Dominici M, Katz AJ, March KL, et al. Stromal cells from the adipose tissue-derived stromal vascular fraction and culture expanded adipose tissue-derived stromal/stem cells: A joint statement of the International Federation for Adipose Therapeutics and Science (IFATS) and the International Society for Cellular Therapy (ISCT). Cytotherapy. 2013;**15**(6):641-648. DOI: 10.1016/j.jcyt.2013.02.006

[65] Hass R, Kasper C, Böhm S, Jacobs R. Different populations and sources of human mesenchymal stem cells (MSC): A comparison of adult and neonatal tissue-derived MSC. Cell Communication and Signaling: CCS. 2011;**9**:12. DOI: 10.1186/1478-811X-9-12

[66] Hoogduijn MJ, de Witte SF, Luk F, van den Hout-van Vroonhoven MC, Ignatowicz L, Catar R, et al. Effects of freeze-thawing and intravenous infusion on mesenchymal stromal cell gene expression. Stem Cells and Development. 2016;**25**(8):586-597. DOI: 10.1089/scd.2015.0329

[67] Jossen V, van den Bos C, Eibl R, Eibl D. Manufacturing human mesenchymal stem cells at clinical scale: Process and regulatory challenges. Applied Microbiology and Biotechnology. 2018;**102**(9):3981-3994. DOI: 10.1007/s00253-018-8912-x

[68] de Soure AM, Fernandes-Platzgummer A, da Silva CL, Cabral JM. Scalable microcarrier-based manufacturing of mesenchymal stem/stromal cells. Journal of Biotechnology. 2016;**236**:88-109. DOI: 10.1016/j.jbiotec.2016.08.007

[69] Giancola R, Bonfini T, Iacone A. Cell therapy: cGMP facilities and manufacturing. Muscle, Ligaments and Tendons Journal 2012;**2**(3):243-247 [Print Jul 2012]

[70] Soleimani M, Nadri S. A protocol for isolation and culture of mesenchymal stem cells from mouse bone marrow. Nature Protocols. 2009;**4**(1):102-106. DOI: 10.1038/nprot.2008.221

[71] Zuk PA, Zhu M, Ashjian P, De Ugarte DA, Huang JI, Mizuno H, et al. Human adipose tissue is a source of multipotent stem cells. Molecular Biology of the Cell. 2002;**13**(12):4279-4295

[72] Nordberg RC, Loboa EG. Our fat future: Translating adipose stem cell therapy. Stem Cells Translational Medicine. 2015;**4**(9):974-979. DOI: 10.5966/sctm.2015-0071

[73] Van den Bos C. Off the beaten track—regulatory changes. European Biopharmaceutical Reviews. 2012;**165**(winter):32-36

[74] Halme DG, Kessler DA. FDA regulation of stem-cell-based therapies. New England Journal of Medicine. 2006;**355**:409-415. DOI: 10.1056/NEJMhpr063086

[75] Yi T, Kim SN, Lee HJ, Kim J, Cho YK, Shin DH, et al. Manufacture of clinical-grade human clonal mesenchymal stem cell products from single colony forming unit-derived colonies based on the subfractionation culturing method. Tissue Engineering. Part C, Methods. 2015;**21**(12):1251-1262. DOI: 10.1089/ten.TEC.2015.0017

[76] Dessels C, Potgieter M, Pepper MS. Making the switch: Alternatives to fetal bovine serum for adipose-derived stromal cell expansion. Frontiers in Cell and Development Biology. 2016;**4**:115

[77] Horwitz EM, Gordon PL, Koo WKK, Marx JC, Neel MD, McNall RY, et al. Isolated allogeneic bone marrow-derived mesenchymal cells engraft and stimulate growth in children with osteogenesis imperfecta: Implications for cell therapy of bone. Proceedings of the National Academy of Sciences of the United States of America. 2002;**99**:8932-8937. DOI: 10.1073/pnas.132252399

[78] Selvaggi TA, Walker RE, Fleisher TA. Development of antibodies to fetal calf serum with arthus-like reactions in human immunodeficiency virus-infected patients given syngeneic lymphocyte infusions. Blood. 1997;**89**:776-779

[79] Mackensen A, Dräger R, Schlesier M, Mertelsmann R, Lindemann A. Presence of IgE antibodies to bovine serum albumin in a patient developing anaphylaxis after vaccination with human peptide-pulsed dendritic cells. Cancer Immunology, Immunotherapy. 2000;**49**:152-156

[80] Josh F, Kobe K, Tobita M, Tanaka R, Suzuki K, Ono K, et al. Accelerated and safe proliferation of human adipose-derived stem cells in medium supplemented with human serum. Journal of Nippon Medical School. 2012;**79**:444-452. DOI: 10.1272/jnms.79.444

[81] Koellensperger E, Bollinger N, Dexheimer V, Gramley F, Germann G, Leimer U. Choosing the right type of serum for different applications of human adipose tissue-derived stem cells: Influence on proliferation and differentiation abilities. Cytotherapy. 2014;**16**(6): 789-799. DOI: 10.1016/j.jcyt.2014.01.007

[82] Cho H, Lee A, Kim K. The effect of serum types on Chondrogenic differentiation of adipose-derived stem cells. Biomater Research. 2018;**22**:6. DOI: 10.1186/s40824-018-0116-z

[83] Kocaoemer A, Kern S, Klüter H, Bieback K. Human AB serum and thrombin-activated platelet-rich plasma are suitable alternatives to fetal calf serum for the expansion of mesenchymal stem cells from adipose tissue. Stem Cells. 2007;**25**(5):1270-1278

[84] Pierce J, Benedetti E, Preslar A, Jacobson P, Jin P, Stroncek DF, et al. Comparative analyses of industrial-scale human platelet lysate preparations. Transfusion. 2017;**57**(12): 2858-2869. DOI: 10.1111/trf.14324

[85] Saury C, Lardenois A, Schleder C, Leroux I, Lieubeau B, David L, et al. Human serum and platelet lysate are appropriate xeno-free alternatives for clinical-grade production of human MuStem cell batches. Stem Cell Research & Therapy. 2018;**9**(1):128. DOI: 10.1186/s13287-018-0852-y

[86] Bieback K, Hecker A, Kocaömer A, Lannert H, Schallmoser K, Strunk D, et al. Human alternatives to fetal bovine serum for the expansion of mesenchymal stromal cells from bone marrow. Stem Cells. 2009;**27**(9):2331-2341. DOI: 10.1002/stem.139

[87] Tang H, Xiang Y, Jiang X, Ke Y, Xiao Z, Guo Y, et al. Dual expression of hTERT and VEGF prolongs life span and enhances angiogenic ability of aged BMSCs. Biochemical and Biophysical Research Communications. 2013;**440**:502-508. DOI: 10.1016/j.bbrc.2013.09.053

[88] Turinetto V, Vitale E, Giachino C. Senescence in human mesenchymal stem cells: functional changes and implications in stem cell-based therapy. International Journal of Molecular Sciences. 2016;**17**(7). pii: E1164. DOI: 10.3390/ijms17071164

[89] Peng Y, Huang S, Wu Y, Cheng B, Nie X, Liu H, et al. Platelet rich plasma clot releasate preconditioning induced PI3K/AKT/NFκB signaling enhances survival and regenerative function of rat bone marrow mesenchymal stem cells in hostile microenvironments. Stem Cells and Development. 2013;**22**(24):3236-3251. DOI: 10.1089/scd.2013.0064

[90] Mellado-López M, Griffeth RJ, Meseguer-Ripolles J, Cugat R, García M, Moreno-Manzano V. Plasma rich in growth factors induces cell proliferation, migration, differentiation, and cell survival of adipose-derived stem cells. Stem Cells International. 2017;**2017**:5946527. DOI: 10.1155/2017/5946527

[91] Usta SN, Scharer CD, Xu J, Frey TK, Nash RJ. Chemically defined serum-free and xeno-free media for multiple cell lineages. Annals of Translational Medicine. 2014;**2**:97. DOI: 10.3978/j.issn.2305-5839.2014.09.05

[92] Patrikoski M, Juntunen M, Boucher S, Campbell A, Vemuri MC, Mannerström B, et al. Development of fully defined xeno-free culture system for the preparation and propagation of cell therapy compliant human adipose stem cells. Stem Cell Research & Therapy. 2013;**4**:27. DOI: 10.1186/scrt175

[93] Oikonomopoulos A, vanDeen WK, Manansala AR, Lacey PN, Tomakili TA, Ziman A, et al. Optimization of human mesenchymal stem cell manufacturing: The effects of animal/xeno-free media. Scientific Reports. 2015;**5**:16570. DOI: 10.1038/srep16570

[94] Lindroos B, Boucher S, Chase L, Kuokkanen H, Huhtala H, Haataja R, et al. Serum-free, xeno-free culture media maintain the proliferation rate and multipotentiality of adipose stem cells *in vitro*. Cytotherapy. 2009;**11**(7):958-972. DOI: 10.3109/14653240903233081

[95] Chen C, Tang Q, Zhang Y, Yu M, Jing W, Tian W. Physioxia: A more effective approach for culturing human adipose-derived stem cells for cell transplantation. Stem Cell Research & Therapy. 2018;**9**(1):148. DOI: 10.1186/s13287-018-0891-4

[96] Okolicsanyi RK, Camilleri ET, Oikari LE, Yu C, Cool SM, van Wijnen AJ, et al. Human mesenchymal stem cells retain multilineage differentiation capacity including neural

marker expression after extended *In vivo* expansion. PLoS One. 2015;**10**(9):e0137255. DOI: 10.1371/journal.pone.0137255 (eCollection 2015)

[97] Zachar L, Bačenková D, Rosocha J. Activation, homing, and role of the mesenchymal stem cells in the inflammatory environment. Journal of Inflammation Research. 2016;**9**:231-240. DOI: 10.2147/JIR.S121994 (eCollection 2016)

[98] Ma OK, Chan KH. Immunomodulation by mesenchymal stem cells: Interplay between mesenchymal stem cells and regulatory lymphocytes. World Journal of Stem Cells. 2016;**8**(9):268-278. DOI: 10.4252/wjsc.v8.i9.268

[99] Li W, Ren G, Huang Y, Su J, Han Y, Li J, et al. Mesenchymal stem cells: A double-edged sword in regulating immune responses. Cell Death and Differentiation. 2012;**19**(9): 1505-1513. DOI: 10.1038/cdd.2012.26

[100] Prockop DJ, Oh JY. Mesenchymal stem/stromal cells (MSCs): Role as guardians of inflammation. Molecular Therapy. 2012;**20**(1):14-20. DOI: 10.1038/mt.2011.211

[101] Le Blanc K, Davies LC. Mesenchymal stromal cells and the innate immune response. Immunology Letters. 2015;**168**(2):140-146. DOI: 10.1016/j.imlet.2015.05.004

[102] Hoogduijn MJ, Roemeling-van Rhijn M, Engela AU, Korevaar SS, Mensah FK, Franquesa M, et al. Mesenchymal stem cells induce an inflammatory response after intravenous infusion. Stem Cells and Development. 2013;**22**(21):2825-2835. DOI: 10.1089/scd.2013.0193

[103] Ulivi V, Tasso R, Cancedda R, Descalzi F. Mesenchymal stem cell paracrine activity is modulated by platelet lysate: Induction of an inflammatory response and secretion of factors maintaining macrophages in a proinflammatory phenotype. Stem Cells and Development. 2014;**23**(16):1858-1869. DOI: 10.1089/scd.2013.0567

[104] de Witte SF, Franquesa M, Baan CC, Hoogduijn MJ. Toward development of mesenchymal stem cells for immunomodulatory therapy. Frontiers in Immunology. 2016;**6**:648. DOI: 10.3389/fimmu.2015.00648

[105] Moll G, Jitschin R, von Bahr L, Rasmusson-Duprez I, Sundberg B, Lönnies L, et al. Mesenchymal stromal cells engage complement and complement receptor bearing innate effector cells to modulate immune responses. PLoS One. 2011;**6**(7):e21703. DOI: 10.1371/journal.pone.0021703

[106] Tu Z, Li Q, Bu H, Lin F. Mesenchymal stem cells inhibit complement activation by secreting factor H. Stem Cells and Development. 2010;**19**(11):1803-1809. DOI: 10.1089/scd.2009.0418

[107] Kadoya K, Lu P, Nguyen K, Lee-Kubli C, Kumamaru H, Yao L, et al. Spinal cord reconstitution with homologous neural grafts enables robust corticospinal regeneration. Nature Medicine. 2016;**22**(5):479-487

[108] Mothe AJ, Tator CH. Advances in stem cell therapy for spinal cord injury. The Journal of Clinical Investigation. 2012;**122**(11):3824-3834

[109] Matsui T, Akamatsu W, Nakamura M, Okano H. Regeneration of the damaged central nervous system through reprogramming technology: Basic concepts and potential application for cell replacement therapy. Experimental Neurology. 2014;**260**:12-18

[110] Mothe AJ, Zahir T, Santaguida C, Cook D, Tator CH. Neural stem/progenitor cells from the adult human spinal cord are multipotent and self-renewing and differentiate after transplantation. PLoS One. 2011;**6**(11):e27079. DOI: 10.1371/journal.pone.0027079

[111] Lukovic D, Diez Lloret A, Stojkovic P, Rodríguez-Martínez D, Perez Arago MA, Rodriguez-Jimenez FJ, et al. Highly efficient neural conversion of human pluripotent stem cells in adherent and animal-free conditions. Stem Cells Translational Medicine. 2017;**6**(4):1217-1226. DOI: 10.1002/sctm.16-0371

[112] Tsuji O, Miura K, Okada Y, Fujiyoshi K, Mukaino M, Nagoshi N, et al. Therapeutic potential of appropriately evaluated safeinduced pluripotent stem cells for spinal cord injury. Proceedings of the National Academy of Sciences of the United States of America. 2010;**107**(28):12704-12709

[113] Nori S, Okada Y, Yasuda A, Tsuji O, Takahashi Y, Kobayashi Y, et al. Grafted human-induced pluripotent stem cell derived neurospheres promote motor functional recovery after spinal cord injury in mice. Proceedings of the National Academy of Sciences of the United States of America. 2011;**108**(40):16825-16830

[114] Fujimoto Y, Abematsu M, Falk A, Tsujimura K, Sanosaka T, Juliandi B, et al. Treatment of mouse model of spinal cord injury by transplantation of human induced pluripotent stem cell-derived long-term self-renewing neuroepithelial-like stem cells. Stem Cells. 2012;**30**(6):1163-1173

[115] López-Serrano C, Torres-Espín A, Hernández J, Alvarez-Palomo AB, Requena J, Gasull X, et al. Effects of the post-spinal cord injury microenvironment on the differentiation capacity of human neural stem cells derived from induced pluripotent stem cells. Cell Transplantation. 2016;**25**(10):1833-1852

[116] Säljö K, Barone A, Mölne J, Rydberg L, Teneberg S, Breimer ME. HLA and Histo-blood group antigen expression in human pluripotent stem cells and their derivatives. Scientific Reports. 2017;**7**(1):13072

[117] Badja C, Maleeva G, El-Yazidi C, Barruet E, Lasserre M, Tropel P, et al. Efficient and cost-effective generation of mature neurons from human induced pluripotent stem cells. Stem Cells Translational Medicine. 2014;**3**(12):1467-1472. DOI: 10.5966/sctm.2014-0024

[118] Qu X, Liu T, Song K, Li X, Ge D. Induced pluripotent stem cells generated from human adipose-derived stem cells using a non-viral polycistronic plasmid in feeder-free conditions. PLoS One. 2012;**7**(10):e48161. DOI: 10.1371/journal.pone.0048161

[119] Eminli S, Foudi A, Stadtfeld M, Maherali N, Ahfeldt T, Mostoslavsky G, et al. Differentiation stage determines potential of hematopoietic cells for reprogramming into induced pluripotent stem cells. Nature Genetics. 2009;**41**(9):968-976. DOI: 10.1038/ng.428

[120] Kim JB, Greber B, Araúzo-Bravo MJ, Meyer J, Park KI, Zaehres H, et al. Direct reprogramming of human neural stem cells by OCT4. Nature. 2009;**461**(7264):649-643. DOI: 10.1038/nature08436

[121] Anokye-Danso F, Trivedi CM, Juhr D, Gupta M, Cui Z, Tian Y, et al. Highly efficient miRNA-mediated reprogramming of mouse and human somatic cells to pluripotency. Cell Stem Cell. 2011;**8**(4):376-388. DOI: 10.1016/j.stem.2011.03.001 (Erratum in: Cell Stem Cell. 2012;**11**(6):853)

[122] Khazaei M, Ahuja CS, Fehlings MG. Induced pluripotent stem cells for traumatic spinal cord injury. Frontiers in Cell and Development Biology. 2017;**4**:152. DOI: 10.3389/fcell.2016.00152

[123] Ban H, Nishishita N, Fusaki N, Tabata T, Saeki K, Shikamura M, et al. Efficient generation of transgene-free human induced pluripotent stem cells (iPSCs) by temperature-sensitive Sendai virus vectors. Proceedings of the National Academy of Sciences of the United States of America. 2011;**108**(34):14234-14239. DOI: 10.1073/pnas.1103509108

[124] Zhou W, Freed CR. Adenoviral gene delivery can reprogram human fibroblasts to induced pluripotent stem cells. Stem Cells. 2009;**27**(11):2667-2674. DOI: 10.1002/stem.201

[125] Maherali N, Hochedlinger K. Guidelines and techniques for the generation of induced pluripotent stem cells. Cell Stem Cell. 2008;**3**(6):595-605. DOI: 10.1016/j.stem.2008.11.008

[126] González F, Boué S, Izpisúa Belmonte JC. Methods for making induced pluripotent stem cells: Reprogramming à la carte. Nature Reviews. Genetics. 2011;**12**(4):231-242. DOI: 10.1038/nrg2937

[127] Brouwer M, Zhou H, Nadif KN. Choices for induction of pluripotency: Recent developments in human induced pluripotent stem cell reprogramming strategies. Stem Cell Reviews. 2016;**12**(1):54-72. DOI: 10.1007/s12015-015-9622-8

[128] Fandel TM, Trivedi A, Nicholas CR, Zhang H, Chen J, Martinez AF, et al. Transplanted human stem cell-derived interneuron precursors mitigate mouse bladder dysfunction and central neuropathic pain after spinal cord injury. Cell Stem Cell. 2016;**19**(4):544-557. DOI: 10.1016/j.stem.2016.08.020

[129] Orlova VV, Drabsch Y, Freund C, Petrus-Reurer S, van den Hil FE, Muenthaisong S, et al. Functionality of endothelial cells and pericytes from human pluripotent stem cells demonstrated in cultured vascular plexus and zebrafish xenografts. Arteriosclerosis, Thrombosis, and Vascular Biology. 2014;**34**(1):177-186. DOI: 10.1161/ATVBAHA.113.302598

[130] Yu C, Griffiths LR, Haupt LM. Exploiting Heparan Sulfate proteoglycans in human neurogenesis-controlling lineage specification and fate. Frontiers in Integrative Neuroscience. 2017;**11**:28. DOI: 10.3389/fnint.2017.00028

[131] Rodríguez-Jimenez FJ, Alastrue-Agudo A, Erceg S, Stojkovic M, Moreno-Manzano V. FM19G11 favors spinal cord injury regeneration and stem cell self-renewal by mitochondrial uncoupling and glucose metabolism induction. Stem Cells. 2012;**30**(10):2221-2233

[132] Alastrue-Agudo A, Rodriguez-Jimenez FJ, Mocholi EL, De Giorgio F, Erceg S, Moreno-Manzano V. FM19G11 and ependymal progenitor/stem cell combinatory treatment enhances neuronal preservation and oligodendrogenesis after severe spinal cord injury. International Journal of Molecular Sciences. 2018;**19**(1). pii: E200. DOI: 10.3390/ijms19010200

[133] Carrero R, Cerrada I, Lledó E, Dopazo J, García-García F, Rubio MP, et al. IL1β induces mesenchymal stem cells migration and leucocyte chemotaxis through NF-κB. Stem Cell Reviews. 2012;**8**(3):905-916. DOI: 10.1007/s12015-012-9364-9

[134] Hu C, Li L. Preconditioning influences mesenchymal stem cell properties *in vitro* and *in vivo*. Journal of Cellular and Molecular Medicine. 2018;**22**(3):1428-1442. DOI: 10.1111/jcmm.13492

[135] Wang J, Li Z, Zhang Y, Liu X, Chen L, Chen Y. CX43 change in LPS preconditioning against apoptosis of mesenchymal stem cells induced by hypoxia and serum deprivation is associated with ERK signaling pathway. Molecular and Cellular Biochemistry. 2013;**380**(1-2):267-275. DOI: 10.1007/s11010-013-1683-x

[136] Liao W, Pham V, Liu L, et al. Mesenchymal stem cells engineered to express selectin ligands and IL-10 exert enhanced therapeutic efficacy in murine experimental autoimmune encephalomyelitis. Biomaterials. 2016;**77**:87-97. DOI: 10.1016/j.biomaterials.2015.11.005

[137] Berglund AK, Fortier LA, Antczak DF, Schnabel LV. Immunoprivileged no more: Measuring the immunogenicity of allogeneic adult mesenchymal stem cells. Stem Cell Research & Therapy. 2017;**8**(1):288. DOI: 10.1186/s13287-017-0742-8

[138] Lohan P, Treacy O, Griffin MD, Ritter T, Ryan AE. Anti-donor immune responses elicited by allogeneic mesenchymal stem cells and their extracellular vesicles: Are we still learning? Frontiers in Immunology. 2017;**8**:1626. DOI: 10.3389/fimmu.2017.01626

[139] Hofer HR, Tuan RS. Secreted trophic factors of mesenchymal stem cells support neurovascular and musculoskeletal therapies. Stem Cell Research & Therapy. 2016;**7**:131. DOI: 10.1186/s13287-016-0394-0

[140] Cizkova D, Cubinkova V, Smolek T, Murgoci AN, Danko J, Vdoviakova K, et al. Localized intrathecal delivery of mesenchymal stromal cells conditioned medium improves functional recovery in a rat model of spinal cord injury. International Journal of Molecular Sciences. 2018;**19**(3). pii: E870. DOI: 10.3390/ijms19030870 (Erratum in: International Journal of Molecular Sciences. 2018;**19**(7))

[141] Haynesworth SE, Baber MA, Caplan AI. Cytokine expression by human marrow-derived mesenchymal progenitor cells *in vitro*: Effects of dexamethasone and IL-1 alpha. Journal of Cellular Physiology. 1996;**166**(3):585-592

[142] Keshtkar S, Azarpira N, Ghahremani MH. Mesenchymal stem cell-derived extracellular vesicles: Novel frontiers in regenerative medicine. Stem Cell Research & Therapy. 2018;**9**(1):63. DOI: 10.1186/s13287-018-0791-7

[143] Lötvall J, Hill AF, Hochberg F, Buzás EI, Di Vizio D, Gardiner C, et al. Minimal experimental requirements for definition of extracellular vesicles and their functions: A position statement from the International Society for Extracellular Vesicles. Journal of Extracellular Vesicles. 2014;**3**:26913. DOI: 10.3402/jev.v3.26913

[144] Kordelas L, Rebmann V, Ludwig AK, Radtke S, Ruesing J, Doeppner TR, et al. MSC-derived exosomes: A novel tool to treat therapy-refractory graft-versus-host disease. Leukemia. 2014;**28**:970-973. DOI: 10.1038/leu.2014.41

[145] Drommelschmidt K, Serdar M, Bendix I, Herz J, Bertling F, Prager S, et al. Mesenchymal stem cell-derived extracellular vesicles ameliorate inflammation-induced preterm brain injury. Brain, Behavior, and Immunity. 2017;**60**:220-232. DOI: 10.1016/j.bbi.2016.11.011

[146] Ruppert KA, Nguyen TT, Prabhakara KS, Toledano Furman NE, Srivastava AK, et al. Human mesenchymal stromal cell-derived extracellular vesicles modify microglial response and improve clinical outcomes in experimental spinal cord injury. Scientific Reports. 2018;**8**(1):480. DOI: 10.1038/s41598-017-18867-w

[147] Lener T, Gimona M, Aigner L, Borger V, Buzas E, Camussi G, et al. Applying extracellular vesicles based therapeutics in clinical trials—An ISEV position paper. Journal of Extracellular Vesicles. 2015;**4**:30087

[148] Gimona M, Pachler K, Laner-Plamberger S, Schallmoser K, Rohde E. Manufacturing of human extracellular vesicle-based therapeutics for clinical use. International Journal of Molecular Sciences. 2017;**18**(6). pii: E1190. DOI: 10.3390/ijms18061190

[149] Mora EM, Álvarez-Cubela S, Oltra E. Biobanking of exosomes in the era of precision medicine: Are we there yet? International Journal of Molecular Sciences. 2015;**17**(1). pii: E13. DOI: 10.3390/ijms17010013

[150] García-Contreras M, Vera-Donoso CD, Hernández-Andreu JM, García-Verdugo JM, Oltra E. Therapeutic potential of human adipose-derived stem cells (ADSCs) from cancer patients: A pilot study. PLoS One. 2014;**9**(11):e113288. DOI: 10.1371/journal.pone.0113288

9

Erythropoiesis and Megakaryopoiesis in a Dish

Eszter Varga, Marten Hansen,
Emile van den Akker and Marieke von Lindern

Abstract

Erythrocytes and platelets are the major cellular components of blood. Several hereditary diseases affect the production/stability of red blood cells (RBCs) and platelets (Plts) resulting in anemia or bleeding, respectively. Patients with such disorders may require recurrent transfusions, which bear a risk to develop alloantibodies and ultimately may result in transfusion product refractoriness. Cell culture models enable to unravel disease mechanisms, and to screen for alternative therapeutic products. Besides these applications, the ultimate goal is the large-scale production of blood effector cells for transfusion. Cultured RBCs that lack many of the common blood group antigens and Plts-lacking HLA expression would improve transfusion practice. Large numbers of RBCs and Plts can already be generated using hematopoietic stem cells derived from fetal liver, cord blood, peripheral blood, and bone marrow as starting material for cell culture. The recent advances to generate blood cells from induced pluripotent stem cells provide a donor-independent, immortal primary source for cell culture models. This enables us to study developmental switches during erythropoiesis/megakaryopoiesis and provides potential future therapeutic applications. In this review, we will discuss how erythropoiesis and megakaryopoiesis are mimicked in culture systems and how these models relate to the *in vivo* process.

Keywords: transfusion, cell culture model, GMP conditions, defined culture medium, cultured red blood cells, erythropoiesis, megakaryopoiesis, induced pluripotent stem cells, bioreactors, clinical trials

1. Introduction

Blood cells are by far the most abundant cells of which our body is comprised. Red blood cells (RBCs, or erythrocytes) and platelets (Plts, or thrombocytes) circulate in the vascular system, whereas the white blood cells that form our immune system locate both in the vascular system and in the tissues. RBCs are best known for their function as oxygen transporters and for the clearance of CO_2. Plts exert a crucial function in homeostasis upon vascular damage but they also function during angiogenesis, innate immunity, inflammation, wound healing, cancer, and hemostasis [1, 2]. This chapter focuses on erythropoiesis and megakaryopoiesis. RBCs in the periphery have an average life span of 120 days, constituting approximately 45% of the blood volume. To maintain the population of RBCs, humans generate daily ~2×10^{11} reticulocytes [3]. Plts are shed by megakaryocytes (MK) and live approximately 8–9 days in humans, which require a production of ~8.5×10^{10} Plts/day [4, 5]. The generation of RBCs and Plts occurs mainly in the bone marrow (BM) in adults, although the lung has also been found to host megakaryocytic progenitors as well as Plts-shedding MKs [6]. A small population of hematopoietic stem cells (HSCs) ensures the life-long generation of blood cells, although the HSCs themselves divide rarely. Mostly, HSCs that divide give rise to one new HSC and a daughter cell that develops to an actively dividing multipotent progenitor (MPP) (**Figure 1**) [7]. These MPPs undergo specification through reciprocal actions of transcription factors (TF) that enhance or repress expression of lineage-specific TFs and direct the cells to a lineage-specific gene expression program [7]. Erythropoiesis and megakaryopoiesis were long thought to arise from a common progenitor, the megakaryocytic-erythroid progenitor, but recent lineage tracing indicates that MKs can also differentiate directly from HSCs [8–10]. Not only MPPs, also erythroid progenitors (erythroblasts:

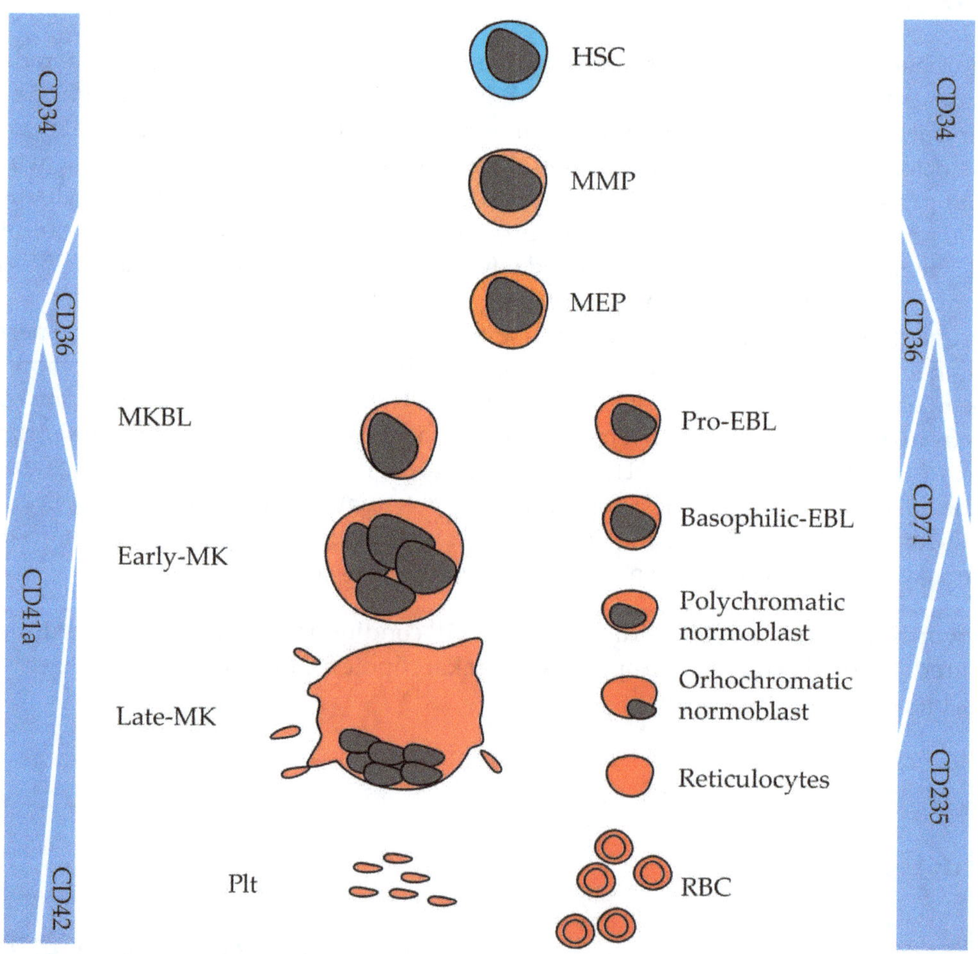

Figure 1. HSC commitment to the erythroid/megakaryocytic lineages with lineage-specific marker expression pattern.

EBLs) and megakaryocytic progenitors (megakaryoblasts: MKBLs) have extensive potential to undergo cell divisions before they commit to the final differentiation program to generate RBCs/MKs. The final differentiation stages of both lineages have unique features. Erythroid progenitors undergo 3–4 additional cell divisions with a short G1 cell cycle phase and without regaining the cell volume (i.e., loss of cell size control) [11–13]. MKBLs, instead, undergo 4–5 cell division cycles without cytokinesis, which results in a single cell with 64–128 genome copies (N = 64–128) [14]. Erythropoiesis and megakaryopoiesis also show spatiotemporal regulation. All blood cell progenitors including erythroid progenitors and MKBLs propagate in close contact with stromal cells that produce membrane-bound factors such as stem cell factor (SCF). Upon terminal differentiation, erythroid progenitors bind to central macrophages that express receptors such as CD163, VCAM1, ICAM4, and CD163 to associate with EBLs [15–19]. Each macrophage binds several progenitors that undergo synchronous differentiation, which ends with phagocytosis of the extruded erythroid nucleus by the macrophage and release of reticulocytes into the circulation. The mature MKs have to interact with the endothelial cells of the vasculature and protrude proplatelets into the capillaries, where shear stress contributes to the shedding of Plts [20].

Whereas steady state erythropoiesis and megakaryopoiesis of adult mammals take place in the BM and lung (MK), distinct anatomic sites of hematopoiesis are employed during development (**Figure 2A**). After gastrulation, in humans, mesodermal precursor cells arise in the primitive streak, migrate to the yolk sac, and develop into blood islands (hemangioblasts), which produce primitive RBCs, primitive MKs, and macrophages [21]. During this process, basic fibroblast growth factor (bFGF) influences the proliferation of the hemangioblast and thereby the production of hematopoietic cells [22]. bFGF is synergistic with vascular endothelial growth factor (VEGF) signaling in this process [23]. The primitive RBCs express embryonic type of hemoglobins (Hbs), retain the size of the early EBLs, and lose their nucleus only after prolonged circulation. Their erythropoietin (EPO)-dependence is unclear at this early stage of development [24, 25]. The primitive MKs are thrombopoietin (TPO) independent, have low ploidy compared to adult MKs, and produce fewer Plts, but contrary to primitive erythroid cells, these cells migrate to the fetal liver, where their polyploidization is TPO dependent [26, 27]. Erythroid-myeloid progenitors (EMPs) arise in the yolk sac from hemogenic endothelium (HE) through endothelial to hematopoietic transition (EHT) and give rise to the first intermediate definitive wave, producing RBCs with fetal type of Hbs, MKs, and other myeloid cells [21, 28]. The EMPs migrate and colonize the developing fetal liver where they transiently produce definitive fetal RBCs and MKs. Permanent definitive hematopoiesis in the fetal liver depends on the "birth" of HSC in the aorta main arteries, and more specifically in the aorta-gonad-mesonephros (AGM) region, where the first $CD34^+$ HSC arises through EHT. These early HSCs are dependent on bone morphogenetic protein 4 (BMP4), VEGF, and bFGF secreted by "feeder cells," which are located near the endothelial cells undergoing EHT, thereby promoting this transition [23, 29]. These HSCs home to the fetal liver to produce definitive fetal blood cells. From the fetal liver, the HSCs migrate to the final site of hematopoiesis; the BM, where they give rise to adult definitive blood cells. Perinatally, hematopoiesis also occurs in the spleen [30]. RBCs and Plts generated at distinct anatomic sites have distinct characteristics; for example, RBCs express different Hb molecules arising from different sites (**Figure 2B**). Hb consists of two α and two β subunits each bound to an iron-containing heme molecule. The α locus expresses ζ and α protein isoforms, the β locus expresses ε, γ (γ1 and 2), and β (β and δ) isoforms. Primitive

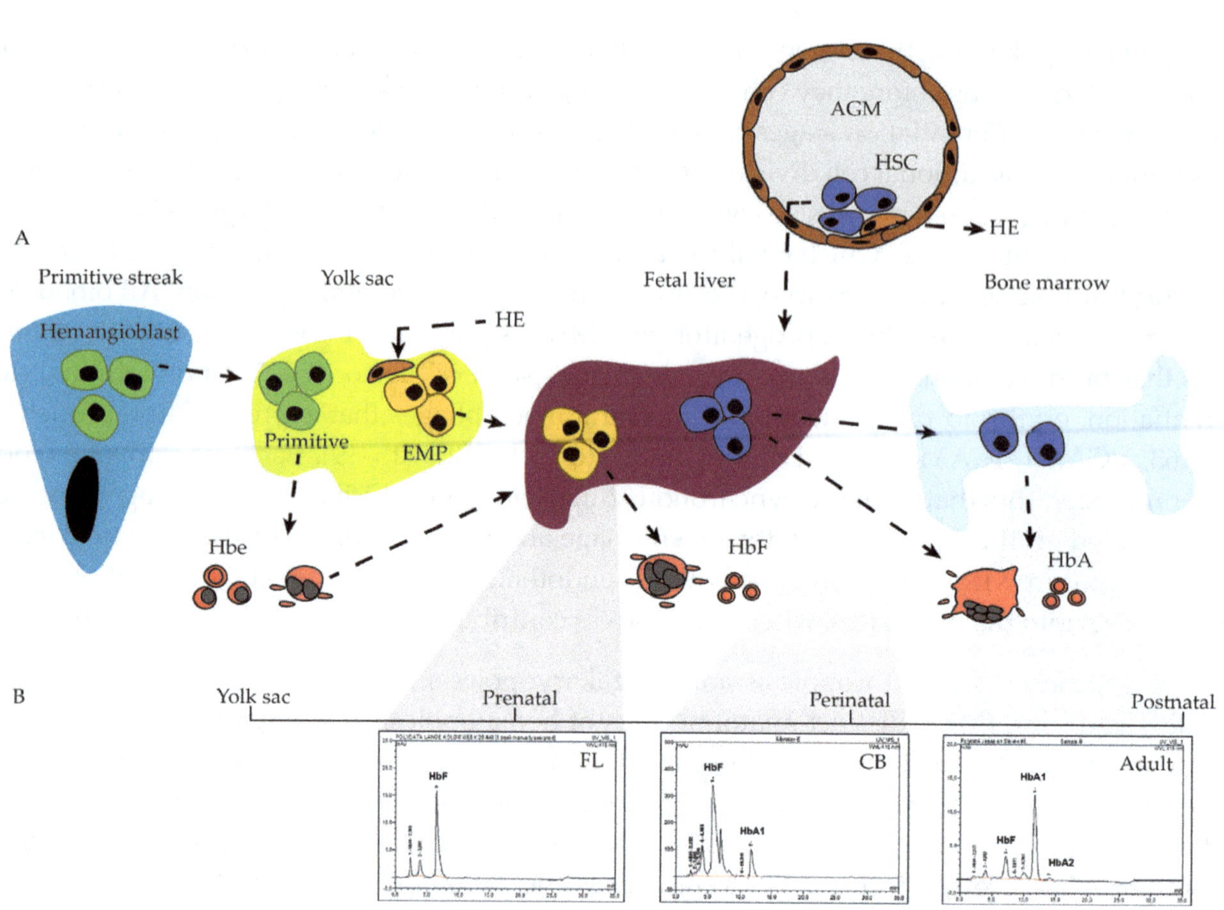

Figure 2. Human erythropoiesis/megakaryopoiesis during development. (A) Schematic depiction of site-specific (yolk sac/fetal liver-AGM/bone marrow) blood production during ontogeny focusing on erythroid and megakaryocytic lineages. (B) Representative HPLC tracks, showing the Hb content of *in vitro*-cultured erythroid cells derived from primary sources originating from distinct anatomical sites. Fetal liver-erythroid cells express fetal Hbs (HbF). Cord blood is obtained at the time of birth when the fetal to adult Hb switch takes place, resulting a mix of HbF and HbA. PBMC/MPB-derived erythroid cells produced by the bone marrow mainly express adult hemoglobins (HbA1, HbA2).

RBCs express Hbe consisting of ζ and ε isoforms (Portland 1: $\zeta_2\gamma_2$; Portland 2: $\zeta_2\beta_2$; Gower 1: $\zeta_2\varepsilon_2$; Gower 2: $\alpha_2\varepsilon_2$); fetal RBCs are characterized by HbF consisting of α and γ isoforms; adult RBCs express HbA consisting of α and β isoforms (HbA1) plus a small amount of HbA2 consisting of α and δ isoforms. Hbs can be used to distinguish RBCs originated from different developmental stages; however, in the megakaryocytic lineages, there is a lack of such markers.

Biochemical and molecular analysis of erythroid/megakaryocytic cells requires large cell numbers. The *in vitro* expansion and differentiation of erythroid and megakaryocytic progenitors from human fetal liver (HFL), cord blood (CB), BM, or peripheral blood enable the production of large cell numbers from distinct ontology and at defined stages of differentiation for basic research, drug testing, disease modeling, or translational purposes (**Figure 3**). Differentiation of pluripotent stem cell types such as embryonic stem cells (ESCs) and induced pluripotent stem cells (iPSCs) toward hematopoietic lineages allows to study early blood ontogeny, which is difficult to study *in vivo* as of ethical issues and availability of human material. The knowledge gained by using the abovementioned culture systems is subsequently of great value to control the expansion and differentiation of erythroid/megakaryocytic cells from ESCs/iPSCs leading to donor-independent blood cell production (**Figure 3**).

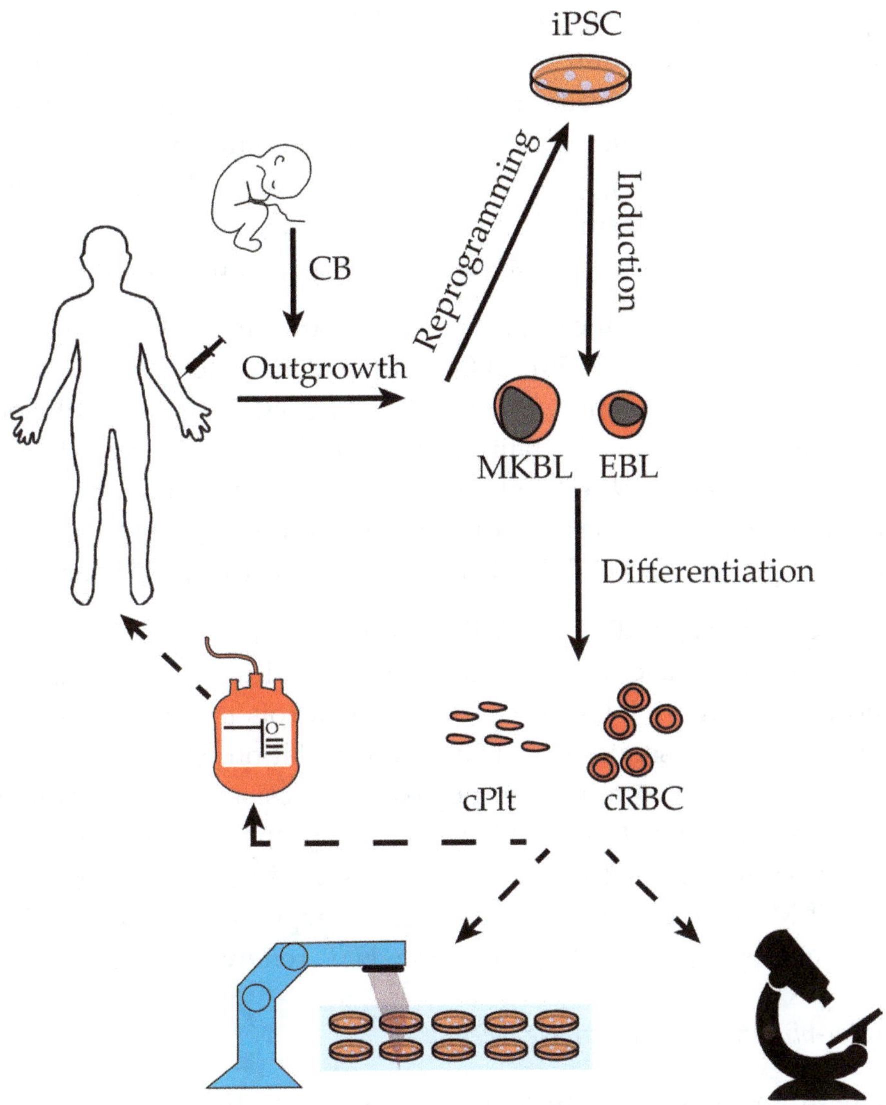

Figure 3. Application of *in vitro*-derived blood products from different primary sources in basic research, drug discovery, and transfusion.

The aim of this chapter is twofold. First, we provide background information of the basic processes of erythropoiesis and megakaryopoiesis that underly the various cell culture models. Second, we provide details, interpret and compare results on current protocols to expand and differentiate erythroid and megakaryocytic progenitors.

2. Erythropoiesis/megakaryopoiesis

2.1. Growth factors and major regulators

2.1.1. Lineage- and stage-specific cytokines

Interleukins (*ILs*) activate cytokine receptors, which do not have enzymatic activity and recruit Janus Kinases (JAK1, JAK2, JAK3, TYK) to phosphorylate tyrosines in their intracellular tail that can subsequently recruit signaling molecules. Cytokine receptors are often expressed on a limited set of cell types/differentiation stages. IL-3 and IL-6 are cytokines that promote survival of hematopoietic stem/progenitor cells (HSPCs), but they do not act on EBL-specific stages. IL-9 synergizes with IL-3 to enhance both erythropoiesis and megakaryopoiesis [31, 32]. IL-1β is an inflammatory cytokine that signals via MyD88 to IRAK and NF-κB to inhibit cell growth, to induce expression of growth factors and the extracellular matrix, and to cause apoptosis in most tissues. In megakaryopoiesis, it enhances commitment and differentiation [33, 34].

Erythropoietin (*EPO*) is mainly produced by the kidney (80%) and partially in the liver (10–15%; 30% upon stress erythropoiesis) [35, 36]. EPO binds and activates EPO receptor (EPOR) and functions as a survival factor during erythropoiesis [37, 38]. Deficiency of Epo and/or Epor causes embryonic lethality due to the failure of definitive fetal liver erythropoiesis leading to the lack of mature RBCs [24, 39–41]. Fetal livers of EPO-deficient embryos contain normal numbers of erythroid progenitors that can form colony-forming unit-erythroid (CFU-E) *in vitro* in the presence of exogenous Epo [39]. EPO shows cross-reactivity in human and mouse.

Thrombopoietin (*TPO*) is the ligand for the MPL receptor. MPL is well conserved between species and TPO showing cross-reactivity between mouse and human. TPO has a direct effect on self-renewal and expansion of HSCs in the BM, but controls megakaryopoiesis as well [42]. MKs and Plts bind TPO and sequester it from the circulation. Upon activation, Plts release TPO into the plasma, thereby stimulating megakaryopoiesis. Thus, the number of Plts and their activation controls TPO levels [43]. TPO is mainly produced by hepatocytes in the human liver. TPO production is increased when hepatocytes bind damaged and aged desialylated Plts via the Ashwell-Morell receptor. In a JAK2/STAT3-dependent manner, this activates TPO transcription and thereby regulates Plt production, HSC renewal, and expansion [42, 44, 45].

2.1.2. *Growth factor-binding tyrosine kinase receptors*

The tyrosine kinase receptors directly cross-phosphorylate tyrosine residues in their own cytoplasmic tail, and they phosphorylate downstream effector molecules. They are more widely expressed compared to cytokine receptors.

Stem cell factor (*SCF*; first described as the *Steel* locus) signals via the mast/stem cell growth factor receptor KIT (CD117; *White locus*). SCF cooperates with other cytokines in order to maintain the viability of HSPCs, and their proliferation/differentiation ability [46–48]. SCF is produced by the stromal cells in the BM as a secreted soluble factor but also as a membrane-bound factor. A specific mutation in SCF (*Steel-Dickie*) disrupting the membrane form of SCF leads to severe anemia, indicating that the membrane-bound form is crucial at least for erythropoiesis [49]. In addition, Kit-deficient mice (*W/W*) suffer from neonatal lethality due to severe anemia [50, 51]. Besides its function on HSPC, SCF is particularly important upon blood loss, to enhance proliferation and delay differentiation of the erythroid and megakaryocytic progenitors [52–54]. In contrast to TPO and EPO, SCF is not interchangeably reactive between mouse and human—mouse SCF activating both, while human does not.

FLT-3-ligand (*FL*) binds to Fms-like tyrosine kinase 3 (FLT-3 (CD135)) as its receptor. FL is an important growth factor for CD34$^+$ HSPC by controlling their proliferation. Underscoring its function, activating mutations in the FLT-3 receptor are prominent in acute myeloid leukemia [55]. FL, however, has a limited role in terminal MKBL/EBL differentiation because FLT-3 is not expressed on committed cells.

Insulin (Ins) and Insulin-like growth factor-I (IGF-I) bind the Ins receptor (InsR) and IGF-I receptor (IGF1R). Both receptors are very well conserved between species to the extent that Ins even cross-reacts between mammals and birds. The homology between the receptors suggested that Ins may also act through the IGF1R, which is not true. IGF-I has a general cell survival function mediated by PI3K activation [56]. Human erythroid progenitors (pro-EBLs, early basophilic EBLs) express the IGF1R, and not the InsR, which changes during differentiation when the InsR becomes the prominent receptor [57]. InsR signaling is particularly important to control trafficking of GLUT4 glucose transporters to the cell membrane [58]. Late EBLs and mature RBCs depend on glycolysis and EBLs express high levels of glucose transporters [59]. Upregulation of GLUT4, however, is also required for import of glutamine, required for nucleotide synthesis [60]. Because Ins and IGF-I act in physiological concentrations, their effect on *in vitro* cultures is not noticed in the presence of serum or plasma [61, 62].

2.1.3. Ligand-activated TFs

Steroid hormone receptors are the best known nuclear hormone receptors. In addition to these ligand-dependent TFs that bind DNA through a Zinc-finger domain, other ligand-activated TFs exist such as the aryl hydrocarbon receptor (AhR) that binds DNA through a helix-loop-helix domain.

Glucocorticoids bind the glucocorticoid receptor (GR), a nuclear hormone receptor, which translocates to the nucleus upon association with its ligand, where it functions as a transcriptional repressor and activator of gene expression. The GR homodimer binds to glucocorticoid response elements that consists of two inverted repeats. As a heterodimer with for instance STAT5, it only needs a "halfmer" GRE, i.e., a single repeat combined with the STAT5 binding site [63, 64]. The ligand of the GR is glucocorticoid (*in vivo*) produced by the adrenal gland. Several synthetic ligands have been designed to be used as immunosuppressive agents (e.g., dexamethasone—DEX). The GR exerts its immunosuppressive function at pharmaceutical levels of glucocorticoids, and as a monomer that binds and inhibits other TFs such as NF-κB or FOS/Jun dimers [65]. In contrast, stress erythropoiesis *in vivo*, and expansion of EBL cultures *in vitro*, is induced at physiological levels of glucocorticoids and depends on dimerization of the GR and on the ligand-dependent transcription activation domain [66, 67]. Polymorphisms of the GR may alter the expansion potential of EBLs [68].

3,5,3 -triiodothyronine (T3) binds the α and β thyroid hormone receptors (TRα, TRβ). T3 deficiency is associated with anemia, although it is not clear whether this is caused by a direct effect on erythropoiesis [69, 70]. The effect of T3 appears to be highly species and developmental stage specific. T3 has a potent differentiation promoting effect on avian erythropoiesis, but mouse EBLs are only sensitive to T3 during neonatal spleen erythropoiesis [71, 72]. In cultures of human erythroid cells, T3 enhances synchronous differentiation to enucleated reticulocytes [73].

StemRegenin 1 (*SR-1*) binds and inhibits the AhR, which mediates toxicity of environmental pollutants, by binding to specific DNA enhancer sequences. This receptor can be activated by endogenous or exogenous ligands and contributes to several physiological processes, among which cell migration, apoptosis, and cell growth [74–77]. Importantly, AhR-deficient mice have increased numbers of BM HSCs [78]. The AhR functions in complex with the AhR Nuclear Translocator to induce differentiation by regulating genes directly such as *c-MYC* and *C/EBP*. During hematopoietic differentiation, its targets also include *PU.1, β-CATENIN, CXCR4,* and *STAT5* [79]. Currently, the antagonist of AhR (SR-1) is used in clinical trials to expand CB-CD34$^+$ cells prior to transplantation, thereby reducing the number of CB units needed per transplant from 5 to 1 [80]. Importantly, SR-1 also enhances *in vitro* proplatelet formation [81]. To achieve proper Plts production, MKs have to produce an extensive membrane system; the demarcation membrane system. This requirs extensive lipid biosynthesis which can be inhibited by AhR [82]. Thus, blocking the inhibition of AhR with SR-1 may increase the proplatelet formation, through increasing the lipid membrane biosynthesis.

2.1.4. *Erythroid-specific regulation*

EPO is sufficient for steady-state erythropoiesis, when proliferative signals are mediated through the EPOR-associated RON receptor and EPO-induced differentiation is dependent on STAT5 [83, 84]. Increased erythropoiesis during development and upon blood loss requires the cooperative action of the EPOR and KIT [85]. Activation of KIT prevents differentiation and propagates the long-term proliferation of erythroid progenitors through inhibition of *FOXO3a* and activation of mRNA translation via the PI3K/mTOR pathway [86, 87]. Activation of the GR is required for stress erythropoiesis and to inhibit *in vitro* differentiation. Interestingly, glucocorticoids activate largely the same growth inhibitory genes in EBLs and in immune cells, but growth inhibition is counteracted by EPOR/KIT activation [88]. Even in the presence of serum, glucocorticoids promote selective proliferation of erythroid progenitors by both supporting erythroid proliferation and inhibiting proliferation of other myeloid and lymphoid cells [57].

Erythropoiesis is also regulated by the availability of iron, which is imported into the cell as holotransferrin via the transferrin receptor (TfR; CD71), and by selenium through selenoproteins [89, 90]. *In vivo,* erythropoiesis is dependent on the formation of erythropoietic islands that form around a central macrophage [91, 92]. Whereas CD169 macrophages are essential for erythropoiesis *in vivo,* high level of enucleation can be achieved *in vitro* in the absence of macrophages (van den Akker and von Lindern, manuscript in preparation) [93]. The complete understanding of which macrophage signals control enucleation *in vivo* may further optimize the *in vitro* production of erythrocytes for transfusion purposes.

2.1.5. *MK-specific regulation*

TPO is the main regulator of megakaryopoiesis, and multiple other factors can work synergistic with it. IL-6 can, in conjunction with TPO, increase hepatic TPO synthesis [94]. This effect is through the shared usage of gp130 and amplification of the same downstream JAK pathways [95]. The direct admission of IL-6, besides its effect on TPO, results in increased polyploidization and subsequently leads to an enhanced Plts production in patients [96]. IL-11 is not constitutively expressed but was shown to be induced in thrombocytopenia patients undergoing

BM transplantation. Exposure of IL-11 on HSPCs directly leads to an inhibition of NF-κB signaling, which suppresses miR-204-5p, that targets and represses the expression of TPO [97]. Because of this effect on TPO, it has a dual role in MK differentiation, first, in the expansion of HSCs, second, in the terminal differentiation of MKs. In a similar way, IL-9 synergizes with IL-3 as well as with IL-4 and SCF to increase the yield of MKBL [33, 98]. Furthermore, the addition of IL-1β promotes selective megakaryocytic differentiation. IL-1β can increase the production of Plts by enhancing the effects of SDF1 and FGF4 that are produced by BM niche cells. IL-1α was shown to induce MK rupture and is considered as a stress megakaryopoiesis regulator [99].

Inhibition of AhR by SR-1 slows the differentiation program in HSCs, leading to increased expansion, and has a direct effect on MK differentiation, by slowing and/or conditioning cells to have a more synchronized maturation [74, 79, 81]. Through its effect on PU.1, it can influence RUNX1 which is one of the TF-regulating megakaryopoiesis. The introduction of SR-1 into *in vitro* MK cultures results in increased cell size, higher polyploidization, and proplatelet production. In these cultures, a specific MK precursor population is identified that has an increased potential to produce proplatelets [81].

2.2. Cell culture models

2.2.1. Erythroid immortalized cell lines

2.2.1.1. Mouse cell lines

The most widely used immortalized mouse erythroid cell lines are MEL cells, which are EBLs transformed by Friend Leukemia virus. The viral gp55 protein activates the Epor to sustain cell growth, whereas integration of the virus upstream of *PU.1* or Friend leukemia integration 1 (*Fli-1*) induces constitutive high expression and inhibits differentiation. Leukemogenesis and the establishment of cell lines require additional inactivation of tumor suppressor protein p53 [100]. Under conditions that enable transient expansion of primary mouse cultures (serum-free medium supplemented with Epo, SCF, and glucocorticoids), lack of p53 is sufficient to establish immortalized cell lines with full differentiation potential in the absence of SCF and glucocorticoids [13].

2.2.1.2. Human cell lines

Deletion of p53 is not sufficient in human to establish immortalized erythroid cell lines. However, expression of the human papilloma virus E6/E7 proteins in EBLs differentiated from iPSC gave rise to the HiDEP cell line [101]. The E6/E7 proteins inactivate both p53 and retinoblastoma tumor suppressor proteins [102]. They are expressed from a doxycycline (dox)-inducible vector allowing for unlimited growth in the presence of EPO, SCF, and dox, and differentiation in the presence of EPO but without dox and SCF. Lack of p53 does not affect differentiation, but retinoblastoma is required for terminal erythroid differentiation [103–105]. Dox-inducible expression was also used to establish immortalized erythroid cells lines from CB (HUDEP) and from adult EBLs cultured from $CD34^+$ HSPC (BEL-A) [106, 107]. The HiDEP, HUDEP, and BEL-A express embryonic, fetal, and adult Hbs, respectively [101, 107].

2.2.2. MK-cell lines

Multiple cell lines have been generated to study megakaryopoiesis with among them; MEG-01 (suspension cells) and DAMI (adherent/suspension cells) both megakaryoblastic leukemia cell lines [108, 109]. These cell lines are mostly positive for MK-specific markers (see in Section 2.3.3) but are not a homogenous population. Although they proliferate in a MKBL-like state with some spontaneous differentiation and limited terminal differentiation, they can be induced to differentiate by the addition of phorbol myristate acetate. Under these conditions, the cells can become polyploid, increase their expression of MK-associated proteins, like von Willebrand factor, and are able to produce proplatelets although with low efficiency. Mechanistic insights were uncovered with these lines; for example, the formation of long, beaded cytoplasmic extensions of MKs that yield platelets upon shear stress. This process was also observed in normal healthy MKs *in vivo*, showing the usefulness of this artificial system to study fundamental processes [110]. Despite the usefulness of these lines, their genetic background (patients with genetic abnormalities) can influence megakaryopoiesis, resulting in incomplete differentiation and potential abnormalities that could possibly be linked to their immortalization process. As such, they are suboptimal models to study megakaryopoiesis and specifically MK polyploidization, synthesis of granules, and proplatelet formation.

2.3. Primary cell culture

2.3.1. Cell source and media

RBCs and MKs can be cultured for research and clinical applications from multiple primary tissue sources including HFL, CB, BM, mobilized peripheral blood (MPB), and peripheral blood mononuclear cells (PBMC). HFL is obtained from abortions on medical indication. HFL-derived erythroid cells express HbF and can be expanded to large numbers and differentiated to hemoglobinized enucleated RBCs. For MK culture, this source is less ideal, mainly because of the harshness of the isolation method. Ethical concerns rule out HFL as a general source for transfusion, but with proper consent allows research into fetal hematopoiesis development. A widely available and ethically accepted source is CB which is commonly used for production of both erythroid cells and MKs. CB is obtained at birth when Hb-switch occurs from HbF to HbA1 (γ to β switch) and both Hb types are expressed in CB-derived cultures. The presence of HSPCs with a fetal hematopoietic program in CB has notable effects in the MK cultures. MKBL expansion is high; MK polyploidization and proplatelet formation are decreased compared to cultures of adult cells. As adult hematopoietic source, either BM or PBMC can be applied. BM is a limited source that can be more difficult to obtain, but does yield large quantities of HSPCs that can differentiate to erythroid and MK lineages. HSPCs can also be isolated from PBMCs, which is less invasive therefore a less limited source. This makes PBMC an ideal source to scale-up RBC production. The HSPC percentage in PBMC is significantly lower compared to BM, which can be enhanced by leukophoresis and by mobilizing BM HSPCs using G-CSF (10 μg/kg) alone or in combination with CXCR4/CXCL12 inhibition [19, 111]. G-CSF alone leads to 5–30% mobilization [15]. HSPC mobilization is caused by the downregulation of adhesion and chemokine processes and by the loss of BM macrophages [15, 112–115]. The mobilization can cause side effects for the donor, including headaches, fatigue, vomiting, muscle pain, bone pain, thrombocytopenia, citrate toxicity, etc., in which females experience the most adverse events [116]. These MPBs are ideal for RBC/Plt production because of their HSPC richness.

The MKs can be cultured in Cellgro (Corning), and both cell types can be cultured in StemSpan (Stem Cell Technologies) or other Iscove's modified Dulbecco's medium (IMDM)-based media. We generated a completely defined GMP-grade medium called Cell-Quin (Migliaccio et al. with minor modifications) that is highly efficient in expanding and differentiating EBLs/RBCs and MKs/Plts, with the ability to culture other hematopoietic progenitors and blood cell types [117].

2.3.2. *Erythroid-specific culture system*

Several parameters characterize the differentiation stage of erythroid cells. Expansion of EBL cultures is only possible when they maintain cell size control during their cell cycle, which is achieved by the cooperative action of SCF and glucocorticoids [12, 13, 118]. Terminal differentiation in the presence of EPO involves 3–4 cell divisions during which cells' surface marker expression changes and gets smaller due to loss of cell size control until cell cycle arrest and extrusion of their nuclei, concurrently, accumulating Hb [119]. Thus, surface marker expression pattern, cell size and morphology (enucleation), Hb content, and cumulative cell numbers are a measure of differentiation (**Figure 4A–C**). Morphological features of the cells (nuclei-cytoplasm ratio, hemoglobinization, nuclei condensation, and polarization) are commonly assessed by cytospins coupled with Giemsa/benzidine stainings (**Figure 4A**). The purity of the erythroid population and its distribution over different maturation stages can be assessed by monitoring the progression of various cell surface markers. Commonly used markers are CD36, CD71 (transferrin receptor), CD117, and the erythroid-specific markers band 3 (SLC4A1) and CD235 (glycophorin A). The generally accepted dynamics of these markers during erythroid differentiation: pro-EBLs (immature EBL stage) are characterized by $CD34^-/CD36^+/CD117^+/CD71^{high}/CD235^{low/-}$, while during expansion phase, EBLs gain CD235 expression and become $CD117^+/CD71^+/CD235^+$. In terminal differentiation phase, EBLs remain positive for CD235 and lose their expression of CD117 followed by the gradual loss of CD71, which is associated with reticulocyte formation [120].

The first human erythroid culture systems utilized the knowledge obtained from genetics, e.g., discoveries in the field of cytokines, growth factors, and their receptors. In these first protocols, HSPCs were expanded in the presence of IL-3, SCF plus or minus IL-6. It is followed by a step in which the resulting erythroid progenitors were further expanded and differentiated in the presence of EPO [121]. This protocol was modified, using low EPO concentrations in step 1 (0.5 U/ml) and high concentrations (>3 U/ml) in step 2 [122]. Others used high concentrations of EPO throughout step 1 and step 2 [123, 124]. These two-step protocols are based on the original protocol of Fibach and coworkers who employ IMDM supplemented with serum or plasma [121]. Serum and plasma contain factors that support erythropoiesis in these cultures. The major factor in the serum is transforming growth factor β (TGFβ), which is a potent differentiation factor for erythropoiesis [57, 125]. These cultures

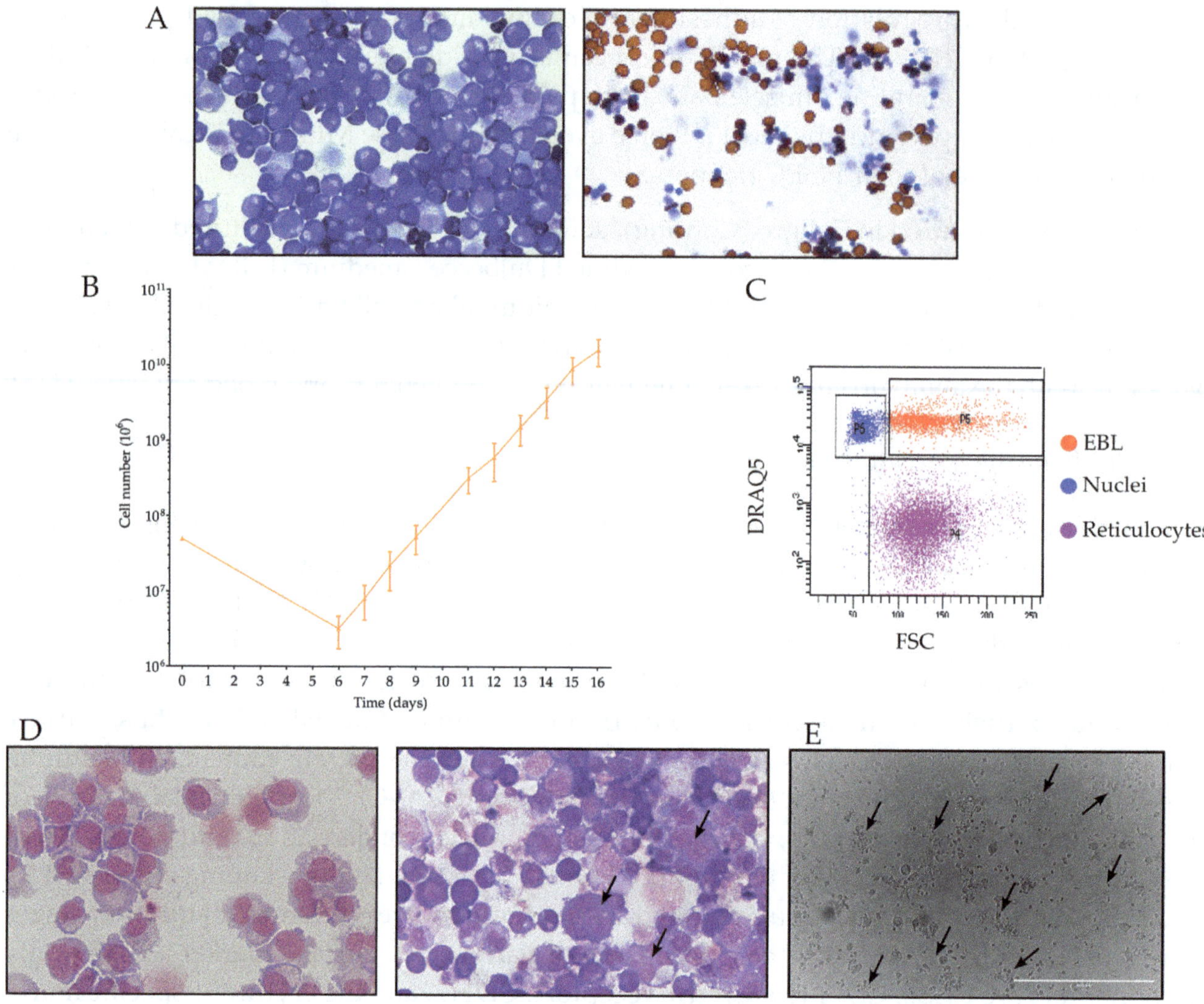

Figure 4. Characteristics of erythroid and megakaryocytic cultures. (A) Erythroid-specific morphology by cytospin with Giemsa/benzidine staining. Left: pro-EBL, right reticulocytes. (B) Erythroid expansion growth curve from PBMCs (n = 4). (C) Flow cytometry of terminal erythroid differentiation, DNA staining by DRAQ5 resulting in three distinct populations: DRAQ5$^+$ big cells: nucleated EBLs (red); DRAQ5$^+$ small cells: nuclei (blue); DRAQ5$^-$ cells: enucleated reticulocytes (purple). (D) MK-specific morphology by cytospin with MGG-staining. Left: MKBL; right: polyploid MK (arrows). (E) Proplatelet-forming MK (arrows (beads on a string)).

show a high degree of spontaneous differentiation, which is often used to study expansion and differentiation of EBLs carrying a genetic defect. The quality differences between serum batches and the use of different cytokines make these culture protocols difficult to compare. Two major changes increase the yield of these erythroid cultures and enable synchronous differentiation. First, glucocorticoids cooperate with SCF to retain pro-EBLs and early basophilic EBLs in their undifferentiated state [57, 126]. Second, serum-free medium avoids the differentiation promoting effect of TGFβ. However, the available serum-free media are suboptimal and require complementation with lipids [73]. Even better expansion is achieved with a serum-free medium optimized for expansion of EBLs [117]. The differentiation arrest in the presence of glucocorticoids and the absence of serum enables the expansion of a homogeneous early EBLs culture that can undergo up to 20 cell divisions to achieve a million-fold expansion [73, 119]. Using Cell-Quin, we can obtain 2×10^{10} EBLs within 16 days, starting from 5×10^{7} PBMCs (**Figure 4B**). Expansion of EBLs in the presence of serum and in the absence of glucocorticoids irrevocably results in differentiation and transfer of the cells to differentiation conditions. Of note, addition of glucocorticoid agonists in a serum-based culture will still induce spontaneous differentiation due to the presence of TGFβ [57]. At any moment during expansion phase, cells can be transferred to differentiation conditions in which the medium is supplemented with EPO, Ins, and low level of plasma/serum [119, 120]. Although expansion of EBL cultures is achieved in serum-free medium, terminal differentiation to enucleated cells requires at least 2% serum or plasma [119, 120]. Using Cell-Quin medium, we currently obtain >90% enucleation, a deformability that corresponds to values between freshly isolated reticulocytes and erythrocytes, and normal oxygen association and dissociation values (van den Akker and von Lindern, manuscript in preparation). We use DRAQ5 staining coupled with flow cytometry analyses to quantify reticulocyte/nuclei/nucleated cell ratio (**Figure 4C**). Flexibility is measured on a ARCA, and oxygen binding by the Hemox analyzer [127, 128].

2.3.3. MK-specific culture system

Commitment of MKBL and differentiation of MKs can be monitored by the expression of cell surface markers and by the morphological features of the cells (**Figure 4D** and **E**). MKBLs are characterized by $CD34^{+}/CD41a^{+}$ expression and blast-like morphology. In terminal differentiation, MKs gradually lose their expression of $CD34^{+}$, leading to a subdivision of stages: early MKs are $CD34^{+}/CD41a^{+}/CD42^{+}$ and late MKs are $CD34^{-}/CD41a^{+}/CD42^{+}$.

To obtain large numbers of MKBLs, SCF/FL and TPO are used during the first 4–7 days of cultures started from $CD34^{+}$-HSPCs. TPO without SCF and FL allows terminal differentiation to proplatelet-forming cells. To increase the expansion potential, IL-3 can be included only in the initial phase as its prolonged exposure directs the HSPC toward the monocyte/granulocyte lineage. With the addition of IL-6, the MK specification and TPO signaling can be enhanced. With the addition of either IL-1β, IL-9, or IL-11 during the first phase of $CD34^{+}$ differentiation, MK commitment is enhanced instead of progenitor proliferation. It is important to determine the main goal of an experiment before starting the culture: does the experiment require large numbers of MKBLs, or should MK enrichment be maximal, because a good expansion of MKBL tends to compromise terminal differentiation and vice versa (Hansen and van den Akker, unpublished results). Factors such as IL-1β and IL-9 increase polyploidization, formation of proplatelets, and Plts shedding. There is some concern about using IL-1β, because of its proinflammatory nature. Particularly, as it is closely related to IL-1α, and the increased Plt shedding may cause rupture of MKs [33, 99]. To introduce proplatelet formation, IL-6 can be used in high concentrations (>100 ng/ml), by itself or in combination with TPO. SR-1 influences megakaryopoiesis on an early and late stage of the culture, as described above, having a positive effect on the expansion of HSC and terminal differentiation of MK [80, 81]. During the terminal stages of MK cultures (during proplatelet formation), it becomes increasingly essential to prevent the activation of the MKs and produced proplatelets. The addition of heparin prevents the coagulation of plasma added to the media but cells are still able to clump together, thereby having a negative impact on the differentiation and proplatelet production. To prevent activation, signaling via the GPIIb/IIIa (ITGA2B) receptor can be blocked with tirofiban hydrochloride monohydrate. Whereas an MK sheds thousands of Plts *in vivo*, shedding large amounts *in vitro* from a single MK has not yet been achieved. *In vivo* of this process requires that proplatelets extrude between the endothelial cells of the blood vessel wall into the capillaries. This increases level of SP1 among others, combined with shear stress of the blood flow is required for the Plts to be released. To mimic this *in vitro*, several specialized bioreactors are being tested (see Section 2.5).

2.4. Erythropoiesis/megakaryopoiesis from iPSC

Pluripotent stem cells offer a novel approach for developmental studies, drug screening/ discovery, disease modeling, and regenerative medicine. ESCs originate from the inner cell mass of a blastocyst stage embryo, while iPSCs are somatic cells that are reprogrammed back to this embryonic stage [129–132]. Hematopoietic differentiation of ESC/iPSC cells follows the various stages of blood development from early embryonic stages (**Figure 2A**). This offers a valuable tool to study early human hematopoiesis which is difficult because of ethical issues and tissue availability. Besides, differentiation of iPSCs opens opportunities for large-scale manufacture of blood products with the expectancy of clinical application [133]. Several groups showed the potential of ESCs in blood cell production, the source which was later replaced by iPSCs with similar outcome including our group (**Figure 5**) [134–140]. The published protocols generally include four culture phases: (1) mesoderm induction, (2) hematopoietic/erythroid/megakaryocytic commitment, (3) expansion of the specific cell pool, and (4) terminal maturation. The hematopoietic differentiation phases *in vitro* are directed by stepwise addition of cytokines. This is commonly achieved by BMP4, bFGF, and

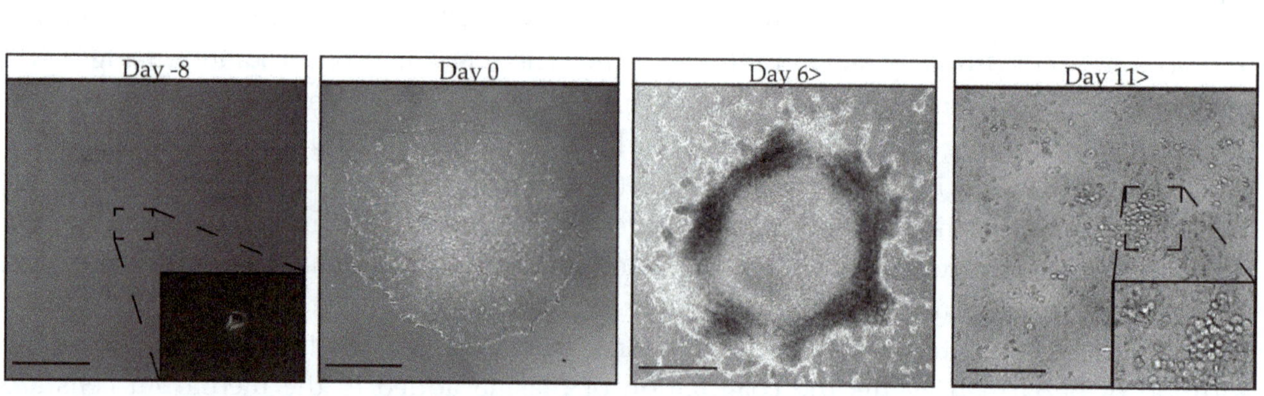

Figure 5. Erythroid/MK differentiation of iPSC according to Hansen et al. showing the different phases of differentiation, with their corresponding growth factor combination and morphological changes [136].

VEGF that drive the cells toward mesoderm, followed by the addition of IL-3, IL-6, SCF, and TPO stimulating hematopoietic specification. Lineage-committed progenitors can be further differentiated toward mature cell types, which is achieved by the combination of medium, growth factors, and hormones. Although most of the differentiation protocols are following the abovementioned scheme, including the listed growth factors, there are multiple technical variations during iPSCs differentiation toward hematopoietic lineages. Two main technical details underlie the major differences in the applied protocols: (i) the induction of differentiation as a 2D monolayer versus 3D embryoid body formation and (ii) the use of coculture with feeder versus feeder-free systems [141–144]. The choice of the differentiation system depends on the application need. 3D systems more closely resemble the *in vivo* process in comparison to 2D systems, offering a tool to study embryogenesis [141, 144–147]. 2D systems, however, are relatively simple, more reproducible, and therefore suitable to scale up production, enabling clinical application [77, 148–150]. The choice of feeder-based or feeder-free differentiation similar to 2D/3D systems also depends on the purpose of the specific research question. Feeder-based coculture systems more resemble the niche, including secretome and cell-cell contact; however, for future clinical application, feeder-free systems are imperative [142, 143, 151, 152]. Protocols can also differ in timing, in the applied media and cytokine cocktails used, which makes comparisons between research groups and methods difficult. Stemline II (SIGMA) is a widely applied base medium during the first and, in some cases, the second phase in feeder-free settings [138, 149, 153]. From the second/third phase onward, the same basic medium are applied that are generally used for other definitive blood cell types such as IMDM with serum/holotransferrin/lipid/Ins supplementation or StemSpan (Stem Cell Technologies) [137, 138, 150, 153]. From the second-step onward, we apply Cell-Quin which in comparison to StemSpan was more efficient in their iPSC-erythroid expansion potential (**Figure 6A**). Some methods still include BMP4, VEGF, and bFGF (or either of them) at this second stage, typically in embryoid body-based system because the 3D structures are less homogenous. Therefore, the transition between phases is not entirely uniform and clear [140, 149, 154, 155]. The few feeder-free 2D systems that have been published mostly do not rely on these three additional factors during commitment phase [148, 149]. FL is also used in erythroid/MK commitment cytokine mixes to improve progenitor expansion [140, 154]. MK commitment/expansion is always based on TPO with or without the addition of multiple other cytokines (e.g., IL-1β, IL-9, IL-11) [99, 136, 156–158].

As pointed out before, early erythropoiesis/megakaryopoiesis (yolk sac) in humans is not well studied, resulting in a lack of knowledge on the regulatory program at these developmental stages. Therefore, the generally applied cytokines might not ideally mimick the *in vivo* situation or the iPSC-driven hematopoietic program. This could underlie inefficient iPSC to RBC/MK differentiation. For example, EPO is applied in all systems to induce erythropoiesis; however, the role of EPO during primitive wave is not entirely clear. Disruption of *EPO* and/or *EPOR* causes embryonic lethality in mice due to the failure of the definitive fetal liver erythropoiesis with reduced primitive erythropoiesis, suggesting that *EPO* and *EPOR* are already functional in early yolk sac [24, 25, 39–41]. However, others showed that additional *EPO* did not affect heme synthesis in early mouse embryos [159]. Furthermore, Malik et al. found that *EPOR*-null embryos have normal number of primitive, early stage progenitors but subsequently develop anemia with loss of primitive EBL [24]. In line with these findings, the

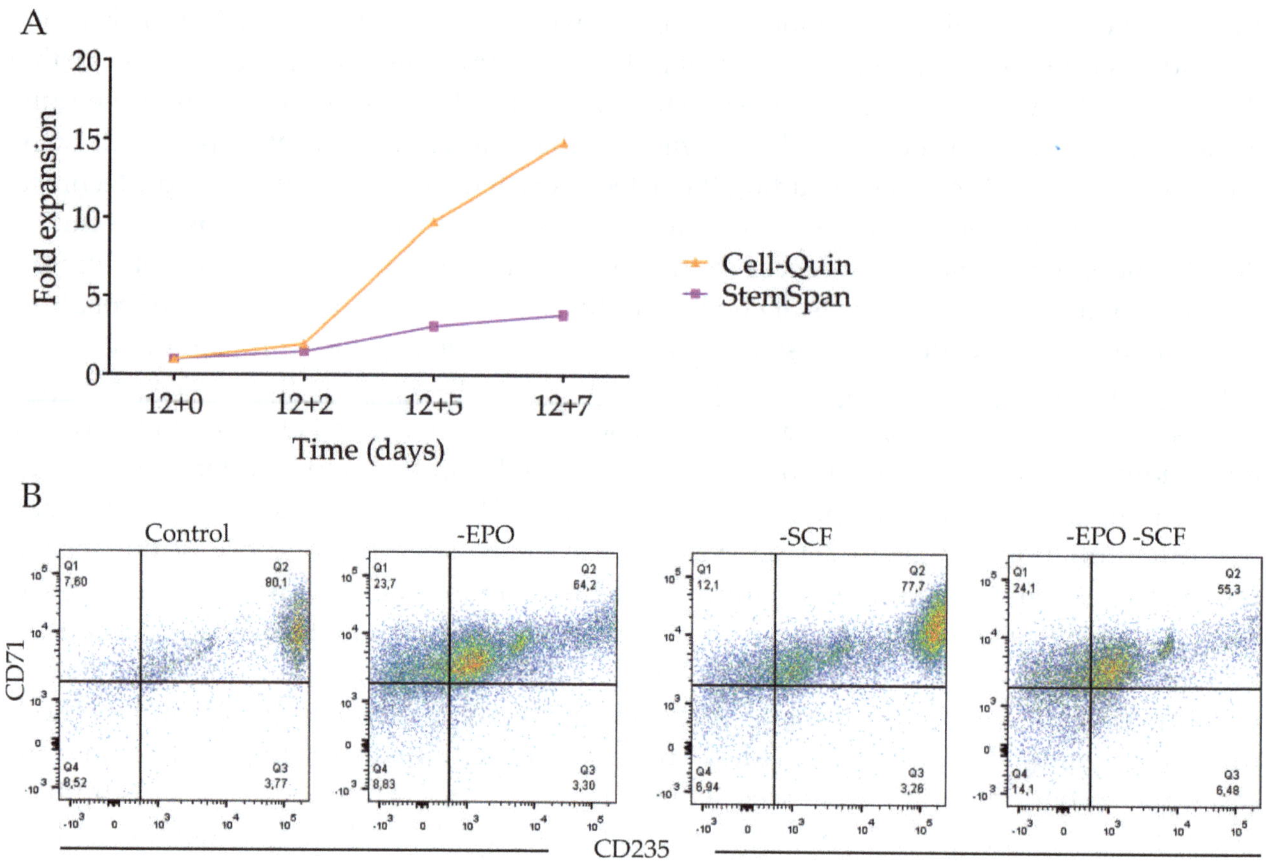

Figure 6. Optimization of iPSC-erythroid differentiation cultures. (A) iPSC-derived erythroid cells arose (D12+0) and expanded in Cell-Quin or StemSpan media. (B) Representative FACS-plots of iPSC-derived erythroid cell (D12 harvest), with or without EPO and SCF from day 6 onward.

same group concluded that EPO signaling is not critical for the survival of human primitive erythroid progenitors, but have a less understood role to promote proliferation and maturation of these cells. Since the role of EPO is controversial in yolk sac erythropoiesis, we tested whether iPSC-erythroid commitment is EPO-dependent. We have tested the requirement of the two most important growth factors for definitive erythropoiesis in various combinations; EPO and SCF (Control), without EPO (-EPO), without SCF (-SCF), and without both cytokine (-EPO and -SCF) [136]. Without EPO, we noticed lower harvest rate/colony number and the loss of CD71/CD235 population. The erythroid commitment was not affected by the deprivation of SCF; however, the addition of SCF together with EPO resulted in a more pure CD71/CD235 population (**Figure 6B**). These data suggest that EPO is required to allow early erythroid commitment while the role of SCF is not entirely clear.

Introduction of erythroid/MK-specific TFs into iPSC-derived hematopoietic cells, often named "forward reprogramming", is being pursued as an approach to improve differentiation outcome. HOXA9, ERG, RORA, SOX4, and MYB have been introduced into human pluripotent stem cells. Engraftment into NSG mice resulted in erythroid cells, which were more skewed to definitive erythropoiesis (lack of embryonic Hbs, mainly HbF and some HbA, some enucleation) compared to TF-free counterparts [160, 161]. These results suggest the possibility of more mature erythroid cell production from iPSCs if certain TFs are included; however, the *in vitro* feasibility is not provided presently. To direct the cells more lineage-specific, other lineage instructive TFs may be used. GATA1, FLI-1, and TAL1 (MK-specific TFs) can be overexpressed to direct iPSC to MKs, thereby achieving ~100% MK yield within 15 days [158]. Besides this set of genes, there are other combinations that are used to achieve a similar aim [156, 162].

The technical differences between published differentiation methods are leading to slight discrepancies in marker expression pattern, purity, yield, and stage of development. However, currently all published methods are limited by technical pitfalls, including the production of developmentally immature (nonadult) cell types which may be the cause of low yield and difficulty to terminally differentiate toward functional end stage blood cell types (e.g., low enucleation potential of iPSC-erythroid cells and low efficiency of iPSC-Plt formation).

2.4.1. Marker expression pattern of differentiating iPSC cells

The purity of the iPSC-derived erythroid population, and its distribution over different maturation stages can be assessed by the erythroid-specific markers used for definitive erythroid culture systems (Section 2.3.2); however, their progression differs in some aspects. Based on our differentiation scheme (**Figure 5**), we recognize three maturation stages: (i) an early erythroid population (harvest at day 10–14) is $CD71^{high}/CD235^{high}/CD36^{med/high}$, which is not yet hemoglobinized and displays big nuclei [136]. Furthermore, the cells are negative for CD18 (myeloid lineage marker) confirming specification toward the erythroid lineage; (ii) a 100% pure erythroid population (day 7–9 expansion) is $CD71/CD235/CD36^{med}$ with some spontaneous differentiation, which is recognized by hemoglobinization and condensation of the nuclei; (iii) a mature erythroid population (D7-14 terminal differentiation) gives rise to $CD235^{high}/CD71^{high/med}/CD36^{low}$ cells. However, there is a slight CD71 decrease associated with reticulocyte formation, and iPSC-derived erythroid cells do not become CD71 negative. Morphologically, these cells were somewhat different from their definitive counterparts. Despite hemoglobinization, nuclear condensation, and polarization, we do not observe a decrease of cytoplasm size and the enucleation potential is poor. Technical variations in the published methods (timing, added growth factors) cause notable differences in the erythroid marker expression pattern; therefore, it is hard to compare and/or draw general conclusions. The emergence of $CD71^{+}/CD235^{+}$ population is generally reported with purity discrepancies. For example, Yang et al. [163] reported 80% CD71/CD235 purity (with $CD34^{+}/CD43^{+}$ preselection and OP9 coculture), Salvagiotto et al. [148] by a feeder-free monolayer system reached 40% pure population, while Kobari et al. [135] with EB-based induction reached 98–99% comparable to our findings. The pattern of CD36 expression is not entirely clear. Mao et al., for example, used a four-step differentiation scheme, including an AGM coculture induction step, and defined the following gene expression profile: early definitive EBLs derived from $CD235^{+}/CD34^{low}/CD36^{-}$, and they develop to $CD235^{+}/CD34^{-}/CD36^{-}$, $CD235^{+}/CD34^{-}/CD36^{-/low}$, $CD235^{+}/CD34^{-}/CD36^{-}$ cells in sequence [164]. Others including us found high CD36 expression during the early erythroid stage [136].

The kinetics during differentiation/maturation of MK from iPSC follow the same steps as from definitive $CD34^+$ cells, namely MKBL ($CD34^+/CD41a^+$), early MK ($CD34^+/CD41a^+/CD42^+$), and late MK ($CD34^-/CD41a^+/CD42^+$). The MKs can undergo some polyploidization albeit not in similar level as *in vivo* or primary $CD34^+$ cultures. MKs derived from mouse iPSC can form proplatelets but also in low numbers [157, 165–168]. These Plts can be activated and contribute to clot formation and wound healing when transfused to injured mice, showing that iPSC megakaryopoiesis, although still inefficient, leads to functional Plts. The MK-specific cell surface markers can be also used for iPSC-MKs with the exception of CD41a. This marker is also an early endothelial/hematopoietic marker. Therefore, it is essential to always use it in a combination, for example, with CD42 to confirm the specific MK commitment. Despite this, most groups report MK percentage only based on CD41a expression (~30–80% MK induction) [136, 150, 168]. With our method, we are able to achieve an average purity of 78% (CD41a) that can be easily used in scaleup production [136].

2.4.2. Developmental stage of iPSC-derived erythroid/MK cells

Human ESC/iPSC-derived erythropoiesis/megakaryopoiesis, with the current knowledge, do not reach the adult definitive stage, but give rise to a mixture of primitive and definitive fetal/adult cells. Very little is known about human erythropoiesis/megakaryopoiesis in the early stages, between days 17 and 23 of embryogenesis (yolk-sac, AGM region) due to the fact that abortions are primarily performed at later fetal stages and in addition have serious ethical concerns. Hbs are commonly used to distinguish between developmental waves; however, HbF expressing RBCs both arises from yolk sac and later from fetal liver, and momentarily, there is a lack of markers, which can clearly distinguish these two waves (**Figure 2A** and **B**). The iPSC-derived erythroid cells predominantly express HbF, in addition to embryonic types of Hb. We and others also showed the presence of a small portion of adult Hb [135, 136, 153–155]. From the embryonic type of globins, both the presence of Gower 1 and Gower 2 Hb has been reported, but the ratio greatly differs between methods [135]. The presence of Gower 2 Hb indicates that the cells are capable of the first globin switch (ζ to α) to provide more mature primitive-state RBCs. Interestingly, the presence of adult types of Hb also differs between protocols as some group were able to show HbA1 or the presence of β chain, whereas others, including us, observed mainly HbA2 [135, 155]. It is unknown whether primitive erythropoiesis *in vivo* goes through a stage that corresponds to iPSC-derived erythroid cells (Hbe^{mid}, HbF^{high}, $HbA^{low/-}$) or whether globin synthesis is impaired in these cells. The globins, however, may not be the most accurate markers to define the waves. Altogether, these findings make it difficult to define the state of iPSC-RBCs with respect to their developmental stage. From a technical point of view, these cells are able to produce definitive RBCs; therefore, they potentially can give rise to the required therapeutic product. Better understanding of the underlying regulatory mechanisms such as the site where erythropoiesis/megakaryopoiesis takes place, and Hb switches during development might pave the way for the necessary improvement of the current methods.

Unlike in the erythroid lineages where the expression of stage-specific Hbs can be used to determine the ontogeny phase, this type of readout is not available in the MK lineage. There are, however, intrinsic differences between megakaryocytic cells derived at different sites during ontogeny. For instance, polyploidy is a measure of ontogeny *in vivo* as yolk sack MK exhibits low ploidy (4–8N), whereas adult MK reach >64N [14, 169, 170]. Besides the lower polyploidization, the number of Plts that are produced follows the same trend from a low number per MK in embryonic/fetal tissues too high numbers of in adult MK, linking polyploidization and MK size to Plts production [170]. However, ploidy levels and Plt production *in vitro* are generally lower compared to their *in vivo* counterparts; therefore, it is not a good marker to access iPSC-MK development stages. The current best approach would be to use data from erythroid cultures and their ontogeny stage/wave and extrapolated this to the MK development because of their close relationship. However, investigation of purified megakaryocytic cultures of defined ontogeny stages through, e.g., RNAseq or mass spectrometry could yield specific makers.

2.4.3. Expansion potential/yield of iPSC-erythroid/MK cells

The final yield of our method is relatively high; however, the comparison with other methods is difficult due to technical discrepancies. There are different ways to calculate the final yield, which also depends on the iPSC maintenance system (single cell vs. clumps) and on the induction system (2D or 3D), resulting various ways to report the final yield. Single cell-seeded iPSC cultures can be normalized both to the initial number of seeded cells or to colony number. Furthermore, the comparison of absolute cell number produced (harvested) between 2D and 3D systems is not entirely realistic because of the different nature of the two cultures. We use single cell-seeding, which allows to calculate the yield/iPSC and yield/colony number to represent differentiation efficiency. In our hands, one iPSC colony can give rise on average 5.6×10^6 erythroid cells after 9 days of expansion. While a single iPSC gives rise to $\sim 8 \times 10^3$ erythroid cells (harvest) and subsequent 9 days expansion results in $\sim 2 \times 10^5$ erythroid cells/iPSC on average [136]. The expansion potential (from harvest day) compared to definitive cell types remains relatively low, in line with other methods irrespective of culture condition (**Figure 4B**). The terminal maturation of iPSC-erythroid cells toward enucleating reticulocytes is inefficient with the existing methods and is currently one of the major hurdles to overcome. In our hands, matured iPSC-erythroid cultures had 30–40% enucleation rate based on their nuclei count; however, the resulting reticulocytes appeared to be instable. Altogether, iPSC-derived erythroid cells are able to expand but for limited time and length, with suboptimal enucleation capacity. Probably, this is also coming from the fact that iPSC-derived erythroid cells, based on their globin expression, do not entirely correspond to a fetal/adult definitive wave.

MK yield from iPSCs is higher than $CD34^+$ differentiation (6.9×10^3 cells/iPSC). Unlike the erythroid system in which differentiation can be inhibited for several days by the addition of glucocorticoid analogues, the megakaryocytic system currently lacks such a specific expansion advantage. As a result, iPSC-MK yield and purity is currently low compared to the iPSC-erythroid yield. Even though the yield of MK is low, a small number of MK still could produce significantly large amount of Plts. The *in vivo* production of 2000–8000 Plts/MK is not achieved by far (1–50 Plts/MK in static condition) [4, 5, 81, 168]. Initiatives to increase this yield are needed but will require increasing our knowledge on the later stages of proplatelet formation, which is a current hurdle to overcome in culture conditions and further discussed in the next paragraph devoted to bioreactors (Section 2.5).

2.5. Bioreactors

The production of cultured red blood cells for transfusion purposes has been the holy grail for transfusion medicine. However, a main challenge is the well-described limitation in cell density during the expansion phase [171]. A single transfusion unit contains 2×10^{12} RBCs. Conventional cultivation systems using dishes or flasks can reach up to 10×10^6 cells/mL, meaning that more than 1000 L of culture would be required for the manufacture of a single transfusion unit [172]. Handling of such large volumes is impractical if static culture conditions are maintained. Thus, multiple bioreactor designs have been proposed to improve the volumetric productivity of the process (produced cells/volume of medium). A static culturing system mimicking BM tissue has been proposed, in which cells are grown in a porous scaffold and nutrients are continuously fed through hollow fibers, while used media containing waste from metabolism are removed [173]. This system allows to have continuous replenishment of spent components in the media, while it is possible to re-use some of the most expensive components such as growth factors and transferrin. Also, it separates cells from large shear stress sources. Although an optimal design of this system could lead to the production of transfusion units at competitive costs compared to the price of rare blood units, it would require significant improvements in transfer of nutrients and matured RBCs between the scaffold and the inflow/outflow streams [174]. Mass transfer limitations in diffusion-governed systems can be tackled with agitation. It is relevant to note that conflicting reports have been made on the effect of shear due to agitation in *in vitro* erythropoiesis [172, 175]. Nevertheless, important advances have been made toward culture of EBLs in stirred reactors. Enucleated RBCs were produced in microbioreactors (<20 mL) in which the effect of shear stress on expansion and enucleation of EBLs was evaluated. Gas sparging caused cell death, whereas stirring enhanced enucleation [172]. Expansion can also be performed in shake flasks (40 mL) and stirred tank bioreactors of larger volumes (500 mL) with similar growth kinetics [176]. Hybrid systems combining mechanical agitation and growth of cells in porous materials have also been proposed. This type of systems could improve mass transfer while protecting cells from shear, but harvesting of mature cells from these carriers is still a challenge that must be addressed.

The high sensitivity of MK to shear stress renders culturing and flow cytometry assays challenging. However, it can also be exploited to generate *in vitro* Plts. Multiple techniques can be used to induce the culturing and shedding of Plts like pipetting or the use of flow chambers [157, 165–168]. These approaches mimic more the *in vivo* situation of Plts formation where they are shed in the vasculature, in contrast to static cultures. Plt shedding can be induced by repetitive pipetting of the cultured cells with a hand pipette (p1000) [81]. There is a risk that too much pressure is enforced on the cells, causing them to lyse instead of shed. A more sophisticated method of *in vitro* Plts production is the use of flow chambers, which come in different versions [167, 177]. The most basic one has a linear flow with a surface that is coated to which MKs can attach [165, 167]. These versions were further developed to chambers that include small gaps where MKs are trapped and can extend their cytoplasmic extensions through the gap where they start forming Plts [165, 177]. Another flow chamber setup is the use of pillars in the chamber instead of gaps where the MKs can attach. Coating with a various matrices like fibronectin, von Willebrand factor, or TPO can enhance the efficiency of the Plts generation. In addition, the flow rates that are produced in these chambers can be regulated.

The produced Plts in these systems can be harvested and used for functional test. The aforementioned approaches can also be combined in a way that the culturing and differentiation of iPSC toward MK is performed on coated beads in a stirred spinner flask bioreactor, where subsequently also the Plt production is induced and enhanced, although with a relatively low Plt yield (1 Plt/MK) [168]. Recently, Ito et al. showed that in the vasculature, MKs are not only exposed to shear stress but also to turbulence during Plt sheading, where the process was implemented in their bioreactor [178]. By using their iPSC lines with inducible c-MYC-BMI-1 and BCL-XL expression, they were able to produce a thrombocyte transfusion unit in an 8 L system. This approach proves the possibility of large-scale Plt production but still with a discrepancy between *in vitro* (80–100 Plts) and *in vivo* (2000–8000 Plts) Plt production capacity.

2.6. Clinical trial

The ability to produce large numbers of enucleated, hemoglobinized RBCs opens the perspective of producing cultured red blood cells (cRBC) for transfusion purposes. The feasibility to do so has been demonstrated by the team of Luc Douay who cultured 1 mL of packed cRBC from $CD34^+$ HSCs and transfused it to a healthy volunteer, with a cRBCs half-life of 26 days after injection [133, 179]. Donor-derived RBC transfusion is a cornerstone of modern medicine in the treatment of trauma, chronic anemia, and in surgery. The existence of 30 blood group systems, such as the ABO and Rhesus system, generates at least 300 distinct blood group antigens [180]. Recurrent transfusions carry an inherent risk on alloimmunization to non-identical blood group antigens, which complicate further transfusions. Besides, this cellular therapy is dependent on donor availability with a potential risk of blood-borne diseases. *In vitro* derivation of cRBCs allows their thorough characterization, therefore providing access to better matched product. Improved cell culture protocols may eventually enable us to generate cRBCs for transfusion purposes from iPSCs that were selected and/or modified to lack most blood group antigens with the advance of donor-independency offered by iPSCs source.

In a recent publication, Ito et al. [178] were able to generate Plts in transfusion quantities, where the functionality of these Plts was shown in *in vivo* mouse experiments. This work gives a feasible prospect of clinical trials, which still requires the system to be converted to GMP grade [178].

Our culture protocol from human PBMC to RBC

- day 0 expansion: purify PBMCs by Ficoll and seed in StemSpan or Cell-Quin supplemented with 1 ng/ml human IL-3, 2 U/ml EPO, 10 ng/ml hSCF, 10^{-6} M DEX at 10×10^6 cells/ml, 37°C, and 5% CO_2.
    - Optional: to remove remaining RBCs after Ficoll purification, the use of RBC lysis buffer is suggested.
- day 2 and 4: replace half of the medium; add all factors to the medium except for IL-3.
- day 5: EBLs appear as large (nongranulated) blasts.
    - Optional: to remove remaining lymphocytes, purify the culture by density centrifugation on a 1.075 g/ml Percoll gradient.

- day 6–20: put the cells daily or every second day to 0.5–0.7 × 10^6/ml medium supplemented with EPO, SCF, and DEX (concentration is same as day 0).
- day 0 terminal differentiation: wash the cells twice with PBS and re-seed in medium supplemented with 10 U/ml EPO, 1 mg/ml holotransferrin, 2–5% plasma, 5 U/ml heparin at 2 × 10^6/ml.
- day 9–14: let the cells differentiate with half media change every 2–3 days.

Our culture protocol from CD34+ to MK

- day 0: collect $CD34^+$ cells by MACS isolation from CB, BM, PBMC, or MPB and seed in Cellgro or Cell-Quin with 100 ng/ml FL, 50 ng/ml hSCF, 50 ng/ml TPO, 20 ng/ml IL-6 at 1 × 10^6 cells/ml, 37°C, and 5% CO_2.
  - Re-seed cells when concentration exceeds 2.5 × 10^6 cells/ml, otherwise keep cells as undisturbed as possible at stable CO_2 levels.
- day 4: cells start to commit to the MK lineage (~10–20%), cells are collected and spun down at 200 g to start terminal differentiation.
- day 0 terminal differentiation: re-seed cells in media (Cellgro or Cell-Quin) supplemented with: 50 ng/ml TPO and 10 ng/ml IL-1β at 0.5 × 10^6 cells/ml.
  - During terminal differentiation, the addition of tirofiban hydrochloride monohydrate and heparin is recommended.
  - Pipetting should be kept to minimal.
  - 1 μM of SR-1 can be used to increase polyploidization and Plt production.
- day 11: cells will be committed to the MK lineage (80–100%) and consist of MKBLs and early MKs.
  - Cell collection from these days onward should be performed using 2 mL or larger pipettes and avoid the usage of hand pipettes to circumvent cell lysis and shear stress [157, 166].
  - Centrifugation steps are on 150 g, low ramp, and brake.
  - Flow cytometry techniques on unfixed MKs at this stage will induce granule release and proplatelet formation.
- day 12–16: MK cells will mature to late MKs and start producing proplatelets.
  - Centrifugation steps are 100 g, low ramp, and brake.
  - CB starts proplatelet production earlier than adult sources.
  - Proplatelets can be harvested using Plts isolation protocols.
  - Proplatelets are easily activated, treat them as regular Plts.

3. Conclusions

As described in this chapter, there are multiple protocols to culture RBCs and MKs from a variety of hematopoietic tissues. Depending on the goal (fundamental research, drug screening, or clinical applications), one should consider beforehand which source can be used. For clinical applications, the use of a fully defined unlimited source would be preferred. For this, iPSC (generated with nonintegrating method) hold great potential but differentiation toward RBCs and Plts has to be improved.

Acknowledgements

We would like to thank Esther Heideveld, Joan Gallego Murillo, Department of Hematopoiesis and Laboratory for Cell Therapy (Sanquin, Amsterdam) for their experimental and scientific input.

We are grateful for the support by the Ministry of Health (PPOC: 11-035, 15-2089), the Landsteiner Foundation for Blood Transfusion Research (LSBR1141), the European Union (FA H2020-MSCA ITN-2015, RELEVANCE), and the Netherlands Organization for Scientific Research (NWO/ZonMw 40-41400-98-1327; 40-00812-98-12128).

Conflict of interest

None.

Abbreviations

AGM	aorta-gonad-mesonephros
AhR	aryl hydrocarbon receptor
bFGF	basic fibroblast growth factor
BM	bone marrow
BMP4	bone morphogenetic protein 4
CB	cord blood
cRBC	cultured red blood cell
DEX	dexamethasone
dox	doxycycline

EBL	erythroblast
EHT	endothelial to hematopoietic transition
EMP	erythroid-myeloid progenitor
EPO	erythropoietin
EPOR	EPO receptor
ESC	embryonic stem cell
FL	FLT-3 ligand
FLI-1	friend leukemia integration 1
FLT-3	Fms-like tyrosine kinase 3
GR	glucocorticoid receptor
Hb	hemoglobin
HE	hemogenic endothelium
HFL	human fetal liver
HSC	hematopoietic stem cell
HSPC	hematopoietic stem/progenitor cell
IL	interleukin
Ins	insulin
InsR	insulin receptor
iPSC	induced pluripotent stem cell
MK	megakaryocyte
MKBL	megakaryoblast
MPB	mobilized peripheral blood
MPP	multipotent progenitor
PBMC	peripheral blood mononuclear cell
Plt	platelet
RBC	red blood cell
SCF	stem cell factor
SR-1	StemRegenin 1
TF	transcription factor

TPO	thrombopoietin
T3	3,5,3′-triiodothyronine
VEGF	vascular endothelial growth factor

Author details

Eszter Varga[†], Marten Hansen[†], Emile van den Akker and Marieke von Lindern*

*Address all correspondence to: m.vonlindern@sanquin.nl

Department of Hematopoiesis, Sanquin Research, and Landsteiner Laboratory Amsterdam UMC, Amsterdam, The Netherlands

[†] These authors contributed equally.

References

[1] Machlus KR, Italiano JE Jr. The incredible journey: From megakaryocyte development to platelet formation. The Journal of Cell Biology. 2013;**201**:785-796

[2] Franco AT, Corken A, Ware J. Platelets at the interface of thrombosis, inflammation, and cancer. Blood. 2015;**126**:582-588

[3] Frick PG. The erythrocyte as an example of biological fitness. Schweizerische Medizinische Wochenschrift. 1961;**91**:1245-1249

[4] Trowbridge EA, Martin JF, Slater DN, Kishk YT, Warren CW, Harley PJ, et al. The origin of platelet count and volume. Clinical Physics and Physiological Measurement: An Official Journal of the Hospital Physicists' Association, Deutsche Gesellschaft fur Medizinische Physik and the European Federation of Organisations for Medical Physics. 1984;**5**:145-170

[5] Kaufman RM, Airo R, Pollack S, Crosby WH, Doberneck R. Origin of Pulmonary Megakaryocytes, Blood. 1965;**25**:767-775

[6] Lefrancais E, Ortiz-Munoz G, Caudrillier A, Mallavia B, Liu F, Sayah DM, et al. The lung is a site of platelet biogenesis and a reservoir for haematopoietic progenitors. Nature. 2017;**544**:105-109

[7] Weiskopf K, Schnorr PJ, Pang WW, Chao MP, Chhabra A, Seita J, et al. Myeloid cell origins, differentiation, and clinical implications. Microbiology Spectrum. 2016;**4**:10.1128

[8] Roch A, Trachsel V, Lutolf MP. Brief report: Single-cell analysis reveals cell division-independent emergence of megakaryocytes from phenotypic hematopoietic stem cells. Stem Cells. 2015;**33**:3152-3157

[9] Nishikii H, Kurita N, Chiba S. The road map for megakaryopoietic lineage from hematopoietic stem/progenitor cells. Stem Cells Translational Medicine. 2017;**6**:1661-1665

[10] Debili N, Coulombel L, Croisille L, Katz A, Guichard J, Breton-Gorius J, et al. Characterization of a bipotent erythro-megakaryocytic progenitor in human bone marrow. Blood. 1996;**88**:1284-1296

[11] Fraser ST, Isern J, Baron MH. Maturation and enucleation of primitive erythroblasts during mouse embryogenesis is accompanied by changes in cell-surface antigen expression. Blood. 2007;**109**:343-352

[12] Dolznig H, Bartunek P, Nasmyth K, Mullner EW, Beug H. Terminal differentiation of normal chicken erythroid progenitors: Shortening of G1 correlates with loss of D-cyclin/cdk4 expression and altered cell size control. Cell Growth & Differentiation: The Molecular Biology Journal of the American Association for Cancer Research. 1995;**6**:1341-1352

[13] von Lindern M, Deiner EM, Dolznig H, Parren-Van Amelsvoort M, Hayman MJ, Mullner EW, et al. Leukemic transformation of normal murine erythroid progenitors: V- and c-ErbB act through signaling pathways activated by the EpoR and c-Kit in stress erythropoiesis. Oncogene. 2001;**20**:3651-3664

[14] Ma DC, Sun YH, Chang KZ, Zuo W. Developmental change of megakaryocyte maturation and DNA ploidy in human fetus. European Journal of Haematology. 1996;**57**:121-127

[15] Chow A, Lucas D, Hidalgo A, Mendez-Ferrer S, Hashimoto D, Scheiermann C, et al. Bone marrow CD169+ macrophages promote the retention of hematopoietic stem and progenitor cells in the mesenchymal stem cell niche. The Journal of Experimental Medicine. 2011;**208**:261-271

[16] Fabriek BO, Polfliet MM, Vloet RP, van der Schors RC, Ligtenberg AJ, Weaver LK, et al. The macrophage CD163 surface glycoprotein is an erythroblast adhesion receptor. Blood. 2007;**109**:5223-5229

[17] Sadahira Y, Yoshino T, Monobe Y. Very late activation antigen 4-vascular cell adhesion molecule 1 interaction is involved in the formation of erythroblastic islands. The Journal of Experimental Medicine. 1995;**181**:411-415

[18] Ulyanova T, Scott LM, Priestley GV, Jiang Y, Nakamoto B, Koni PA, et al. VCAM-1 expression in adult hematopoietic and nonhematopoietic cells is controlled by tissue-inductive signals and reflects their developmental origin. Blood. 2005;**106**:86-94

[19] Heideveld E, Masiello F, Marra M, Esteghamat F, Yagci N, von Lindern M, et al. CD14+ cells from peripheral blood positively regulate hematopoietic stem and progenitor cell survival resulting in increased erythroid yield. Haematologica. 2015;**100**:1396-1406

[20] Junt T, Schulze H, Chen Z, Massberg S, Goerge T, Krueger A, et al. Dynamic visualization of thrombopoiesis within bone marrow. Science. 2007;**317**:1767-1770

[21] Palis J. Hematopoietic stem cell-independent hematopoiesis: Emergence of erythroid, megakaryocyte, and myeloid potential in the mammalian embryo. FEBS Letters. 2016;**590**:3965-3974

[22] Faloon P, Arentson E, Kazarov A, Deng CX, Porcher C, Orkin S, et al. Basic fibroblast growth factor positively regulates hematopoietic development. Development. 2000;**127**: 1931-1941

[23] Wang C, Tang X, Sun X, Miao Z, Lv Y, Yang Y, et al. TGFbeta inhibition enhances the generation of hematopoietic progenitors from human ES cell-derived hemogenic endothelial cells using a stepwise strategy. Cell Research. 2012;**22**:194-207

[24] Malik J, Kim AR, Tyre KA, Cherukuri AR, Palis J. Erythropoietin critically regulates the terminal maturation of murine and human primitive erythroblasts. Haematologica. 2013;**98**:1778-1787

[25] McGann JK, Silver L, Liesveld J, Palis J. Erythropoietin-receptor expression and function during the initiation of murine yolk sac erythropoiesis. Experimental Hematology. 1997;**25**:1149-1157

[26] Potts KS, Sargeant TJ, Markham JF, Shi W, Biben C, Josefsson EC, et al. A lineage of diploid platelet-forming cells precedes polyploid megakaryocyte formation in the mouse embryo. Blood. 2014;**124**:2725-2729

[27] Li W, Johnson SA, Shelley WC, Ferkowicz M, Morrison P, Li Y, et al. Primary endothelial cells isolated from the yolk sac and para-aortic splanchnopleura support the expansion of adult marrow stem cells *in vitro*. Blood. 2003;**102**:4345-4353

[28] McGrath KE, Frame JM, Fegan KH, Bowen JR, Conway SJ, Catherman SC, et al. Distinct sources of hematopoietic progenitors emerge before HSCs and provide functional blood cells in the mammalian embryo. Cell Reports. 2015;**11**:1892-1904

[29] Crisan M, Solaimani Kartalaei P, Neagu A, Karkanpouna S, Yamada-Inagawa T, Purini C, et al. BMP and hedgehog regulate distinct AGM hematopoietic stem cells ex vivo. Stem Cell Reports. 2016;**6**:383-395

[30] Tavassoli M. Embryonic and fetal hemopoiesis: An overview. Blood Cells. 1991;**17**:269-281 discussion 282-266

[31] Fujiki H, Kimura T, Minamiguchi H, Harada S, Wang J, Nakao M, et al. Role of human interleukin-9 as a megakaryocyte potentiator in culture. Experimental Hematology. 2002; **30**:1373-1380

[32] Goswami R, Kaplan MH. A brief history of IL-9. Journal of Immunology. 2011; **186**:3283-3288

[33] Beaulieu LM, Lin E, Mick E, Koupenova M, Weinberg EO, Kramer CD, et al. Interleukin 1 receptor 1 and interleukin 1beta regulate megakaryocyte maturation, platelet activation, and transcript profile during inflammation in mice and humans. Arteriosclerosis, Thrombosis, and Vascular Biology. 2014;**34**:552-564

[34] Ratajczak J, Marlicz W, Keidel A, Machalinski B, Ratajczak MZ, Gewirtz AM. Effect of Interleukin-1 alpha and Interleukin-1 beta on erythroid progenitor cell growth in serum free cultures: An *in vitro* study relevant to the pathogenesis of the anemia of chronic disease. Hematology. 1997;**2**:21-28

[35] Jacobson LO, Goldwasser E, Fried W, Plzak L. Role of the kidney in erythropoiesis. Nature. 1957;**179**:633-634

[36] Jelkmann W. Physiology and pharmacology of erythropoietin, transfusion medicine and hemotherapy: Offizielles. Organ der Deutschen Gesellschaft fur Transfusionsmedizin und Immunhamatologie. 2013;**40**:302-309

[37] Haase VH. Regulation of erythropoiesis by hypoxia-inducible factors. Blood Reviews. 2013;**27**:41-53

[38] Kapur R, Zhang L. A novel mechanism of cooperation between c-Kit and erythropoietin receptor. Stem cell factor induces the expression of Stat5 and erythropoietin receptor, resulting in efficient proliferation and survival by erythropoietin. The Journal of Biological Chemistry. 2001;**276**:1099-1106

[39] Wu H, Liu X, Jaenisch R, Lodish HF. Generation of committed erythroid BFU-E and CFU-E progenitors does not require erythropoietin or the erythropoietin receptor. Cell. 1995;**83**:59-67

[40] Kieran MW, Perkins AC, Orkin SH, Zon LI. Thrombopoietin rescues *in vitro* erythroid colony formation from mouse embryos lacking the erythropoietin receptor. Proceedings of the National Academy of Sciences of the United States of America. 1996;**93**:9126-9131

[41] Lin CS, Lim SK, D'Agati V, Costantini F. Differential effects of an erythropoietin receptor gene disruption on primitive and definitive erythropoiesis. Genes & Development. 1996;**10**:154-164

[42] de Graaf CA, Metcalf D. Thrombopoietin and hematopoietic stem cells. Cell Cycle. 2011;**10**: 1582-1589

[43] Klinger MH, Jelkmann W. Subcellular localization of thrombopoietin in human blood platelets and its release upon thrombin stimulation. British Journal of Haematology. 2001;**115**:421-427

[44] Fielder PJ, Gurney AL, Stefanich E, Marian M, Moore MW, Carver-Moore K, et al. Regulation of thrombopoietin levels by c-mpl-mediated binding to platelets. Blood. 1996;**87**: 2154-2161

[45] Grozovsky R, Begonja AJ, Liu K, Visner G, Hartwig JH, Falet H, et al. The Ashwell-Morell receptor regulates hepatic thrombopoietin production via JAK2-STAT3 signaling. Nature Medicine. 2015;**21**:47-54

[46] Hoffman R, Tong J, Brandt J, Traycoff C, Bruno E, McGuire BW, et al. The *in vitro* and *in vivo* effects of stem cell factor on human hematopoiesis. Stem Cells. 1993;**11**(Suppl 2): 76-82

[47] Li K, Yang M, Lam AC, Yau FW, Yuen PM. Effects of flt-3 ligand in combination with TPO on the expansion of megakaryocytic progenitors. Cell Transplantation. 2000;**9**:125-131

[48] Hassan HT, Zander A. Stem cell factor as a survival and growth factor in human normal and malignant hematopoiesis. Acta Haematologica. 1996;**95**:257-262

[49] Kapur R, Majumdar M, Xiao X, McAndrews-Hill M, Schindler K, Williams DA. Signaling through the interaction of membrane-restricted stem cell factor and c-kit receptor tyrosine kinase: Genetic evidence for a differential role in erythropoiesis. Blood. 1998;**91**:879-889

[50] Fleischman RA, Mintz B. Prevention of genetic anemias in mice by microinjection of normal hematopoietic stem cells into the fetal placenta. Proceedings of the National Academy of Sciences of the United States of America. 1979;**76**:5736-5740

[51] Bernstein A, Chabot B, Dubreuil P, Reith A, Nocka K, Majumder S, et al. The mouse W/c-kit locus. Ciba Foundation Symposium. 1990;**148**:158-166 discussion 166-172

[52] Avraham H, Vannier E, Cowley S, Jiang SX, Chi S, Dinarello CA, et al. Effects of the stem cell factor, c-kit ligand, on human megakaryocytic cells. Blood. 1992;**79**:365-371

[53] Muta K, Krantz SB, Bondurant MC, Dai CH. Stem cell factor retards differentiation of normal human erythroid progenitor cells while stimulating proliferation. Blood. 1995; **86**:572-580

[54] Broudy VC, Lin NL, Priestley GV, Nocka K, Wolf NS. Interaction of stem cell factor and its receptor c-kit mediates lodgment and acute expansion of hematopoietic cells in the murine spleen. Blood. 1996;**88**:75-81

[55] Tsapogas P, Mooney CJ, Brown G, Rolink A. The cytokine Flt3-ligand in normal and malignant hematopoiesis. International Journal of Molecular Sciences. 2017;**18**:1115

[56] Valentinis B, Baserga R. IGF-I receptor signalling in transformation and differentiation. Molecular Pathology: MP. 2001;**54**:133-137

[57] von Lindern M, Zauner W, Mellitzer G, Steinlein P, Fritsch G, Huber K, et al. The glucocorticoid receptor cooperates with the erythropoietin receptor and c-Kit to enhance and sustain proliferation of erythroid progenitors *in vitro*. Blood. 1999;**94**:550-559

[58] Tunduguru R, Thurmond DC. Promoting glucose transporter-4 vesicle trafficking along cytoskeletal tracks: PAK-Ing them out. Frontiers in Endocrinology. 2017;**8**:329

[59] Montel-Hagen A, Sitbon M, Taylor N. Erythroid glucose transporters. Current Opinion in Hematology. 2009;**16**:165-172

[60] Montel-Hagen A, Blanc L, Boyer-Clavel M, Jacquet C, Vidal M, Sitbon M, et al. The Glut1 and Glut4 glucose transporters are differentially expressed during perinatal and postnatal erythropoiesis. Blood. 2008;**112**:4729-4738

[61] Correa PN, Axelrad AA. Production of erythropoietic bursts by progenitor cells from adult human peripheral blood in an improved serum-free medium: Role of insulinlike growth factor 1. Blood. 1991;**78**:2823-2833

[62] Correa PN, Eskinazi D, Axelrad AA. Circulating erythroid progenitors in polycythemia vera are hypersensitive to insulin-like growth factor-1 *in vitro*: Studies in an improved serum-free medium. Blood. 1994;**83**:99-112

[63] Stocklin E, Wissler M, Gouilleux F, Groner B. Functional interactions between Stat5 and the glucocorticoid receptor. Nature. 1996;**383**:726-728

[64] Luisi BF, Xu WX, Otwinowski Z, Freedman LP, Yamamoto KR, Sigler PB. Crystallographic analysis of the interaction of the glucocorticoid receptor with DNA. Nature. 1991; **352**:497-505

[65] Karin M, Chang L. AP-1—Glucocorticoid receptor crosstalk taken to a higher level. The Journal of Endocrinology. 2001;**169**:447-451

[66] Wessely O, Deiner EM, Beug H, von Lindern M. The glucocorticoid receptor is a key regulator of the decision between self-renewal and differentiation in erythroid progenitors. The EMBO Journal. 1997;**16**:267-280

[67] Bauer A, Tronche F, Wessely O, Kellendonk C, Reichardt HM, Steinlein P, et al. The glucocorticoid receptor is required for stress erythropoiesis. Genes & Development. 1999; **13**:2996-3002

[68] Varricchio L, Migliaccio AR. The role of glucocorticoid receptor (GR) polymorphisms in human erythropoiesis. American Journal of Blood Research. 2014;**4**:53-72

[69] Das KC, Mukherjee M, Sarkar TK, Dash RJ, Rastogi GK. Erythropoiesis and erythropoietin in hypo- and hyperthyroidism. The Journal of Clinical Endocrinology and Metabolism. 1975;**40**:211-220

[70] M'Rabet-Bensalah K, Aubert CE, Coslovsky M, Collet TH, Baumgartner C, den Elzen WP, et al. Thyroid dysfunction and anaemia in a large population-based study. Clinical Endocrinology. 2016;**84**:627-631

[71] Angelin-Duclos C, Domenget C, Kolbus A, Beug H, Jurdic P, Samarut J. Thyroid hormone T3 acting through the thyroid hormone alpha receptor is necessary for implementation of erythropoiesis in the neonatal spleen environment in the mouse. Development. 2005;**132**:925-934

[72] Schroeder C, Gibson L, Zenke M, Beug H. Modulation of normal erythroid differentiation by the endogenous thyroid hormone and retinoic acid receptors: A possible target for v-erbA oncogene action. Oncogene. 1992;**7**:217-227

[73] Leberbauer C, Boulme F, Unfried G, Huber J, Beug H, Mullner EW. Different steroids co-regulate long-term expansion versus terminal differentiation in primary human erythroid progenitors. Blood. 2005;**105**:85-94

[74] Jackson CS, Durandt C, Janse van Rensburg I, Praloran V, Brunet de la Grange P, Pepper MS. Targeting the aryl hydrocarbon receptor nuclear translocator complex with DMOG and Stemregenin 1 improves primitive hematopoietic stem cell expansion. Stem Cell Research. 2017;**21**:124-131

[75] Abdelrahim M, Smith R 3rd, Safe S. Aryl hydrocarbon receptor gene silencing with small inhibitory RNA differentially modulates Ah-responsiveness in MCF-7 and HepG2 cancer cells. Molecular Pharmacology. 2003;**63**:1373-1381

[76] Casado FL, Singh KP, Gasiewicz TA. Aryl hydrocarbon receptor activation in hematopoietic stem/progenitor cells alters cell function and pathway-specific gene modulation reflecting changes in cellular trafficking and migration. Molecular Pharmacology. 2011;**80**:673-682

[77] Smith BW, Rozelle SS, Leung A, Ubellacker J, Parks A, Nah SK, et al. The aryl hydrocarbon receptor directs hematopoietic progenitor cell expansion and differentiation. Blood. 2013;**122**:376-385

[78] Singh KP, Garrett RW, Casado FL, Gasiewicz TA. Aryl hydrocarbon receptor-null allele mice have hematopoietic stem/progenitor cells with abnormal characteristics and functions. Stem Cells and Development. 2011;**20**:769-784

[79] Singh KP, Casado FL, Opanashuk LA, Gasiewicz TA. The aryl hydrocarbon receptor has a normal function in the regulation of hematopoietic and other stem/progenitor cell populations. Biochemical Pharmacology. 2009;**77**:577-587

[80] Wagner JE Jr, Brunstein CG, Boitano AE, DeFor TE, McKenna D, Sumstad D, et al. Phase I/II trial of StemRegenin-1 expanded umbilical cord blood hematopoietic stem cells supports testing as a stand-alone graft. Cell Stem Cell. 2016;**18**:144-155

[81] Strassel C, Brouard N, Mallo L, Receveur N, Mangin P, Eckly A, et al. Aryl hydrocarbon receptor-dependent enrichment of a megakaryocytic precursor with a high potential to produce proplatelets. Blood. 2016;**127**:2231-2240

[82] Hu T, Wang D, Yu Q, Li L, Mo X, Pan Z, et al. Aryl hydrocarbon receptor negatively regulates lipid synthesis and involves in cell differentiation of SZ95 sebocytes *in vitro*. Chemico-Biological Interactions. 2016;**258**:52-58

[83] van den Akker E, van Dijk T, Parren-van Amelsvoort M, Grossmann KS, Schaeper U, Toney-Earley K, et al. Tyrosine kinase receptor RON functions downstream of the erythropoietin receptor to induce expansion of erythroid progenitors. Blood. 2004;**103**: 4457-4465

[84] Grebien F, Kerenyi MA, Kovacic B, Kolbe T, Becker V, Dolznig H, et al. Stat5 activation enables erythropoiesis in the absence of EpoR and Jak2. Blood. 2008;**111**:4511-4522

[85] Wessely O, Bauer A, Quang CT, Deiner EM, von Lindern M, Mellitzer G, et al. A novel way to induce erythroid progenitor self renewal: Cooperation of c-Kit with the erythropoietin receptor. Biological Chemistry. 1999;**380**:187-202

[86] Grech G, Blazquez-Domingo M, Kolbus A, Bakker WJ, Mullner EW, Beug H, et al. Igbp1 is part of a positive feedback loop in stem cell factor-dependent, selective mRNA translation initiation inhibiting erythroid differentiation. Blood. 2008;**112**:2750-2760

[87] Bakker WJ, Blazquez-Domingo M, Kolbus A, Besooyen J, Steinlein P, Beug H, et al. FoxO3a regulates erythroid differentiation and induces BTG1, an activator of protein arginine methyl transferase 1. The Journal of Cell Biology. 2004;**164**:175-184

[88] Kolbus A, Blazquez-Domingo M, Carotta S, Bakker W, Luedemann S, von Lindern M, et al. Cooperative signaling between cytokine receptors and the glucocorticoid receptor in the expansion of erythroid progenitors: Molecular analysis by expression profiling. Blood. 2003;**102**:3136-3146

[89] Muckenthaler MU, Rivella S, Hentze MW, Galy B. A red carpet for iron metabolism. Cell. 2017;**168**:344-361

[90] Liao C, Carlson BA, Paulson RF, Prabhu KS. The intricate role of selenium and selenoproteins in erythropoiesis. Free Radical Biology & Medicine. 2018. In press

[91] Heideveld E, Hampton-O'Neil LA, Cross SJ, van Alphen FPJ, van den Biggelaar M, Toye AM, et al. Glucocorticoids induce differentiation of monocytes towards macrophages that share functional and phenotypical aspects with erythroblastic island macrophages. Haematologica. 2018;**103**:395-405

[92] Lee SH, Crocker PR, Westaby S, Key N, Mason DY, Gordon S, et al. Isolation and immunocytochemical characterization of human bone marrow stromal macrophages in hemopoietic clusters. The Journal of Experimental Medicine. 1988;**168**:1193-1198

[93] Chow A, Huggins M, Ahmed J, Hashimoto D, Lucas D, Kunisaki Y, et al. CD169(+) macrophages provide a niche promoting erythropoiesis under homeostasis and stress. Nature Medicine. 2013;**19**:429-436

[94] Kaser A, Brandacher G, Steurer W, Kaser S, Offner FA, Zoller H, et al. Interleukin-6 stimulates thrombopoiesis through thrombopoietin: Role in inflammatory thrombocytosis. Blood. 2001;**98**:2720-2725

[95] Nakashima K, Taga T. gp130 and the IL-6 family of cytokines: Signaling mechanisms and thrombopoietic activities. Seminars in Hematology. 1998;**35**:210-221

[96] Stahl CP, Zucker-Franklin D, Evatt BL, Winton EF. Effects of human interleukin-6 on megakaryocyte development and thrombocytopoiesis in primates. Blood. 1991; **78**:1467-1475

[97] Wang Y, Niu ZY, Guo YJ, Wang LH, Lin FR, Zhang JY. IL-11 promotes the treatment efficacy of hematopoietic stem cell transplant therapy in aplastic anemia model mice through a NF-kappaB/microRNA-204/thrombopoietin regulatory axis. Experimental & Molecular Medicine. 2017;**49**:e410

[98] Tsuji K, Lyman SD, Sudo T, Clark SC, Ogawa M. Enhancement of murine hematopoiesis by synergistic interactions between steel factor (ligand for c-kit), interleukin-11, and other early acting factors in culture. Blood. 1992;**79**:2855-2860

[99] Nishimura S, Nagasaki M, Kunishima S, Sawaguchi A, Sakata A, Sakaguchi H, et al. IL-1alpha induces thrombopoiesis through megakaryocyte rupture in response to acute platelet needs. The Journal of Cell Biology. 2015;**209**:453-466

[100] Lee CR, Cervi D, Truong AH, Li YJ, Sarkar A, Ben-David Y. Friend virus-induced erythroleukemias: A unique and well-defined mouse model for the development of leukemia. Anticancer Research. 2003;**23**:2159-2166

[101] Kurita R, Suda N, Sudo K, Miharada K, Hiroyama T, Miyoshi H, et al. Establishment of immortalized human erythroid progenitor cell lines able to produce enucleated red blood cells. PLoS One. 2013;**8**:e59890

[102] Roecklein BA, Torok-Storb B. Functionally distinct human marrow stromal cell lines immortalized by transduction with the human papilloma virus E6/E7 genes. Blood. 1995;**85**:997-1005

[103] Spike BT, Dirlam A, Dibling BC, Marvin J, Williams BO, Jacks T, et al. The Rb tumor suppressor is required for stress erythropoiesis. The EMBO Journal. 2004;**23**:4319-4329

[104] Clark AJ, Doyle KM, Humbert PO. Cell-intrinsic requirement for pRb in erythropoiesis. Blood. 2004;**104**:1324-1326

[105] Zhang J, Lee EY, Liu Y, Berman SD, Lodish HF, Lees JA. pRB and E2F4 play distinct cell-intrinsic roles in fetal erythropoiesis. Cell Cycle. 2010;**9**:371-376

[106] Vinjamur DS, Bauer DE. Growing and genetically manipulating human umbilical cord blood-derived erythroid progenitor (HUDEP) cell lines. Methods in Molecular Biology. 2018;**1698**:275-284

[107] Trakarnsanga K, Griffiths RE, Wilson MC, Blair A, Satchwell TJ, Meinders M, et al. An immortalized adult human erythroid line facilitates sustainable and scalable generation of functional red cells. Nature Communications. 2017;**8**:14750

[108] Greenberg SM, Rosenthal DS, Greeley TA, Tantravahi R, Handin RI. Characterization of a new megakaryocytic cell line: The Dami cell. Blood. 1988;**72**:1968-1977

[109] Ogura M, Morishima Y, Ohno R, Kato Y, Hirabayashi N, Nagura H, et al. Establishment of a novel human megakaryoblastic leukemia cell line, MEG-01, with positive Philadelphia chromosome. Blood. 1985;**66**:1384-1392

[110] Saito H. Megakaryocytic cell lines. Bailliere's Clinical Haematology. 1997;**10**:47-63

[111] To LB, Levesque JP, Herbert KE. How I treat patients who mobilize hematopoietic stem cells poorly. Blood. 2011;**118**:4530-4540

[112] Winkler IG, Sims NA, Pettit AR, Barbier V, Nowlan B, Helwani F, et al. Bone marrow macrophages maintain hematopoietic stem cell (HSC) niches and their depletion mobilizes HSCs. Blood. 2010;**116**:4815-4828

[113] Liles WC, Broxmeyer HE, Rodger E, Wood B, Hubel K, Cooper S, et al. Mobilization of hematopoietic progenitor cells in healthy volunteers by AMD3100, a CXCR4 antagonist. Blood. 2003;**102**:2728-2730

[114] Levesque JP, Hendy J, Winkler IG, Takamatsu Y, Simmons PJ. Granulocyte colony-stimulating factor induces the release in the bone marrow of proteases that cleave c-KIT receptor (CD117) from the surface of hematopoietic progenitor cells. Experimental Hematology. 2003;**31**:109-117

[115] Levesque JP, Takamatsu Y, Nilsson SK, Haylock DN, Simmons PJ. Vascular cell adhesion molecule-1 (CD106) is cleaved by neutrophil proteases in the bone marrow following hematopoietic progenitor cell mobilization by granulocyte colony-stimulating factor. Blood. 2001;**98**:1289-1297

[116] Pulsipher MA, Chitphakdithai P, Miller JP, Logan BR, King RJ, Rizzo JD, et al. Adverse events among 2408 unrelated donors of peripheral blood stem cells: Results of a prospective trial from the National Marrow Donor Program. Blood. 2009;**113**:3604-3611

[117] Migliaccio G, Sanchez M, Masiello F, Tirelli V, Varricchio L, Whitsett C, et al. Humanized culture medium for clinical expansion of human erythroblasts. Cell Transplantation. 2010;**19**:453-469

[118] Tusi BK, Wolock SL, Weinreb C, Hwang Y, Hidalgo D, Zilionis R, et al. Population snapshots predict early haematopoietic and erythroid hierarchies. Nature. 2018;**555**:54-60

[119] Dolznig H, Kolbus A, Leberbauer C, Schmidt U, Deiner EM, Mullner EW, et al. Expansion and differentiation of immature mouse and human hematopoietic progenitors. Methods in Molecular Medicine. 2005;**105**:323-344

[120] van den Akker E, Satchwell TJ, Pellegrin S, Daniels G, Toye AM. The majority of the *in vitro* erythroid expansion potential resides in CD34(−) cells, outweighing the contribution of CD34(+) cells and significantly increasing the erythroblast yield from peripheral blood samples. Haematologica. 2010;**95**:1594-1598

[121] Fibach E, Rachmilewitz EA. The two-step liquid culture: A novel procedure for studying maturation of human normal and pathological erythroid precursors. Stem Cells. 1993;**11**(Suppl 1):36-41

[122] Ebert BL, Lee MM, Pretz JL, Subramanian A, Mak R, Golub TR, et al. An RNA interference model of RPS19 deficiency in Diamond-Blackfan anemia recapitulates defective hematopoiesis and rescue by dexamethasone: Identification of dexamethasone-responsive genes by microarray. Blood. 2005;**105**:4620-4626

[123] Neildez-Nguyen TM, Wajcman H, Marden MC, Bensidhoum M, Moncollin V, Giarratana MC, et al. Human erythroid cells produced ex vivo at large scale differentiate into red blood cells *in vivo*. Nature Biotechnology. 2002;**20**:467-472

[124] Giarratana MC, Kobari L, Lapillonne H, Chalmers D, Kiger L, Cynober T, et al. Ex vivo generation of fully mature human red blood cells from hematopoietic stem cells. Nature Biotechnology. 2005;**23**:69-74

[125] Gandrillon O, Schmidt U, Beug H, Samarut J. TGF-beta cooperates with TGF-alpha to induce the self-renewal of normal erythrocytic progenitors: Evidence for an autocrine mechanism. The EMBO Journal. 1999;**18**:2764-2781

[126] Migliaccio G, Di Pietro R, di Giacomo V, Di Baldassarre A, Migliaccio AR, Maccioni L, et al. *In vitro* mass production of human erythroid cells from the blood of normal donors and of thalassemic patients. Blood Cells, Molecules & Diseases. 2002;**28**:169-180

[127] van Zwieten R, van Oirschot BA, Veldthuis M, Dobbe JG, Streekstra GJ, van Solinge WW, et al. Partial pyruvate kinase deficiency aggravates the phenotypic expression of band 3 deficiency in a family with hereditary spherocytosis. American Journal of Hematology. 2015;**90**:E35-E39

[128] Vanhille DL, Nussenzveig RH, Glezos C, Perkins S, Agarwal AM. Best practices for use of the HEMOX analyzer in the clinical laboratory: Quality control determination and choice of anticoagulant. Laboratory Hematology : Official Publication of the International Society for Laboratory Hematology. 2012;**18**:17-19

[129] Takahashi K, Yamanaka S. Induction of pluripotent stem cells from mouse embryonic and adult fibroblast cultures by defined factors. Cell. 2006;**126**:663-676

[130] Takahashi K, Tanabe K, Ohnuki M, Narita M, Ichisaka T, Tomoda K, et al. Induction of pluripotent stem cells from adult human fibroblasts by defined factors. Cell. 2007; **131**:861-872

[131] Evans MJ, Kaufman MH. Establishment in culture of pluripotential cells from mouse embryos. Nature. 1981;**292**:154-156

[132] Martin GR. Isolation of a pluripotent cell line from early mouse embryos cultured in medium conditioned by teratocarcinoma stem cells. Proceedings of the National Academy of Sciences of the United States of America. 1981;**78**:7634-7638

[133] Giarratana MC, Rouard H, Dumont A, Kiger L, Safeukui I, Le Pennec PY, et al. Proof of principle for transfusion of *in vitro*-generated red blood cells. Blood. 2011;**118**: 5071-5079

[134] Ma F, Ebihara Y, Umeda K, Sakai H, Hanada S, Zhang H, et al. Generation of functional erythrocytes from human embryonic stem cell-derived definitive hematopoiesis. Proceedings of the National Academy of Sciences of the United States of America. 2008;**105**:13087-13092

[135] Kobari L, Yates F, Oudrhiri N, Francina A, Kiger L, Mazurier C, et al. Human induced pluripotent stem cells can reach complete terminal maturation: *In vivo* and *in vitro* evidence in the erythropoietic differentiation model. Haematologica. 2012;**97**:1795-1803

[136] Hansen M, Varga E, Aarts C, Wust T, Kuijpers T, von Lindern M, et al. Efficient production of erythroid, megakaryocytic and myeloid cells, using single cell-derived iPSC colony differentiation. Stem Cell Research. 2018;**29**:232-244

[137] Qiu C, Olivier EN, Velho M, Bouhassira EE. Globin switches in yolk sac-like primitive and fetal-like definitive red blood cells produced from human embryonic stem cells. Blood. 2008;**111**:2400-2408

[138] Lu SJ, Feng Q, Park JS, Vida L, Lee BS, Strausbauch M, et al. Biologic properties and enucleation of red blood cells from human embryonic stem cells. Blood. 2008;**112**:4475-4484

[139] Dias J, Gumenyuk M, Kang H, Vodyanik M, Yu J, Thomson JA, et al. Generation of red blood cells from human induced pluripotent stem cells. Stem Cells and Development. 2011;**20**:1639-1647

[140] Lapillonne H, Kobari L, Mazurier C, Tropel P, Giarratana MC, Zanella-Cleon I, et al. Red blood cell generation from human induced pluripotent stem cells: Perspectives for transfusion medicine. Haematologica. 2010;**95**:1651-1659

[141] Kennedy M, D'Souza SL, Lynch-Kattman M, Schwantz S, Keller G. Development of the hemangioblast defines the onset of hematopoiesis in human ES cell differentiation cultures. Blood. 2007;**109**:2679-2687

[142] Ledran MH, Krassowska A, Armstrong L, Dimmick I, Renstrom J, Lang R, et al. Efficient hematopoietic differentiation of human embryonic stem cells on stromal cells derived from hematopoietic niches. Cell Stem Cell. 2008;**3**:85-98

[143] Kaufman DS, Hanson ET, Lewis RL, Auerbach R, Thomson JA. Hematopoietic colony-forming cells derived from human embryonic stem cells. Proceedings of the National Academy of Sciences of the United States of America. 2001;**98**:10716-10721

[144] Zambidis ET, Peault B, Park TS, Bunz F, Civin CI. Hematopoietic differentiation of human embryonic stem cells progresses through sequential hematoendothelial, primitive, and definitive stages resembling human yolk sac development. Blood. 2005; **106**:860-870

[145] Ng ES, Davis RP, Azzola L, Stanley EG, Elefanty AG. Forced aggregation of defined numbers of human embryonic stem cells into embryoid bodies fosters robust, reproducible hematopoietic differentiation. Blood. 2005;**106**:1601-1603

[146] Amit M, Itskovitz-Eldor J. Derivation and spontaneous differentiation of human embryonic stem cells. Journal of Anatomy. 2002;**200**:225-232

[147] Itskovitz-Eldor J, Schuldiner M, Karsenti D, Eden A, Yanuka O, Amit M, et al. Differentiation of human embryonic stem cells into embryoid bodies compromising the three embryonic germ layers. Molecular Medicine. 2000;**6**:88-95

[148] Salvagiotto G, Burton S, Daigh CA, Rajesh D, Slukvin II, Seay NJ. A defined, feeder-free, serum-free system to generate *in vitro* hematopoietic progenitors and differentiated blood cells from hESCs and hiPSCs. PLoS One. 2011;**6**:e17829

[149] Niwa A, Heike T, Umeda K, Oshima K, Kato I, Sakai H, et al. A novel serum-free monolayer culture for orderly hematopoietic differentiation of human pluripotent cells via mesodermal progenitors. PLoS One. 2011;**6**:e22261

[150] Feng Q, Shabrani N, Thon JN, Huo H, Thiel A, Machlus KR, et al. Scalable generation of universal platelets from human induced pluripotent stem cells. Stem Cell Reports. 2014;**3**:817-831

[151] Weisel KC, Gao Y, Shieh JH, Moore MA. Stromal cell lines from the aorta-gonado-mesonephros region are potent supporters of murine and human hematopoiesis. Experimental Hematology. 2006;**34**:1505-1516

[152] Choi KD, Yu J, Smuga-Otto K, Salvagiotto G, Rehrauer W, Vodyanik M, et al. Hematopoietic and endothelial differentiation of human induced pluripotent stem cells. Stem Cells. 2009;**27**:559-567

[153] Olivier EN, Marenah L, McCahill A, Condie A, Cowan S, Mountford JC. High-efficiency serum-free feeder-free erythroid differentiation of human pluripotent stem cells using small molecules. Stem Cells Translational Medicine. 2016;**5**:1394-1405

[154] Chang KH, Nelson AM, Cao H, Wang L, Nakamoto B, Ware CB, et al. Definitive-like erythroid cells derived from human embryonic stem cells coexpress high levels of embryonic and fetal globins with little or no adult globin. Blood. 2006;**108**:1515-1523

[155] Dorn I, Klich K, Arauzo-Bravo MJ, Radstaak M, Santourlidis S, Ghanjati F, et al. Erythroid differentiation of human induced pluripotent stem cells is independent of donor cell type of origin. Haematologica. 2015;**100**:32-41

[156] Takayama N, Nishimura S, Nakamura S, Shimizu T, Ohnishi R, Endo H, et al. Transient activation of c-MYC expression is critical for efficient platelet generation from human induced pluripotent stem cells. The Journal of Experimental Medicine. 2010; **207**:2817-2830

[157] Jiang J, Woulfe DS, Papoutsakis ET. Shear enhances thrombopoiesis and formation of microparticles that induce megakaryocytic differentiation of stem cells. Blood. 2014;**124**:2094-2103

[158] Moreau T, Evans AL, Vasquez L, Tijssen MR, Yan Y, Trotter MW, et al. Large-scale production of megakaryocytes from human pluripotent stem cells by chemically defined forward programming. Nature Communications. 2016;**7**:11208

[159] Cole RJ, Paul J. The effects of erythropoietin on haem synthesis in mouse yolk sac and cultured foetal liver cells. Journal of Embryology and Experimental Morphology. 1966;**15**:245-260

[160] Doulatov S, Vo LT, Chou SS, Kim PG, Arora N, Li H, et al. Induction of multipotential hematopoietic progenitors from human pluripotent stem cells via respecification of lineage-restricted precursors. Cell Stem Cell. 2013;**13**:459-470

[161] Sugimura R, Jha DK, Han A, Soria-Valles C, da Rocha EL, Lu YF, et al. Haematopoietic stem and progenitor cells from human pluripotent stem cells. Nature. 2017;**545**:432-438

[162] Nakamura S, Takayama N, Hirata S, Seo H, Endo H, Ochi K, et al. Expandable megakaryocyte cell lines enable clinically applicable generation of platelets from human induced pluripotent stem cells. Cell Stem Cell. 2014;**14**:535-548

[163] Yang CT, French A, Goh PA, Pagnamenta A, Mettananda S, Taylor J, et al. Human induced pluripotent stem cell derived erythroblasts can undergo definitive erythropoiesis and co-express gamma and beta globins. British Journal of Haematology. 2014; **166**:435-448

[164] Mao B, Huang S, Lu X, Sun W, Zhou Y, Pan X, et al. Early development of definitive erythroblasts from human pluripotent stem cells defined by expression of glycophorin A/CD235a, CD34, and CD36. Stem Cell Reports. 2016;**7**:869-883

[165] Sullenbarger B, Bahng JH, Gruner R, Kotov N, Lasky LC. Prolonged continuous *in vitro* human platelet production using three-dimensional scaffolds. Experimental Hematology. 2009;**37**:101-110

[166] Dunois-Larde C, Capron C, Fichelson S, Bauer T, Cramer-Borde E, Baruch D. Exposure of human megakaryocytes to high shear rates accelerates platelet production. Blood. 2009;**114**:1875-1883

[167] Pallotta I, Lovett M, Kaplan DL, Balduini A. Three-dimensional system for the *in vitro* study of megakaryocytes and functional platelet production using silk-based vascular tubes. Tissue Engineering. Part C, Methods. 2011;**17**:1223-1232

[168] Eicke D, Baigger A, Schulze K, Latham SL, Halloin C, Zweigerdt R, et al. Large-scale production of megakaryocytes in microcarrier-supported stirred suspension bioreactors. Scientific Reports. 2018;**8**:10146

[169] Elagib KE, Brock AT, Goldfarb AN. Megakaryocyte ontogeny: Clinical and molecular significance. Experimental Hematology. 2018;**61**:1-9

[170] Bluteau O, Langlois T, Rivera-Munoz P, Favale F, Rameau P, Meurice G, et al. Developmental changes in human megakaryopoiesis. Journal of Thrombosis and Haemostasis: JTH. 2013;**11**:1730-1741

[171] Timmins NE, Nielsen LK. Manufactured RBC—Rivers of blood, or an oasis in the desert? Biotechnology Advances. 2011;**29**:661-666

[172] Bayley R, Ahmed F, Glen K, McCall M, Stacey A, Thomas R. The productivity limit of manufacturing blood cell therapy in scalable stirred bioreactors. Journal of Tissue Engineering and Regenerative Medicine. 2018;**12**:e368-e378

[173] Panoskaltsis N, Magalhães Macedo, HM, Blanco MTM, Mantalaris A, Livingston AG. 3-Dimensional hollow fibre bioreactor systems for the maintenance, expansion, differentiation and harvesting of human stem cells and their rogeny. 2012, patent EP2652119A1

[174] Misener R, Fuentes Garí M, Rende M, Velliou E, Panoskaltsis N, Pistikopoulos EN, et al. Global superstructure optimisation of red blood cell production in a parallelised hollow fibre bioreactor. Computers & Chemical Engineering. 2014;**71**:532-553

[175] Boehm D, Murphy WG, Al-Rubeai M. The effect of mild agitation on *in vitro* erythroid development. Journal of Immunological Methods. 2010;**360**:20-29

[176] Lee E, Han SY, Choi HS, Chun B, Hwang B, Baek EJ. Red blood cell generation by three-dimensional aggregate cultivation of late erythroblasts. Tissue Engineering. Part A. 2015;**21**:817-828

[177] Maeda T, Wakasawa T, Shima Y, Tsuboi I, Aizawa S, Tamai I. Role of polyamines derived from arginine in differentiation and proliferation of human blood cells. Biological & Pharmaceutical Bulletin. 2006;**29**:234-239

[178] Ito Y, Nakamura S, Sugimoto N, Shigemori T, Kato Y, Ohno M, et al. Turbulence activates platelet biogenesis to enable clinical scale ex vivo production. Cell. 2018;**174**: 636-648 e618

[179] Rousseau GF, Giarratana MC, Douay L. Large-scale production of red blood cells from stem cells: What are the technical challenges ahead? Biotechnology Journal. 2014;**9**:28-38

[180] Mitra R, Mishra N, Rath GP. Blood groups systems. Indian Journal of Anaesthesia. 2014;**58**:524-528

10

Optimization of the Self-Assembly Method for the Production of Psoriatic Skin Substitutes

Alexe Grenier, Isabelle Gendreau and
Roxane Pouliot

Abstract

Tissue engineering of the skin is used for various applications. However, to develop treatments for skin pathologies such as psoriasis, robust pathological skin models are needed. The purpose of the work presented in this chapter was to optimize the production of more reproducible psoriatic skin substitutes by modifying the original self-assembly method. Substitutes were produced according to the self-assembly method partially modified. The culture flasks of 25 cm^2 were replaced by 6-well and 12-well plates. Fibroblasts were cultured in 6-well and 12-well plates with ascorbic acid until they form manipulable sheets, which were superimposed and incubated for 7 days to form a dermal layer. Afterwards, keratinocytes were seeded on the dermal layer forming an epidermal layer. Then, the substitutes were raised to the air-liquid interface and cultured 21 days before being analyzed. Analyses demonstrated that psoriatic substitutes have a significantly thicker epidermis than healthy substitutes and the persistence of nuclear structures in corneocytes, with original and both modified methods. Immunofluorescence markers such as filaggrin, loricrin, and keratin 14 have confirmed these results. However, some differences were observed in substitutes produced with 12-well plates. Modifications made to the original method for the production of psoriatic substitutes are effective and lead to highly reproducible substitutes more suitable for pharmacological testing.

Keywords: tissue engineering, cell culture, skin substitutes, psoriasis, self-assembly approach, *in vitro*

1. Introduction

1.1. Skin

The integumentary system is the largest and heaviest organ of the body [1, 2]. This organ is divided into three distinct layers: the epidermis (superficial layer), the dermis (intermediate layer), and the hypodermis (deepest layer). Its main function is to protect the body from external aggressions, such as chemical, mechanical, thermal, microbial, and UV rays [3, 4]. It is therefore a physical, biological, and immunological barrier. The epidermis, the outer layer of the skin, predominantly ensures this barrier function by a constant renewal of keratinocytes, the epidermal cells. Keratinocytes differentiated into five layers: from the *stratum basale* (*stratum germinativum*), in which skin stem cells are found, to the *stratum spinosum, stratum granulosum, stratum lucidum*, and *stratum corneum*, which is the outer layer where keratinocytes have lost their nuclei and are completely keratinized [5]. Keratinocytes from this layer, also called corneocytes, will gradually detach to cause the phenomenon called desquamation. The highly regulated process of differentiation involves specific proteins to maintain this epidermal structure, and deregulation of these proteins expression can induce skin pathology such as psoriasis.

1.2. Psoriasis

Psoriasis is an erythematous-squamous dermatosis touching both men and women. This chronic skin pathology affects 2–3% of the world's population [6, 7], which correspond to approximately 125 M people [8]. This pathology is characterized by a hyperproliferation and an abnormal differentiation of keratinocytes resulting in reddish and whitish plaques [5]. At a cellular level, histopathological characteristics consist of acanthosis, parakeratosis, hyperkeratosis, agranulose, and papillomatosis [9, 10]. The disease's etiology is still unknown. However, environmental and immune factors, as well as genetic predispositions, would act together to trigger psoriasis [10, 11]. This disease seriously affects the quality of life of patients due to the appearance of their skin and the side effects of drugs. Existing treatments cause many severe side effects such as nephrotoxicity, hepatotoxicity, immunosuppression, teratogenicity, and no curable treatments have been found [12–14]. Moreover, several comorbidities may be related to psoriasis, such as major cardiac events, type 2 diabetes, and psoriatic arthritis [6].

1.3. *In vivo* and *in vitro* psoriatic skin models

The skin is a complex organ. Thus, the production of representative and reproducible skin models is a constant challenge. The use of *ex vivo* human skin biopsies would be more convenient, since with skin biopsies, it is possible to observe the mechanisms and interactions of the human skin. However, because of skin donor availability and inter-individual variability, the use of *ex vivo* biopsies is not practical, and thus, the development of new models is important. Over the years, there has been a lot of progress in the field of tissue engineering [15]. Tissue engineering of the skin is used for various clinical applications and in fundamental research such as for drug development. Now, with the development and optimization of *in vivo* and *in vitro* models, it is possible to research new treatments for a skin disease by studying, for example the antioxidant and antiproliferative potentials, and the toxicology of molecules or extracts [16, 17], to study the mechanism of action of compounds [18] and to perform percutaneous absorption studies, and thus study the permeability of the skin, the diffusion rate, and the site of action of compounds [19, 20].

1.3.1. In vivo models

Many approaches are used to obtain animal models as representative as possible to the human pathology. Spontaneous mutations, like the homozygous asebia, xenotransplantation, like the severe combined immunodeficient mice (SCID) and the athymic nude mouse, and genetic models, such as the CD18 hypomorphic mice model, the K14/TGF-α, and the involucrin/INF-γ, have been used over the years to study psoriasis but all of them displayed some limitations [21–23]. Animal models are mostly used to study specific aspects of the pathology. The development of a representative animal model can be expensive.

1.3.2. In vitro models

There are two types of models: monolayer models (dermal or epidermal) and bilayer substitutes. Monolayer models use only one cell type, keratinocytes or fibroblasts, and will be used to study a specific characteristic or to understand the role of a certain cell type in pathologies such as psoriasis. However, these models exclude interactions between different cell types. Bilayer models displayed two layers of skin: dermis and epidermis, which allow the study of skin complexity more representatively. The challenge of skin engineering is to reproduce the complexity and the functionalities of a pathological skin. There are different *in vitro* psoriatic skin models, which include interesting pathological features. Various pathological bilayer skin models were developed using a collagen gel as dermal equivalent. Most of these studies involve pathological keratinocytes seeded on a dermis made of collagen and fibroblasts [24, 25], but there are also studies where a full-thickness psoriatic skin biopsy is incorporated into the dermal equivalent [26]. These models have been useful to better understand the disease and the interactions between fibroblasts and keratinocytes [26, 27]. However, the main disadvantage of these models is the use of an exogenous material, which does not represent exactly the properties of the human dermis. To counter the use of exogenous material such as collagen, some research teams have used de-epidermized dermis to produce their psoriatic skin models [28–30]. Although these equivalents demonstrate several psoriatic features, the use of these models for pharmaceutical studies would require an excessive amount of skin biopsies. Thus, a pathological model free of exogenous material that can generate many samples at a time is still required for pharmaceutical research.

Our team has developed a psoriatic skin model based on a self-assembly method, which is free of exogenous material [31]. This model has been characterized towards its permeability, lipid organization and response to antipsoriatic drugs [32, 33]. This basic model has also been improved by the addition of other cell types such as endothelial cells in order to reproduce the angiogenesis observed *in vivo* [34]. These studies have confirmed that our psoriatic skin substitute model produced according to the self-assembly approach maintained many characteristics of the disease including the presence of a disorganized and thicker epidermis compared with normal skin substitutes [31]. This self-assembly approach allows the understanding of pathological skin complexity through the possibility of: (1) dissecting step by step the mechanisms of skin pathologies according to which kinds of cells are present in the model at that time and/or (2) using various cell combinations such as healthy fibroblasts and healthy keratinocytes, which can be compared with healthy fibroblasts and pathological keratinocytes. Although the self-assembly method is very effective for the reconstruction of substitutes used in basic mechanisms studies, it required an optimization of its original protocol to consider a productive capacity of it in the pharmaceutical industry. Thus, the aim of this work was to improve the original self-assembly method to allow the reconstruction of more reproducible psoriatic skin substitutes that could be used for pharmacological testing.

2. Modified self-assembly methodology

As mentioned previously, our team has developed a model of *in vitro* psoriatic skin substitutes using the self-assembly method. In this present research, the production of the tissue-engineered psoriatic skin substitutes was done according to the self-assembly method partially modified, using 6-well plates and 12-well plates [31, 35] (**Table 1** and **Figure 1**). All methods were also compared to the reconstruction of healthy skin substitutes. Briefly, pathological fibroblasts were cultured 28 days with Dulbecco-Vogt modification of Eagle's medium (DMEM) supplemented with 10% fetal calf serum, 100 UI/ml penicillin G, 25 μg/ml gentamicin, and 50 μg/ml ascorbic acid until they form manipulable sheets. Then, these fibroblast sheets were detached, and two of them were superimposed to form a new dermal equivalent. Subsequently, they were incubated for 7 days to allow the fusion of the two sheets and thus form the new layer. After this period, pathological keratinocytes were seeded on the dermal equivalent to form a new epidermal layer. Seven days later, the substitutes were raised to the air-liquid interface to promote cell differentiation and obtain the different epidermal layers. Skin substitute biopsies were taken at 21 days after being raised to the air-liquid interface and analyzed by histology and immunohistochemistry.

2.1. Results

2.1.1. *Macroscopic results*

Healthy skin substitutes reconstructed using either the original or modified methods (**Figure 2A–C**) showed a uniform and opaque epidermis recovering all the seeding area within the anchoring paper (white contours). For substitutes produced with psoriatic cells,

Self-assembly method	Culture surface area (cm^2)	Final size of substitute (usable) (cm^2)	Volume of medium by fibroblast sheet (ml)	Anchoring papers
Original	25	3.87	5	Day 42
6-well plates	9.6	3.87	2	Day 0
12-well plates	3.8	1.27	2	Day 0

Table 1. Technical characteristics of the original method, the 6-well plate and the 12-well plate modifications.

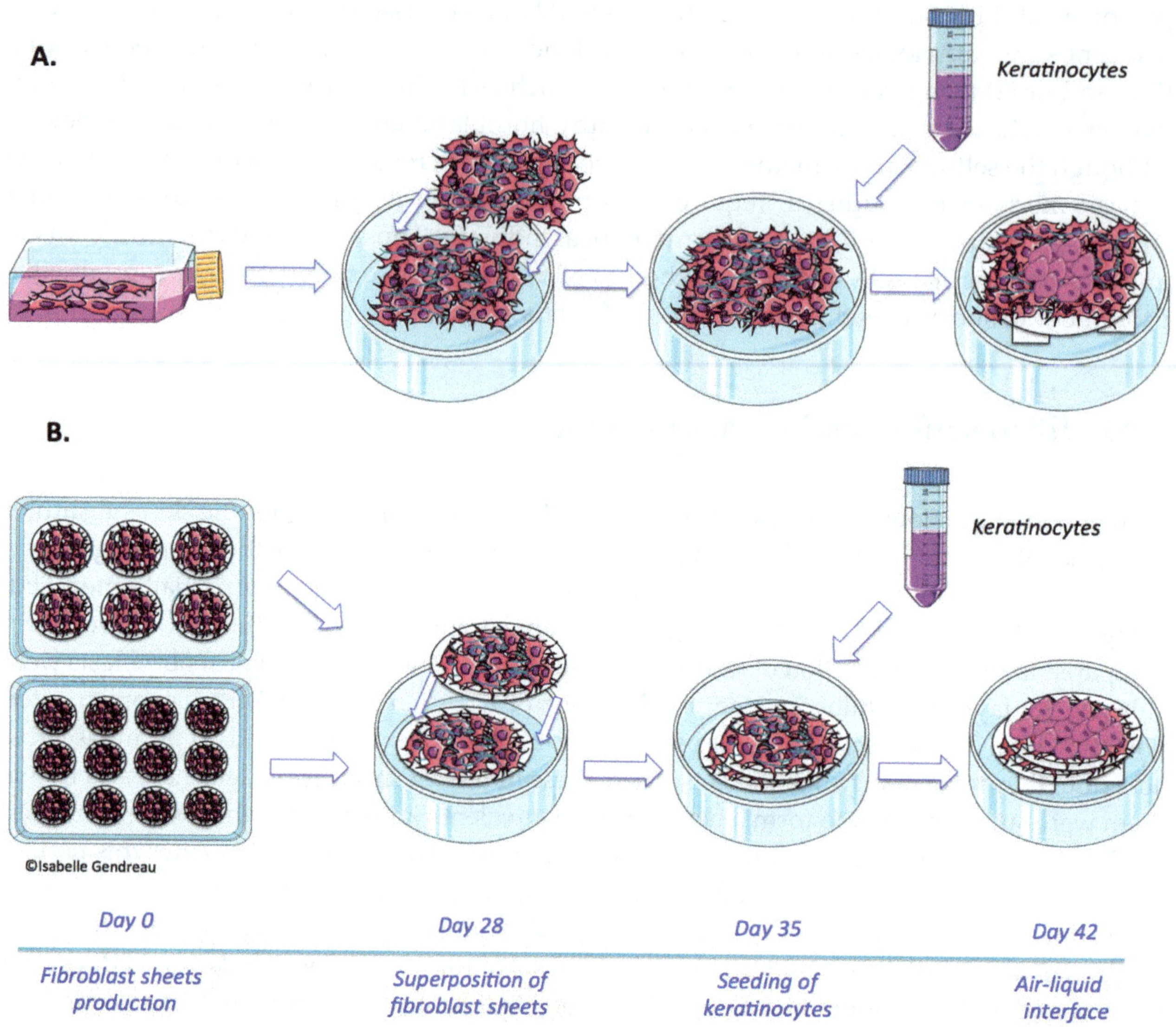

Figure 1. Schematic representation of the self-assembly method. (A) Original self-assembly method: fibroblasts were cultured 28 days in 25 cm^2 flasks for the production of fibroblast sheets. At day 28, flasks were opened with a soldering iron, and two sheets were superimposed to form a dermal equivalent. After one week, a metal ring was deposited on substitutes, and keratinocytes were seeded into the ring. Then, substitutes were cultured one more week in submerged conditions. At day 42, substitutes were placed on an anchoring paper and were raised to the air-liquid interface for three weeks. (B) Modified self-assembly method: fibroblasts were seeded in 6-well or 12-well plates in which anchoring paper have been previously placed. After 28 days of culture, two fibroblast sheets were easily superimposed and solidified with Ligaclip®. Dermal equivalents were cultured seven days before keratinocytes seeding. At day 42, substitutes were raised to the air-liquid interface and cultured for three weeks.

those reconstructed according to the original method and the 6-well plate modifications (**Figure 2D–E**) showed an irregular and contracted epidermis. Psoriatic skin substitutes produced using the 12-well plate modification (**Figure 2F**) demonstrated a more uniform epidermis compared to other methods.

2.1.2. Histology

Healthy skin substitutes reconstructed according to the original method, and the 6-well plate modification (**Figure 3A** and **B**) demonstrated a well-differentiated epidermis and similar characteristics. A thickening of the living epidermis was observed in substitutes produced with psoriatic cells using the original method, as well as in those reconstructed according to the modified protocols (**Figure 3D–F**). A less differentiated epidermis was observed in psoriatic substitutes produced according to the original method and the 6-well plate modification (**Figure 3D–E**), compared with the substitutes produced according to the 12-well plate modification (**Figure 3F**). The substitutes reconstructed according to the 12-well plate modification (**Figure 3C** and **F**) demonstrated a greater cell differentiation which results in a thickening of the *stratum corneum*.

2.1.3. *Epidermal thickness*

Thickness of skin substitutes' living epidermis was measured with the AxioVision software. No significant differences were observed between the skin substitutes reconstructed with healthy cells according to the original method or the two new modifications (65.6 ± 13.9 vs. 69.0 ± 6.8 vs. 78.39 ± 18.69 μm; **Figure 4A**). Measurements of psoriatic substitutes thickness reconstructed according to the original method, and the 6-well plate modification did not show any significant differences between these two methods (93.50 ± 18.9 μm vs. 106.7 ± 23.7 μm; **Figure 4B**), while psoriatic substitutes produced with the 12-well plate modification demonstrated a significant difference compared with the original method (130.8 ± 18.8 vs. 93.50 ± 18.9 μm).

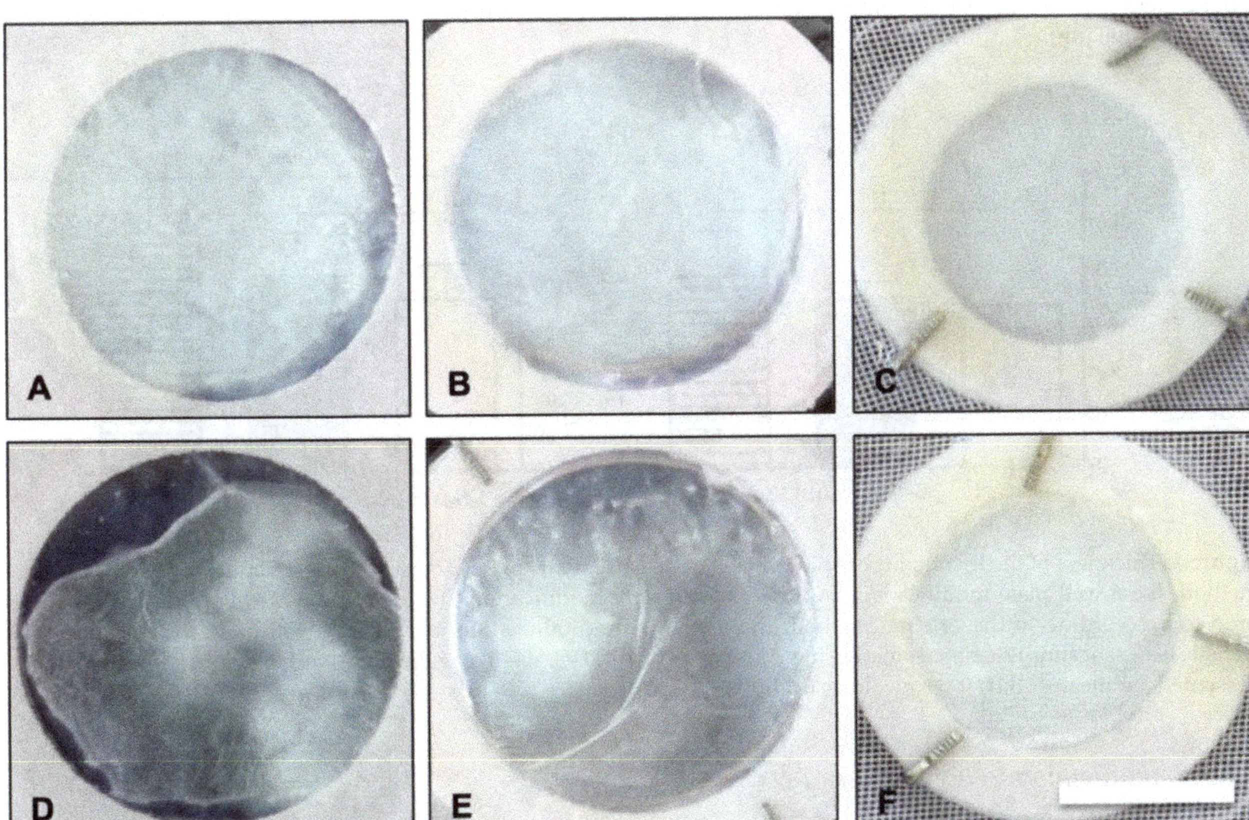

Figure 2. Macroscopic analyses. Macroscopic appearance of healthy substitutes (A–C) and psoriatic substitutes (D–F). Substitutes were produced according to the original method (A and D), 6-well plate modification (B and E), and 12-well plate modification (C and F). Three substitutes of each condition were analyzed, and the results were confirmed with three independent experiments. Cells from two different healthy patients and three different psoriatic patients were used (scale bar = 1 cm).

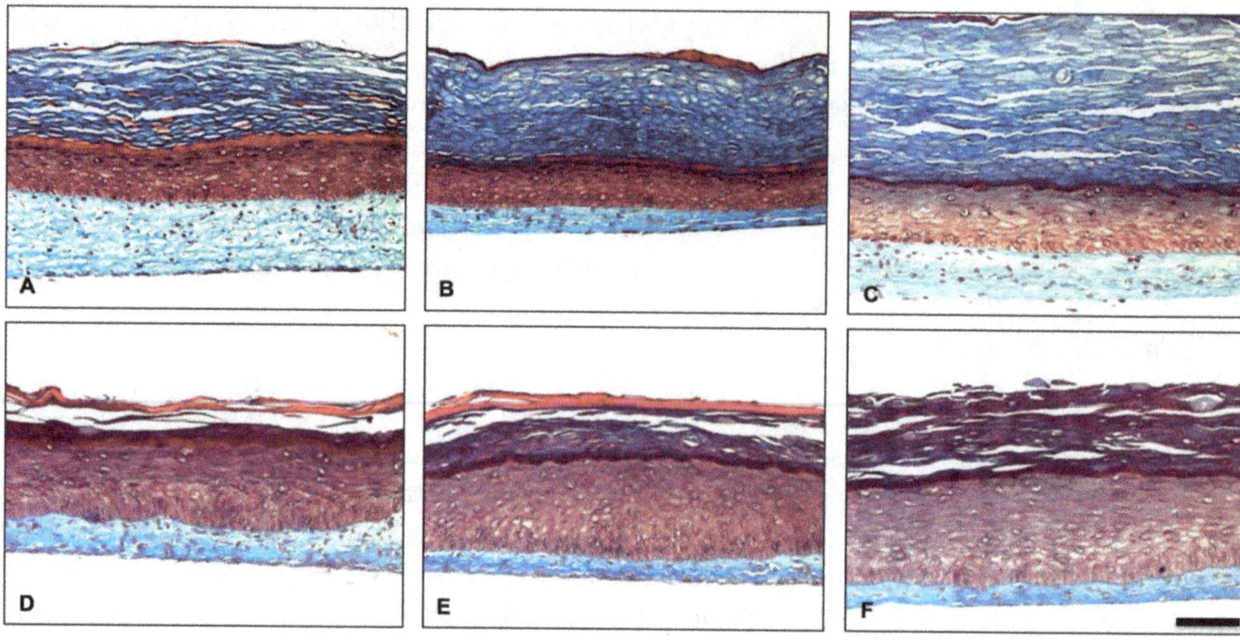

Figure 3. Histological analyses. Masson's trichrome staining of healthy substitutes (A–C), psoriatic substitutes (D–F) and substitutes produced with the original method (A and D), 6-well plate modification (B and E), and 12-well plate modification (C and F). Three substitutes of each condition were analyzed, and the results were confirmed with three independent experiments. Cells from two different healthy patients and three different psoriatic patients were used (scale bar = 100 μm).

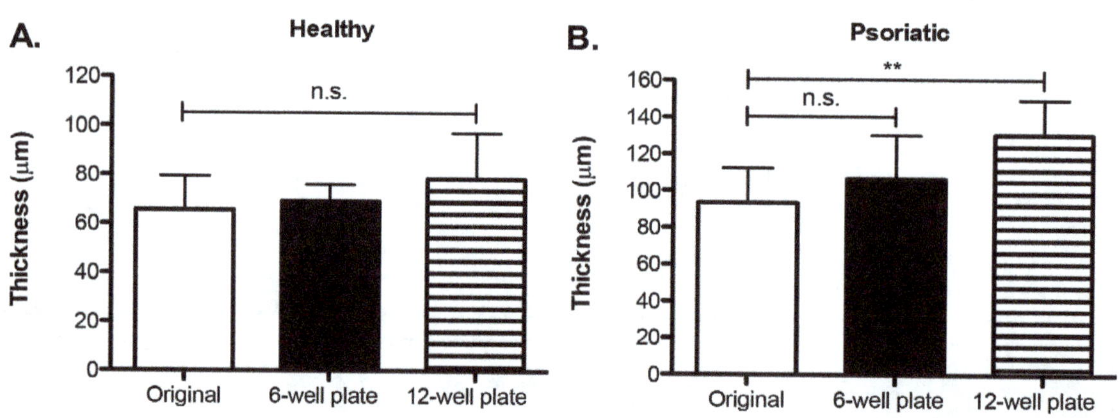

Figure 4. Thickness of the living part of epidermis. (A) Measurements of healthy substitutes produced with the original method, the 6-well plate modification, and the 12-well plate modification. (B) Measurements of psoriatic substitutes produced according to the original method, the 6-well plate modification, and the 12-well plate modification. The statistical significance was determined using ANOVA test ($p < 0.05$, $n = 3$, $N = 3$, 90 measurements by condition). Data presented are means ± S.D. (**) = p-value < 0.001.

2.1.4. *Immunofluorescence analyses*

Immunofluorescent markers were used to compare protein expression. Late differentiation markers, such as filaggrin and loricrin, were observed in order to compare the different skin substitute models. The *stratum granulosum* of *in vivo* skin is characterized by the presence of keratohyalin granules and the expression of late differentiation markers, since their synthesis depends on keratohyalin granules [36, 37]. Moreover, late differentiation markers are found in the *stratum granulosum*. In healthy substitutes, filaggrin was expressed from the last layers of the *stratum granulosum* to the first ones of the *stratum corneum* (**Figure 5A–C**). In healthy substitutes produced according to the 6-well plate and 12-well plate modifications, a slight decrease of filaggrin was observed (**Figure 5B** and **C**). However, since keratohyalin granules are in lesser amount in psoriatic skin than normal skin due to abnormal keratinocytes differentiation [38], expression of late differentiation markers are downregulated. In our psoriatic substitutes, whether the original method was used or the two modifications to protocol were followed, filaggrin is reduced and almost missing (**Figure 5D–F**). Loricrin, another late differentiation marker found in *stratum granulosum*, was observed in healthy substitutes (**Figure 5G–I**). In the psoriatic substitutes, a decrease or an absence of loricrin was observed which is similar to *in vivo* psoriatic skin (**Figure 5J–L**).

Other markers, such as keratin 14, keratin 1, and laminin, were also observed. Keratin 14 (K14) is expressed in the basal layer of the epidermis and is gradually reduced until keratins 1 (K1) and 10 (K10) are synthesized. In healthy substitutes produced according to the original method and the 6-well plate modification, keratin 14 was normally expressed (**Figure 6A** and **B**), whereas in healthy substitutes produced according to the 12-well plate modification, this keratin was still present in the *stratum corneum* (**Figure 6C**). In psoriatic skin, this protein is expressed in all epidermis such as observed regardless of the method used in the production of psoriatic substitutes (**Figure 6D–F**). Keratin 1 staining showed no difference between healthy substitutes reconstructed according to the original method or the 6-well plate modification (**Figure 6G–H**), while this keratin was overexpressed in *stratum corneum* of healthy substitutes produced according to the 12-well plate modification (**Figure 6I**). In psoriasis, keratin 1 is decreased such as observed in the psoriatic substitutes produced according to the original method or the 6-well plate modification (**Figure 6J–K**), whereas psoriatic substitutes produced according to the 12-well plate modification showed a higher expression of this keratin in the *stratum corneum* compared to the other two (**Figure 6L**). Laminin expression was similar for all reconstructed healthy substitutes (**Figure 7A–C**), while for psoriatic substitutes, laminin expression seems more distributed through the dermis using the original method and the 6-well plate modification (**Figure 7D–E**). Laminin staining of psoriatic substitutes produced according to the 12-well plate modification (**Figure 7F**) was similar to the healthy substitute expression.

2.2. Discussion

With psoriasis etiology still unknown, several *in vivo* models of psoriasis have been generated for a better understanding of the pathology. Genetically modified, spontaneous mutations, cytokine injections, xenografts, and gene knockout mouse models are all examples [21, 39–41]. Although these models showed psoriasis-like lesions, psoriasis is a specific human disease, and mouse skin does not represent characteristics of human skin such as epidermal thickness and structure, rete ridges and differences in certain immune cells [40, 41]. Due to the advances that have been made in the field of tissue engineering in the past years, it has been possible to develop *in vitro* psoriatic skin models to understand the disease, improve the development of new treatments, and limit the use of animals [24, 25, 28, 42]. Our team has developed a representative psoriatic skin model reconstructed according to a self-assembly method showing *in vivo* features of psoriasis such as hyperproliferation and abnormal differentiation of

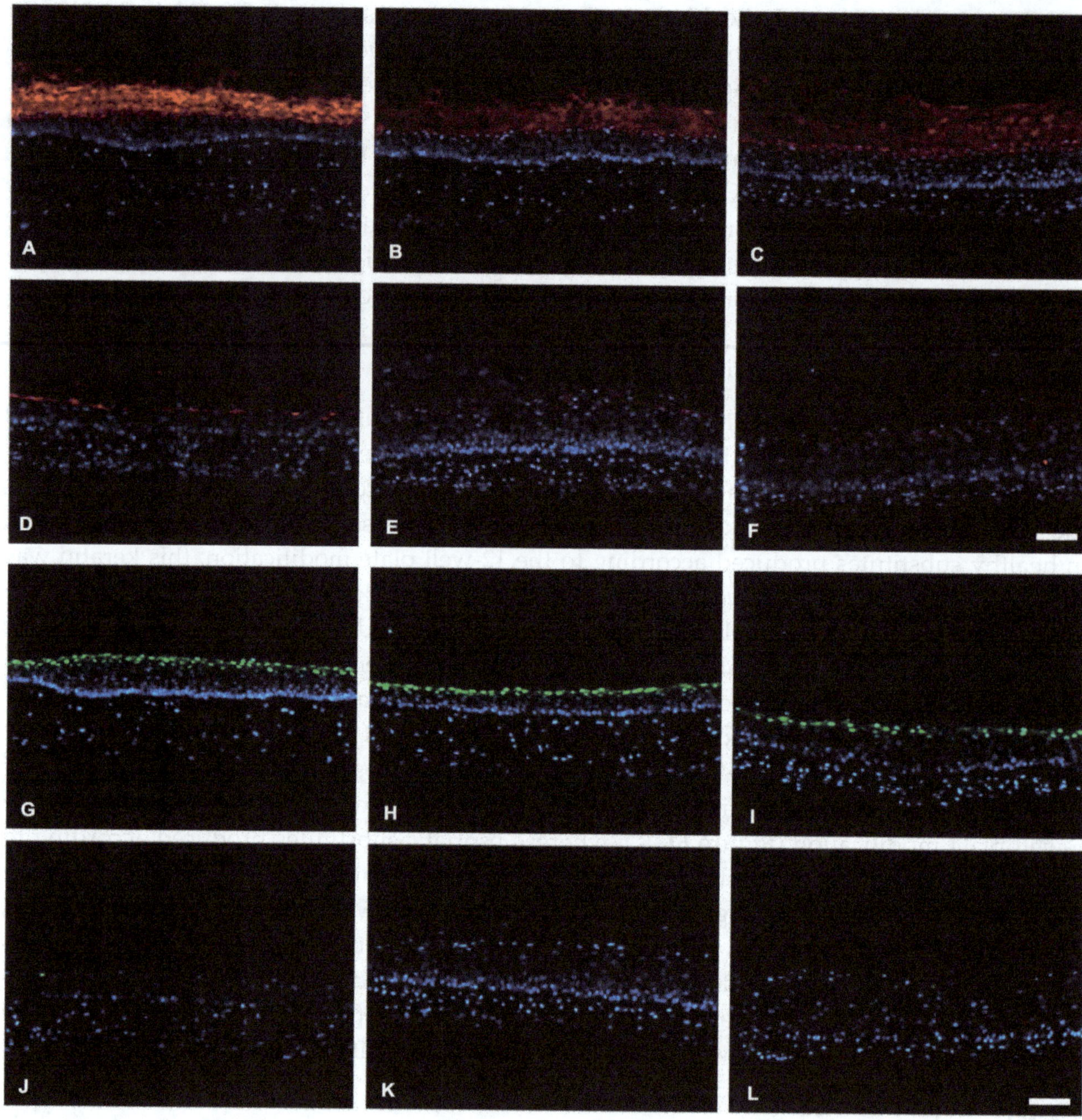

Figure 5. Filaggrin and loricrin staining. Expression of filaggrin (red) in healthy substitutes (A–C), psoriatic substitutes (D–F) and substitutes produced according to the original method (A and D), the 6-well plate modification (B and E), and the 12-well plate modification (C and F). Expression of loricrin (green) in healthy substitutes (G–I), psoriatic substitutes (J–L) and substitutes produced according to the original method (G and J), the 6-well plate modification (H and K), and the 12-well plate modification (I and L). The nuclei were stained with Hoechst (blue). Three substitutes of each condition were analyzed, and the results were confirmed with three independent experiments. Cells from two different healthy patients and three different psoriatic patients were used (scale bar = 100 μm).

keratinocytes [31]. However, each model has limitations, and some of the model developed in our team is the large number of cells required for the production of skin substitutes, in addition to complex manipulations that may generate less reproducibility. Therefore, the optimization of the production of psoriatic skin substitutes with the aim of making them suitable models for their use in pharmacological testing remains a challenge.

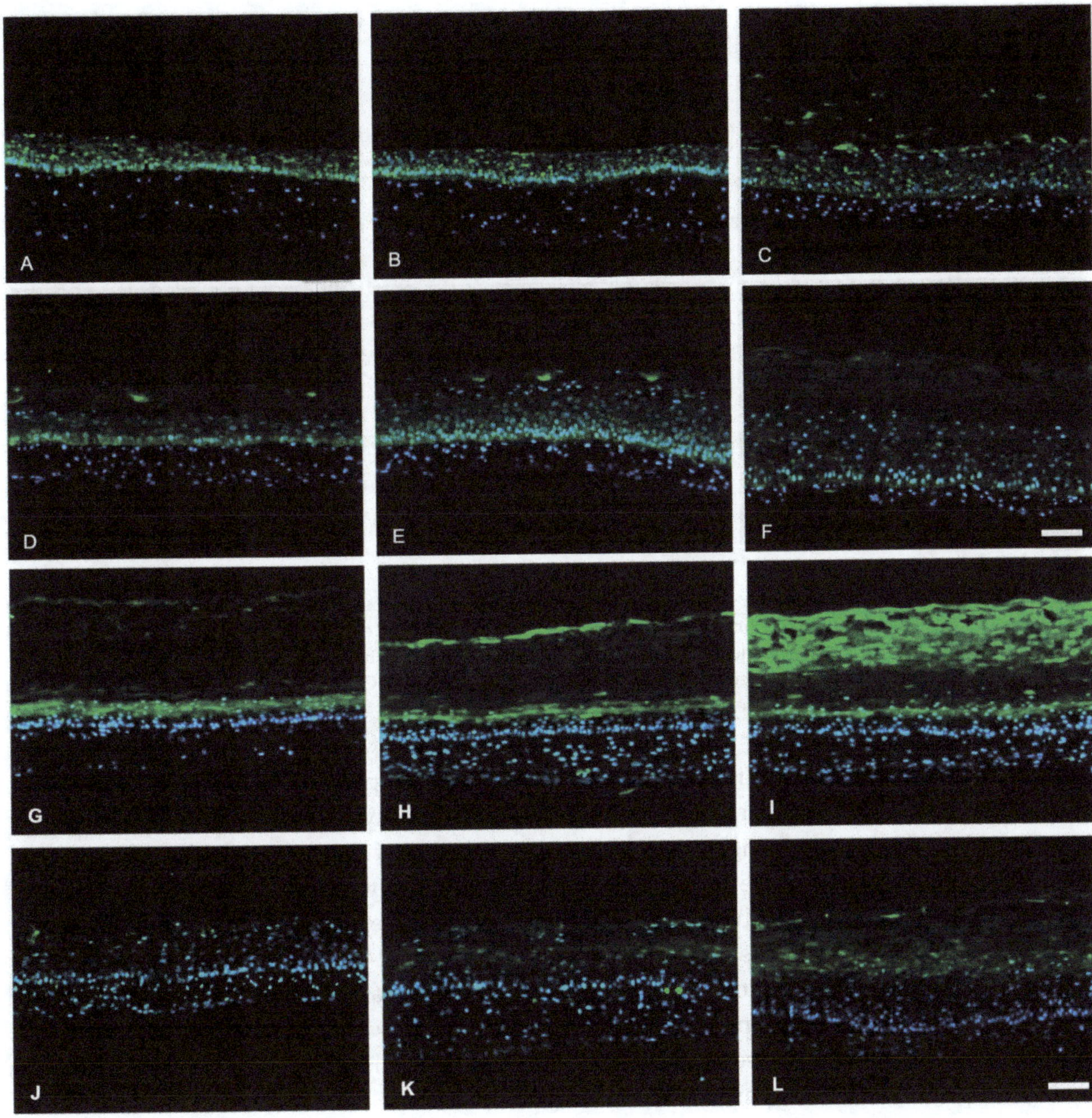

Figure 6. Keratin 14 and keratin 1 staining. Expression of keratin 14 (green) in healthy substitutes (A–C), psoriatic substitutes (D–F) and substitutes produced with the original method (A and D), the 6-well plate modification (B and E), and the 12-well plate modification (C and F). Expression of keratin 1 (green) in healthy substitutes (G–I), psoriatic substitutes (J–L) and substitutes produced according to the original method (G and J), the 6-well plate modification (H and K), and the 12-well plate modification (I and L). The nuclei were stained with Hoechst (blue). Three substitutes of each condition were analyzed, and the results were confirmed with three independent experiments. Cells from two different healthy patients and three different psoriatic patients were used (scale bar = 100 μm).

Previous studies of our group have shown that this psoriatic skin model demonstrated phenotypic characteristics of *in vivo* psoriasis [31–33]. In the present study, it is demonstrated that the improvements made to the original self-assembly method did not affect the psoriatic phenotype of the substitutes. *In vivo* psoriatic features are still expressed in the substitutes reconstructed according to the 6-well plate and 12-well plate modifications to protocol. Indeed,

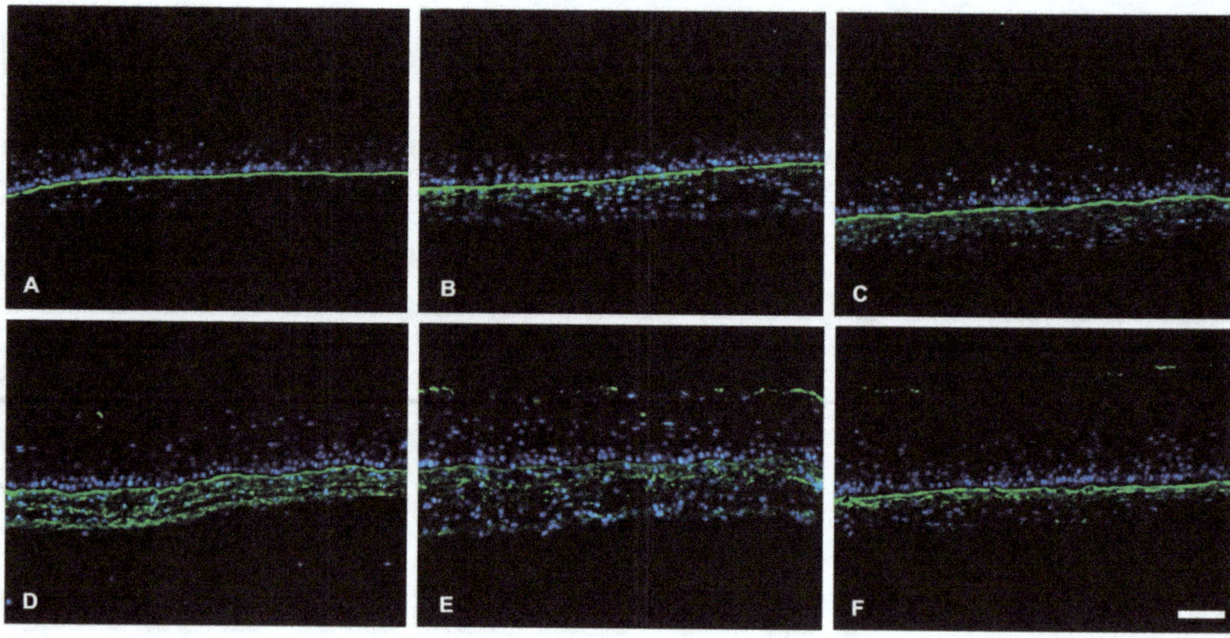

Figure 7. Laminin staining. Expression of laminin (green) in healthy substitutes (A–C), psoriatic substitutes (D–F) and substitutes produced according to the original method (A and D), the 6-well plate modification (B and E), and the 12-well plate modification (C and F). The nuclei were stained with Hoechst (blue). Three substitutes of each condition were analyzed, and the results were confirmed with three independent experiments. Cells from two different healthy patients and three different psoriatic patients were used (scale bar = 100 μm).

in both new protocols, a significant increase in the living epidermis thickness of psoriatic substitutes is observed compared with their respective healthy substitute counterparts as for the original method. Interestingly, the epidermal thickness of psoriatic substitutes produced according to the 12-well plate modification was significantly thicker than those produced according to the original method. This difference could be explained by the same seeding number of keratinocytes in a smaller seeding area for the 12-well plate protocol. Thus, this induces an increase of acanthosis. Considering histological analysis, our model stands out from other models of *in vitro* psoriatic skin that did not show acanthosis [24, 25]. However, based on these histological observations, the 6-well plate modification to protocol seems to be the most representative compared with the original protocol.

Differentiation of psoriatic skin is characterized by the altered expression of several epidermal proteins [43]. In a normal differentiation process (approximately 28 days), the basal layer transit amplifying cells differentiate and migrate into upper epidermal layers and synthesize important proteins involved in the differentiation and the skin barrier function such as filaggrin, loricrin, and keratins [36, 37, 44]. Filaggrin is normally synthesized from a precursor, profilaggrin, found in the granular layer. This protein is a key role in the formation of the cornified envelope [37]. In skin disease, such as psoriasis, expression of filaggrin is decreased, such that it is sometimes even absent due to an altered differentiation process and a reduction or an absence of the granular layer (agranulose) [38, 45, 46]. Loricrin, a major component of the cornified envelope, is stored in granules of the *stratum granulosum*, and its expression is also decreased or absent in psoriatic skin [36, 47]. In agreements with these observations, these features were observed with the substitutes reconstructed according to the original method and the two modifications to protocol (6-well plate and 12-well plate modifications). Indeed, filaggrin and loricrin were detected in healthy skin substitutes, whereas their absence was observed in psoriatic skin substitutes. This therefore confirms that the characteristics associated with the psoriatic phenotype are preserved with these new methods.

Keratins are intermediate filaments highly involved in epidermal structure and different types are expressed in the varying differentiation stages [48]. K5 and K14 are normally found in the basal layer of the epidermis, and they are progressively replaced by K1 and K10 in suprabasal layers [49]. However, *in vivo* psoriatic skin shows K14 in all layers of the epidermis, including the *stratum corneum*. This therefore suggests that the degradation mechanism of this keratin is altered in psoriasis [48, 50]. Moreover, in such hyperproliferative diseases, a new pair of keratins, K6 and K16, is appearing, causing a decrease in the expression of K1 and K10 [48, 50–52]. In the present work, healthy and psoriatic skin substitutes reconstructed according to the original method and the 6-well plate modification to protocol have demonstrated the same K1 and K14 expression than *in vivo*. Thus, taking together, K1, K14, filaggrin, and loricrin results validated the conservation of psoriatic skin differentiation in the new 6-well plate modification. This suggests that the 6-well method would be a great alternative to the original method. In healthy substitutes produced following the 12-well plate modification, an abnormal presence of K14 and K1 was observed in the *stratum corneum*, showing a less efficient differentiation of keratinocytes. This can probably be explained by the higher number of cells seeded in the culture area. These observations demonstrated that the 6-well plate modification is more effective for the production of healthy substitutes than the 12-well plate modification.

Some studies suggested that alterations in the basal membrane of psoriatic skin play an important role in the abnormal proliferation and differentiation of psoriatic keratinocytes [53–55]. Indeed, the expression of proteins such as laminin, which is one of the main proteins that forms the basal membrane, is decreased and disrupted in psoriatic skin unlike in healthy skin. In this last, laminin forms a linear and continuous structure [54]. Laminin expression in our skin substitutes produced with healthy cells regardless of the method (original method and two new modifications) was intense, continuous and more restricted to the basal lamina, demonstrating a good structure of the basal membrane. For psoriatic skin substitutes (original method and 6-well plate modification), the expression of laminin was more distributed through the dermis compared to healthy substitutes, showing disorganization in the basal membrane. For the psoriatic skin substitutes reconstructed according to the 12-well plate modification, laminin staining was more compact and similar to healthy substitute expression. Interestingly, this observation is showing that the basal membrane was more organized, thus less similar to the psoriatic phenotype. These results showed that the 6-well plate modification is more representative of the *in vivo* psoriatic skin than the 12-well plate modification.

2.3. Conclusion

These new modifications to protocol provide several advantages in the production of skin substitutes. Indeed, the 6-well and 12-well plate modifications require almost 3 times fewer fibroblasts and culture medium than the original method for the production of fibroblast sheets, which favorably reduce production costs. In addition, the use of plates and anchoring papers at the beginning of the production greatly facilitate the handling and the superposition of fibroblast sheets increasing the quality of the dermal equivalent. The anchoring papers allow the production of more reproducible and uniform size substitutes within the same experiment and between the different studies, which are essential for pharmaceutical studies. In conclusion, the modifications made to the original self-assembly method for the production of psoriatic substitutes are effective and demonstrate a comparable phenotype. However, the 6-well method is the one that leads to reconstructed substitutes with characteristics more similar to those seen *in vivo*. These modifications make it possible to obtain substitutes that are distinguished by better reproducibility making them new tools of choice for pharmacological analyses.

Acknowledgements

The authors acknowledge financial support from the Natural Sciences and Engineering Research Council of Canada (NSERC) and the Canadian Institutes of Health Research (CIHR) through their joint Collaborative Health Research Projects (CHRP) program. The "Fonds d'Enseignement et de Recherche (FER)" of the Faculty of Pharmacy, Université Laval, Québec, QC, Canada (Isabelle Gendreau scholarship), the "Fonds de Recherche du Québec—Santé (FRQS)" (Alexe Grenier scholarship) and the support of the "Réseau ThéCell du Québec" are also acknowledged. Moreover, Dr. Pouliot is a FRQS career award scholar.

Conflict of interest

The authors state no conflict of interest. The authors have no relationship with a for-profit or a not-for-profit organization to disclose. There is no financial conflict with the subject or the materials discussed in the manuscript apart from those disclosed.

Author details

Alexe Grenier[1,2,3†], Isabelle Gendreau[1,2,3†] and Roxane Pouliot[1,2,3*]

*Address all correspondence to: roxane.pouliot@pha.ulaval.ca

1 Centre de Recherche en Organogénèse Expérimentale de l'Université Laval/LOEX, Québec, QC, Canada

2 Centre de Recherche du CHU de Québec-Université Laval, Québec, QC, Canada

3 Faculté de Pharmacie, Université Laval, Québec, QC, Canada

† These two authors contributed equally to the work and therefore should be considered equivalent first authors.

References

[1] Mélissopoulos A, Levacher C. La peau: Structure et physiologie. 2nd ed. Paris: Éditions Tec & doc; 2012

[2] Schaefer H, Redelmeier TE. Skin Barrier: Principles of Percutaneous Absorption. Basel: Karger; 1996

[3] Lee SH, Jeong SK, Ahn SK. An update of the defensive barrier function of skin. Yonsei Medical Journal. 2006;**47**(3):293-306

[4] Marieb EN. Human Anatomy & Physiology. 5th ed. San Francisco: Benjamin Cummings; 2001

[5] Betts JG, Desaix P, Johnson E, et al. Anatomy and Physiology. Houston: OpenStax; 2013

[6] Warren R, Menter A. Handbook of Psoriasis and Psoriatic Arthritis. Cham: Adis; 2016

[7] Raychaudhuri S, Farber E. The prevalence of psoriasis in the world. Journal of the European Academy of Dermatology and Venereology. 2001;**15**(1):16-17

[8] Thomas J. Textbook of Psoriasis. 1st ed. New Delhi: Jaypee Brothers Medical Pub; 2016

[9] Camisa C. Handbook of Psoriasis. Malden: Blackwell Pub; 2004

[10] van de Kerkhof PCM. Textbook of Psoriasis. Malden, Mass: Blackwell; 2003

[11] Ortonne JP. Aetiology and pathogenesis of psoriasis. British Journal of Dermatology. 1996;**135**(s49):1-5

[12] García-Pérez ME, Jean J, Pouliot R. Antipsoriatic drug development: Challenges and new emerging therapies. Recent Patents on Inflammation & Allergy Drug Discovery. 2012;**6**:3-21

[13] Warren RB, Griffiths CEM. Systemic therapies for psoriasis: Methotrexate, retinoids, and cyclosporine. Clinics in Dermatology. 2008;**26**(5):438-447

[14] Winterfield L, Menter A, Gordon K, Gottlieb A. Psoriasis treatment: Current and emerging directed therapies. Annals of the Rheumatic Diseases. 2005;**64**(Suppl 2):ii87-ii90

[15] Jean J, García-Pérez ME, Pouliot R. Bioengineered skin: The self-assembly approach. Journal of Tissue Science & Engineering. 2011;**S5**:001

[16] García-Pérez ME, Royer M, Duque-Fernandez A, Diouf PN, Stevanovic T, Pouliot R. Antioxidant, toxicological and antiproliferative properties of Canadian polyphenolic extracts on normal and psoriatic keratinocytes. Journal of Ethnopharmacology. 2010;**132**(1):251-258

[17] García-Pérez ME, Royer M, Herbette G, Desjardins Y, Pouliot R, Stevanovic T. Picea mariana bark: A new source of trans-resveratrol and other bioactive polyphenols. Food Chemistry. 2012;**135**(3):1173-1182

[18] García-Pérez ME, Allaeys I, Rusu D, Pouliot R, Janezic TS, Poubelle PE. Picea mariana polyphenolic extract inhibits phlogogenic mediators produced by TNF-α-activated psoriatic keratinocytes: Impact on NF-κB pathway. Journal of Ethnopharmacology. 2014;**151**(1): 265-278

[19] Abd E, Yousef SA, Pastore MN, et al. Skin models for the testing of transdermal drugs. Clinical Pharmacology: Advances and Applications. 2016;**8**:163-176

[20] Angers L, Dubois-Declercq S, Masson L-C, et al. Effects of freezing on functionality and physicochemical properties of a 3D-human skin model. Journal of Dermatology & Cosmetology. 2017;**1**(2):00007

[21] Jean J, Pouliot R. In vivo and in vitro models of psoriasis. In: Eberli D, editor. Tissue Engineering. Croatia: In-Tech; 2010. pp. 359-382

[22] Jean J, García-Pérez M, Pouliot R. Psoriatic skin models: A need for the pharmaceutical industry. In: Soung J, editor. Psoriasis. Croatia: InTech; 2012. pp. 47-62

[23] Boehncke WH, Schön MP. Animal models of psoriasis. Clinics in Dermatology. 2007; **25**(6):596-605

[24] Barker CL, McHale MT, Gillies AK, et al. The development and characterization of an in vitro model of psoriasis. Journal of Investigative Dermatology. 2004;**123**(5):892-901

[25] Desmet E, Ramadhas A, Lambert J, Van Gele M. In vitro psoriasis models with focus on reconstructed skin models as promising tools in psoriasis research. Experimental Biology and Medicine. 2017;**242**(11):1158-1169

[26] Saiag P, Coulomb B, Lebreton C, Bell E, Dubertret L. Psoriatic fibroblasts induce hyperproliferation of normal keratinocytes in a skin equivalent model in vitro. Science. 1985; **230**(4726):669-672

[27] Konstantinova N, Duong D, Remenyik E, Hazarika P, Chuang A, Duvic M. Interleukin-8 is induced in skin equivalents and is highest in those derived from psoriatic fibroblasts. Journal of Investigative Dermatology. 1996;**107**(4):615-621

[28] Tjabringa G, Bergers M, van Rens D, de Boer R, Lamme E, Schalkwijk J. Development and validation of human psoriatic skin equivalents. The American Journal of Pathology. 2008;**173**(3):815-823

[29] van den Bogaard EH, Tjabringa GS, Joosten I, et al. Crosstalk between keratinocytes and T cells in a 3D microenvironment: A model to study inflammatory skin diseases. Journal of Investigative Dermatology. 2014;**134**(3):719-727

[30] Jansen PAM, Rodijk-Olthuis D, Hollox EJ, et al. β-Defensin-2 protein is a serum biomarker for disease activity in psoriasis and reaches biologically relevant concentrations in Lesional skin. PLoS One. 2009;**4**(3):e4725

[31] Jean J, Lapointe M, Soucy J, Pouliot R. Development of an in vitro psoriatic skin model by tissue engineering. Journal of Dermatological Science. 2009;**53**(1):19-25

[32] Jean J, Leroy M, Duque-Fernandez A, Bernard G, Soucy J, Pouliot R. Characterization of a psoriatic skin model produced with involved or uninvolved cells. Journal of Tissue Engineering and Regenerative Medicine. 2015;**9**(7):789-798

[33] Jean J, Soucy J, Pouliot R. Effects of retinoic acid on keratinocyte proliferation and differentiation in a psoriatic skin model. Tissue Engineering Parts A. 2011;**17**(13-14):1859-1868

[34] Ayata RE, Bouhout S, Auger M, Pouliot R. Study of in vitro capillary-like structures in psoriatic skin substitutes. BioResearch open access. 2014;**3**(5):197-205

[35] Larouche D, Jean J, Berthod F, Germain L, Pouliot R. Markers for an in vitro skin substitute. In: Maguire T, Novik E, editors. Methods in Bioengineering: Alternative Technologies to Animal Testing. Boston: Artech House; 2010. pp. 183-203

[36] Steven AC, Bisher ME, Roop DR, Steinert PM. Biosynthetic pathways of filaggrin and loricrin—Two major proteins expressed by terminally differentiated epidermal keratinocytes. Journal of Structural Biology. 1990;**104**(1):150-162

[37] Sandilands A, Sutherland C, Irvine AD, McLean WHI. Filaggrin in the frontline: Role in skin barrier function and disease. Journal of Cell Science. 2009;**122**(9):1285-1294

[38] Brody I. The ultrastructure of the epidermis in psoriasis vulgaris as revealed by electron microscopy: 3. Stratum intermedium in parakeratosis without keratohyalin. Journal of Ultrastructure Research. 1962;**6**(3):341-353

[39] Wagner EF, Schonthaler HB, Guinea-Viniegra J, Tschachler E. Psoriasis: What we have learned from mouse models. Nature Reviews Rheumatology. 2010;**6**:704

[40] Schön MP. Animal models of psoriasis: A critical appraisal. Experimental Dermatology. 2008;**17**(8):703-712

[41] Gudjonsson JE, Johnston A, Dyson M, Valdimarsson H, Elder JT. Mouse models of psoriasis. Journal of Investigative Dermatology. 2007;**127**(6):1292-1308

[42] Boniface K, Bernard F-X, Garcia M, Gurney AL, Lecron J-C, Morel F. IL-22 inhibits epidermal differentiation and induces proinflammatory gene expression and migration of human keratinocytes. The Journal of Immunology. 2005;**174**(6):3695-3702

[43] McKay IA, Leigh IM. Altered keratinocyte growth and differentiation in psoriasis. Clinics in Dermatology. 1995;**13**(2):105-114

[44] Eckert RL, Rorke EA. Molecular biology of keratinocyte differentiation. Environmental Health Perspectives. 1989;**80**:109-116

[45] Bernard BA, Asselineau D, Schaffar-Deshayes L, Darmon MY. Abnormal sequence of expression of differentiation markers in psoriatic epidermis: Inversion of two steps in the differentiation program? Journal of Investigative Dermatology. 1988;**90**(6):801-805

[46] Lowes MA, Bowcock AM, Krueger JG. Pathogenesis and therapy of psoriasis. Nature. 2007;**445**(7130):866-873

[47] Hohl DMD, Pierard GE. Expression patterns of Loricrin in dermatological disorders. American Journal of Dermatopathology. 1993;**15**(1):20-27

[48] Thewes M, Stadler R, Korge B, Mischke D. Normal psoriatic epidermis expression of hyperproliferation-associated keratins. Archives of Dermatological Research. 1991; **283**(7):465-471

[49] Candi E, Schmidt R, Melino G. The cornified envelope: A model of cell death in the skin. Nature Reviews Molecular Cell Biology. 2005;**6**(4):328-340

[50] Stoler A, Kopan R, Duvic M, Fuchs E. Use of monospecific antisera and cRNA probes to localize the major changes in keratin expression during normal and abnormal epidermal differentiation. The Journal of Cell Biology. 1988;**107**(2):427-446

[51] Weiss RA, Eichner R, Sun TT. Monoclonal antibody analysis of keratin expression in epidermal diseases: A 48- and 56-kdalton keratin as molecular markers for hyperproliferative keratinocytes. The Journal of Cell Biology. 1984;**98**(4):1397-1406

[52] Leigh IM, Navsaria H, Purkis PE, McKay IA, Bowden PE, Riddle PN. Keratins (Kl6 and Kl7) as markers of keratinocyte hyperproliferation in psoriasis in vivo and in vitro. British Journal of Dermatology. 1995;**133**(4):501-511

[53] Fleischmajer R, Kuroda K, Hazan R, et al. Basement membrane alterations in psoriasis are accompanied by epidermal overexpression of MMP-2 and its inhibitor TIMP-2. Journal of Investigative Dermatology. 2000;**115**(5):771-777

[54] Mondello MR, Magaudda L, Pergolizzi S, et al. Behaviour of laminin 1 and type IV collagen in uninvolved psoriatic skin. Immunohistochemical study using confocal laser scanning microscopy. Archives of Dermatological Research. 1996;**288**(9):527

[55] Eşrefoğlu M, Seyhan ME, Aktaş A, Gül M, Öztürk F. Histopathological findings and the distribution of Laminin and Fibronectin in psoriatic skin. İnönü Üniversitesi Tıp Fakültesi Dergisi. 2005;**12**(4):217-222

Permissions

All chapters in this book were first published in CC, by InTech Open; hereby published with permission under the Creative Commons Attribution License or equivalent. Every chapter published in this book has been scrutinized by our experts. Their significance has been extensively debated. The topics covered herein carry significant findings which will fuel the growth of the discipline. They may even be implemented as practical applications or may be referred to as a beginning point for another development.

The contributors of this book come from diverse backgrounds, making this book a truly international effort. This book will bring forth new frontiers with its revolutionizing research information and detailed analysis of the nascent developments around the world.

We would like to thank all the contributing authors for lending their expertise to make the book truly unique. They have played a crucial role in the development of this book. Without their invaluable contributions this book wouldn't have been possible. They have made vital efforts to compile up to date information on the varied aspects of this subject to make this book a valuable addition to the collection of many professionals and students.

This book was conceptualized with the vision of imparting up-to-date information and advanced data in this field. To ensure the same, a matchless editorial board was set up. Every individual on the board went through rigorous rounds of assessment to prove their worth. After which they invested a large part of their time researching and compiling the most relevant data for our readers.

The editorial board has been involved in producing this book since its inception. They have spent rigorous hours researching and exploring the diverse topics which have resulted in the successful publishing of this book. They have passed on their knowledge of decades through this book. To expedite this challenging task, the publisher supported the team at every step. A small team of assistant editors was also appointed to further simplify the editing procedure and attain best results for the readers.

Apart from the editorial board, the designing team has also invested a significant amount of their time in understanding the subject and creating the most relevant covers. They scrutinized every image to scout for the most suitable representation of the subject and create an appropriate cover for the book.

The publishing team has been an ardent support to the editorial, designing and production team. Their endless efforts to recruit the best for this project, has resulted in the accomplishment of this book. They are a veteran in the field of academics and their pool of knowledge is as vast as their experience in printing. Their expertise and guidance has proved useful at every step. Their uncompromising quality standards have made this book an exceptional effort. Their encouragement from time to time has been an inspiration for everyone.

The publisher and the editorial board hope that this book will prove to be a valuable piece of knowledge for researchers, students, practitioners and scholars across the globe.

List of Contributors

Ansie Martin and Angshuman Sarkar
CMBL, Department of Biological Sciences, Birla Institute of Technology and Sciences, K K Birla Goa Campus, Sancoale, South Goa, India

Arezou Teimouri, Pollen Yeung and Remigius Agu
Dalhousie University, Halifax, Canada

Raymond W. Nims
RMC Pharmaceutical Solutions, Inc., Longmont, CO, USA

Amanda Capes-Davis
CellBank Australia, Children's Medical Research Institute, The University of Sydney, Westmead, New South Wales, Australia

Christopher Korch
Division of Medical Oncology, University of Colorado Denver, Aurora, CO, USA

Yvonne A. Reid
ATCC, Manassas, VA, USA

Jitcy Saji Joseph, Sibusiso Tebogo Malindisa and Monde Ntwasa
Department of Life and Consumer Sciences, College of Agriculture and Environmental Sciences, University of South Africa, South Africa

John L. Collins
University of Tennessee at Martin, Martin, TN, USA

Bart van Knippenberg
Cytosmart Technologies BV, Eindhoven, The Netherlands

Kai Ding
Johns Hopkins University School of Medicine, Baltimore, MD, USA

Alexander V. Kofman
Department of Biological & Environmental Sciences, Troy University, Troy, AL, USA

Siew-Wai Pang, Noel Jacques Awi, Hooi-Yeen Yap and Sin-Yeang Teow
Department of Medical Sciences, School of Healthcare and Medical Sciences, Sunway University, Jalan Universiti, Subang Jaya, Selangor Darul Ehsan, Malaysia

Moisés Armides Franco-Molina, Silvia Elena Santana-Krímskaya and Cristina Rodríguez-Padilla
Laboratory of Immunology and Virology of the Faculty of Biological Sciences of the Autonomous University of Nuevo León, San Nicolás de los Garza, México

Victoria Moreno-Manzano
Cellular Reprogramming Laboratory, Centro de Investigación Príncipe Felipe (CIPF), Valencia, Spain

Elisa Oltra García
School of Medicine, Universidad Católica de Valencia San Vicente Mártir, Valencia, Spain
Unidad Mixta CIPF-UCV, Centro de Investigación Príncipe Felipe, Valencia, Spain

Eszter Varga, Marten Hansen, Emile van den Akker and Marieke von Lindern
Department of Hematopoiesis, Sanquin Research, and Landsteiner Laboratory Amsterdam UMC, Amsterdam, The Netherlands

Alexe Grenier, Isabelle Gendreau and Roxane Pouliot
Centre de Recherche en Organogénèse Expérimentale de l'Université Laval/LOEX, Québec, QC, Canada
Centre de Recherche du CHU de Québec-Université Laval, Québec, QC, Canada
Faculté de Pharmacie, Université Laval, Québec, QC, Canada

Index

L

M

N

O

P

S

T

V

W

www.ingramcontent.com/pod-product-compliance
Lightning Source LLC
LaVergne TN
LVHW060539210926
848485LV00005B/1224